Selected Titles in This Series

FIELDS INSTITUTE
MONOGRAPHS

THE FIELDS INSTITUTE FOR RESEARCH IN MATHEMATICAL SCIENCES

Introduction to Homotopy Theory

Paul Selick

American Mathematical Society
Providence, Rhode Island

The Fields Institute
for Research in Mathematical Sciences

The Fields Institute is named in honour of the Canadian mathematician John Charles Fields (1863–1932). Fields was a visionary who received many honours for his scientific work, including election to the Royal Society of Canada in 1909 and to the Royal Society of London in 1913. Among other accomplishments in the service of the international mathematics community, Fields was responsible for establishing the world's most prestigious prize for mathematics research—the Fields Medal.

The Fields Institute for Research in Mathematical Sciences is supported by grants from the Ontario Ministry of Education and Training and the Natural Sciences and Engineering Research Council of Canada. The Institute is sponsored by McMaster University, the University of Toronto, the University of Waterloo, and York University and has affiliated universities in Ontario and across Canada.

The author was supported in part by grants from the Natural Sciences and Engineering Research Council of Canada and The Fields Institute for Research in Mathematical Sciences.

1991 *Mathematics Subject Classification.* Primary 55–01;
Secondary 18–01, 54–01, 57–01.

Library of Congress Cataloging-in-Publication Data

Selick, Paul, 1950–
    Introduction to homotopy theory / Paul Selick.
      p. cm. — (Fields Institute monographs, ISSN 1069-5273 ; 9)
    Includes bibliographical references (p. –) and index.
    ISBN 0-8218-0690-4 (alk. paper)
    1. Homotopy theory. I. Title. II. Series.
QA612.7.S44   1997
514′.24—dc21

                                97-5777
                                CIP

To my parents,

*who made it possible for me
to study mathematics*

Contents

Contents   ix

Preface

These notes are based on a one-semester graduate course I gave at The Fields Institute in Fall, 1995 as part of the homotopy theory program which constituted the institute's major program that year. The purpose of the course was to bring graduate students who had completed a first course in algebraic topology to the point where they could understand research lectures in homotopy theory and to prepare them for the other, more specialized, graduate courses being held in conjunction with the program. The notes are divided into two parts: prerequisites and those which constituted the course proper.

Part I, the prerequisites, contains a review of material which might be taught in a first course in algebraic topology. Although it is probable that each of the topics discussed there has appeared in *some* such course, it seems unlikely that any individual first course would contain all of these topics. The students were expected to use this section to help fill in the gaps in their background knowledge. Chapter 4 (homological algebra) in particular contains some material which is likely to be new to most readers at this level. Not all of the material in Part I is applied in these notes; much is presented for completeness or to give a better overall picture of a topic. The reader may wish to skim or skip sections, referring back to them if needed. It is hoped that Part I will provide a useful summary for students and non-specialists who are interested in learning the basics of algebraic topology.

Experts in homotopy theory will recognize immediately upon seeing the chapter titles for Part II that it is not possible to cover all of those topics in detail in a one semester course; indeed many of them could constitute the material for such a course all by themselves. These notes, which were available in preliminary draft form at the time of the course, contain many details which were referred to during the lectures but not presented in detail.

An attempt has been made to make these notes self contained in an expositional sense. Even in the sections on prerequisites the development contains motivation for the major ideas and exact statements of definitions and theorems although the proofs are usually omitted in those sections. The percentage of statements for which a proof is given increases throughout the notes, starting at nearly 0% in Part I, increasing dramatically in Part II, nearing 100% by Chapter 11. Although time constraints prevented me from doing so in all cases, I was particularly concerned about providing a proof for those statements for which a proof is not readily available in the standard textbooks on the subject, since the original papers are often harder for students at this level to read.

The course consisted of eight major topics which appear as the eight chapters of Part II, numbered 7–14. The first of these, Chapter 7, is a lengthy chapter on "classical homotopy theory". It contains cellular approximation, the Hurewicz theorem, properties of H-spaces and co-H-spaces, Whitehead and Samelson products, and a large amount of space devoted to the manipulations of fibrations and cofibrations. It also introduces the James construction, Hopf-invariant maps, and

Lusternik-Schnirrelmann category. Some of this material appears in many places in the literature while other parts would fall into the category of "folk theorems". For example, many product and wedge decompositions which result from manipulating fibrations and cofibrations are given. These are well known and in everyday use but not easy to find in the literature.

Chapter 8 contains the basics of simplicial sets. It discusses Kan complexes, the singular complex, simplicial groups, simplicial abelian groups, and how the homotopy theory of simplicial sets parallels that of topological spaces. However some material, such as the simplicial classifying construction and simplicial loop construction, is omitted and readers who intend to make large use of simplicial sets will need to consult a book devoted explicitly to that subject for extra details. In these notes, simplicial sets are used primarily in discussing localization (Chapter 12). Although it is not explicitly a statement about simplicial sets. The fact that the quotient map from the singular chain complex of a space to that complex modulo its degenerate subcomplex is a chain homotopy equivalence appears in this chapter as a consequence of a theorem on simplicial abelian groups. This is used later in these notes in the section on Adams-Hilton models.

Chapter 9 is a brief introduction to fibre bundles. Basic definitions, classification theorems, and Milnor's construction are given. A student of algebraic topology would be well advised to read more on this subject, for example in Milnor's "Characteristic Classes" [MS].

The material of Chapter 10 on Hopf algebras is taken mostly from the classic paper by Milnor and Moore [MM] on this subject. The reader will have to refer to that source for proofs which (s)he cannot work out alone.

Chapter 11 is a long chapter on spectral sequences. The first half of the chapter develops the general theory of spectral sequences including a discussion of filtered complexes, exact couples, and the notion of convergence of the spectral sequence of a filtered complex. Then, with one exception, the major spectral sequences used in homotopy theory are discussed individually in the second half of the chapter. It is probably inevitable that parts of this chapter will be heavy going for readers who are seeing the material for the first time and rereading is recommended. The major spectral sequence in homotopy theory omitted from these notes is the Adams spectral sequence. In the context of the homotopy program at The Fields Institute, it would not have made sense to include that in this course since an entire course on the Adams spectral sequence was being given by Stan Kochman alongside this course. Readers are referred to the companion volume by Stan Kochman in this monograph series for a complete discussion of the Adams spectral sequence. The presentations of the two forms of the Eilenberg-Moore spectral sequence, that from base to fibre and that from fibre to base are given from opposite points of view. For the second, we take a geometric approach and obtain the spectral sequence by means of a suitable filtration on a space. This has some obvious advantages including giving us for free the commutativity of the spectral sequence with cohomology operations and ready generalization to generalized (co)homology theories. For the first Eilenberg-Moore spectral sequence we take a very algebraic approach closely related to Eilenberg and Moore's original proof. This allows the demonstration of homological techniques useful in other contexts and also demonstrates clearly how much information is lost when one passes from the singular chain complex on a space to its homology. The reader should be aware however that there are more geometric approaches to deriving this spectral sequence, although it would make sense to first develop the properties of a category of spectra to avoid having to express the statements in an awkward form. See Smith [Sm] for a derivation from this point of view. Also included in this chapter is a discussion of Adams-Hilton models. From a logical point of view, this material would fit better into Chapter 7, but the proof of the main theorem uses some material presented in Chapter 11. The section on Adams-Hilton models also contains a brief introduction to the homotopy theory of differential graded algebras.

Chapter 12 discusses localization and completion. Alternative methods of localization are mentioned, and details are given for the Bousfield-Kan construction with an outline of the proof that it does indeed construct a localization. There is also an introduction to cosimplicial sets which are used in the construction. The reader is referred to the book by Bousfield and Kan [BK] for more information about cosimplicial sets and the Bousfield-Kan construction.

The main purpose of Chapter 13 is to develop the connection between generalized cohomology theory and representing spaces, and in particular the Brown Representability Theorem. This provides the framework for the discussion of cohomology operations in the following chapter. Spectra and stable homotopy concepts are introduced, but there is much more that can be (and has been!) said on these subjects. More information can be found in Stan Kochman's monograph in this series. The reader is also referred to the books by Adams [A2] and Lewis, May, Steinberger [LMS] for detailed construction of categories of spectra and their properties. The reader interested in stable homotopy theory will want to read some works devoted entirely to that subject.

The last chapter applies material from many of the earlier chapters to the discussion of cohomology operations and in particular the construction and properties of the Steenrod algebra, and some sample applications. There is also a glimpse into the subject of secondary cohomology operations.

The topics covered in these notes are, for the most part, those referred to in Adams guide [A1]. An omission from the topics mentioned there is Postnikov systems, but the reader who has understood the main ideas of these notes should be able to find and understand a discussion of this without much difficulty. The main addition to the topics in Adams list is localization and completion (Chapter 13) whose growth into a major subject has taken place during the interval since [A1] appeared.

The reader of these notes should watch for occasional out-of-sequence numbers. In the interest of collecting together material on a given subject, occasionally the statement of a theorem whose proof requires later techniques will appear in an earlier section if it fits with the material at hand and the reader can be reasonably expected to understand the statement at that time. In these cases the fact that the theorem's number shows that it belongs to a later section will serve to alert the reader that the statement is being given only for reference and intuitional purposes, and that the statement will reappear in the appropriate place and will not be used in the interval in the proof of other theorems.

Another convention used in these notes is in connection with the symbol $\square$. This symbol is used to mean "end of discussion" and so in addition to appearing at the end of proofs may also appear at the end of examples or at the conclusion of the statement of a theorem whose proof has either been omitted or given previously.

I would like to thank those who attended the course both for their contributions to the course and for pointing out the mistakes and misprints which they observed in earlier drafts of these notes. In particular I mention Kasper Anderson, Wojcieck Chacholsky, Arleigh Crawford, Jasper Grodal, Sadok Kallel, Michael Mather, Jerome Scheerer, and Bjorn Schuster. Above all I want to thank Matvei Libine for his invaluable assistance in the preparation of these notes, ranging from correction of misprints and mathematical errors to stylistic suggestions and advice on which additional proofs or examples would be most useful to a reader seeing this material for the first time. Finally I would like to thank The Fields Institute for providing an atmosphere and facilities so conducive to the discussion of mathematics.

Paul Selick
October 1996

Summary of Global Notation

For each notation, we give a brief description and/or the nearest result or section number to the spot where the notation is first defined. Some notations appear more than once in the following list when the same symbol is used for different things in different contexts.

$\mathcal{A}$	14.2.1	the Steenrod algebra when the prime is understood
$\mathcal{A}_*$	14.5.1	the dual of the Steenrod algebra when the prime is understood
$\mathcal{A}_{(p)}$	14.2.1	the mod p Steenrod algebra
$\mathcal{A}(-)$	11.10.7	Adams-Hilton model of a space
$\mathcal{AB}$		the category of abelian groups
AW	5.5.2	Alexander-Whitney homomorphism
$B(-)$	9.2.2	classifying space of a topological group
$B_n(-)$	4.1.1	boundaries of a chain complex
$B_n(-)$	9.2.2	nth Milnor filtration on the classifying space of a topological group
$\mathbb{C}$		complex numbers
$C(-)$	7.1.16	reduced cone on a pointed space
$C_*(-)$	5.4.1	cellular chain complex of a space
C_f	7.1.16	reduced mapping cone of a function
CC	4.2.1	the category of chain complexes
CoTor	11.11.1	derived functor of $\square$
$\mathbb{C}P^n$		n-dimensional complex projective space
d		often used for the differential in a chain complex
$\hat{D}$	11.3.2	elements infinitely divisibly by i in the D-term of an exact couple
$\check{D}$	11.3.2	direct limit of D-term in an exact couple
d_i	8.1.1	boundary map of a simplicial set
d^r	11.3.2	differential in the E^r-term of a spectral sequence or the composition $j^r k^r$ in an exact couple
D^n		n-dimensional disk
D^r	11.3.2	term in an exact couple
$D(-)$	8.5.1	degenerate sub-chain-complex of a simplicial set
DGA	11.6.1	differential graded algebra
$E(-)$	9.2.2	total space of the universal bundle of a topological group
$E_n(-)$	9.2.2	nth Milnor filtration on the total space the universal bundle of a topological group
E^r	11.2.1	term in a spectral sequence or exact couple
$E(V)$		exterior algebra on a vector space or free module

$E[S]$		exterior algebra on the vector space or free module generated by a set
Ext	4.3.5	derived functor of Hom
EZ	5.5.2	Eilenberg-Zilber homomorphism
$\mathcal{F}_{(-)}$	11.1.1	a filtered object
$F_n(-)$	11.1.1	nth filtration of a filtered object
$\mathrm{FW}_n(-)$	7.9.1	nth fat wedge of a pointed space
$\mathrm{Gr}(-)$	11.1.1	associated graded object to a filtered object
h_X	7.4.3	Hurewicz homomorphism
$h_{(X,A)}$	7.4.3	relative Hurewicz homomorphism
$\mathrm{Hom}_{\underline{C}}(-,-)$	1.1.1	set of morphisms in a category. Equivalent to $\underline{C}(-,-)$
$H_*(-)$	4.1.1	homology of a chain complex or space
$\tilde{H}_*(-)$	4.1.1	reduced homology of a space
$H_*(-;G)$		homology of a chain complex or space with coefficients in a module
$H_*(-;{}^tL)$	11.9.1	homology of a space with local coefficients
$\tilde{H}_*(-;G)$	4.1.1	reduced homology of a space with coefficients in a module
$H^*(-)$	4.1.1	cohomology of a cochain complex or space
$\tilde{H}^*(-)$	4.1.1	reduced cohomology of a space
$H^*(-;G)$		cohomology of a cochain complex or space with coefficients in a module
$H^*(-;{}^tL)$	11.9.1	cohomology of a space with local coefficients
$\tilde{H}^*(-;G)$	4.1.1	reduced cohomology of a space with coefficients in a module
$H_*^{\mathrm{alg}}(-)$	4.3.8	algebraic group homology
$H^*_{\mathrm{alg}}(-)$	4.3.8	algebraic group cohomology
H_t	3.1.1	function obtained from a homotopy by fixing a value of t
I		the interval $[0,1]$ or its simplicial equivalent. Sometimes also used for the cellular chain complex of the interval
$I(-)$	10.1.2	augmentation ideal of a graded algebra
i_0		inclusion $x \mapsto (x,0)$ of a space into its product with $[0,1]$
i_1		inclusion $x \mapsto (x,1)$ of a space into its product with $[0,1]$
i^r	11.3.2	morphism between D^r terms in an exact couple
$\mathrm{Int}\,(-)$		interior of a set. Equivalent to $(\overset{\circ}{-})$
$J(-)$	10.1.2	cokernel of the augmentation in a graded coalgebra
j^r	11.3.2	morphism from D^r to E^r in an exact couple
$J(-)$	7.9.1	James construction on a pointed space
$J_n(-)$	7.9.1	nth stage of James construction on a pointed space
k^r	11.3.2	morphism from E^r to D^r in an exact couple
$(-)_\kappa$	2.6.1	compactly generated topology of a space

$K(\pi, n)$	13.4.3	space with one nonvanishing homotopy group
$L(V)$		free Lie algebra on a vector space or free module
$L[S]$		free Lie algebra on the vector space or free module generated by a set
$L_n(-)$	4.3.4	left derived functor
$\varinjlim$	1.2	direct limit
$\varprojlim$	1.2	inverse limit
$\varprojlim^1$	4.7	derived functor of $\varprojlim$
M_f	7.1.16	reduced mapping cylinder of a function
$\mathrm{Map}\,(X, Y)$	2.3.8	space of continuous functions or function space of simplicial sets. Equivalent to Y^X
$\mathrm{Map}_*(X, Y)$	2.3.8	basepoint preserving functions between pointed spaces. Equivalent to Y^X when X and Y are pointed spaces
$(-)$-mods		category of modules over a ring
$\mathbb{N}$		natural numbers
N	4.3.8	sum within the group ring of all elements of a group
$N(-)$	8.5.1	normalized sub-chain-complex of a simplicial set
NDR	7.1.10	neighbourhood deformation retract
$P(-)$	10.4.1	primitive elements in an augmented coalgebra
$P(-)$	7.1.4	path space
$P'(-)$	7.9.1	Moore path space
P^f	7.1.16	mapping path space of a function
$\mathcal{P}$	14.3.9	Steenrod reduced power operation
$\mathcal{P}^{\Delta_k}$	14.5.3	Milnor primitive in the Steenrod algebra
PID		principal ideal domain
$\mathbb{Q}$		rationals
$Q(-)$	10.4.1	indecomposable quotient in an augmented algebra
$Q(-)$	13.3.1	infinite loop space formed by taking the direct limit of $\Omega^n S^n(-)$
$\mathcal{Q}^{\Delta_k}$	14.5.3	Milnor primitive in the Steenrod algebra
$\mathbb{R}$		real numbers
$\mathbb{R}P^n$		n-dimensional real projective space
$S(-)$		reduced suspension of a chain complex or a space
s		often used for a chain homotopy
s_i	8.1.1	degeneracy map of a simplicial set
S^n		n-dimensional sphere
S_n		symmetric group on n-symbols
$S_*(-)$	5.2	singular chain complex of a space
$S^*(-)$	5.6	singular cochain complex of a space
$S_*^{(k,A)}(-)$	7.4.1	Eilenberg subcomplex of a space
$S(V)$		free commutative algebra on a vector space or free module
$S[X]$		free commutative algebra on the vector space or free module generated by a set
sd	5.2.4	barycentric subdivision operator

$\mathcal{S}et_*$		the category of pointed sets
sh	7.5.14	shearing map
$\mathrm{Sing}\,(-)$	8.4.1	singular complex of a simplicial set
$\mathcal{S}pecS$		the category of spectral sequences
Sq	14.3.9	Steenrod square
SS		the category of simplicial sets
T	4.3.8	generator of a cyclic group
T	10.1.1	map which performs the graded interchange of the two factors of a tensor product
$T(-)$		tangent space of a manifold
$T(V)$		tensor algebra on a vector space or free module
$T[S]$		tensor algebra on the vector space or free module generated by a set
$\mathcal{T}op$		the category of topological spaces and continuous functions
$\mathcal{T}op_*$		the category of pointed topological spaces and basepoint preserving continuous functions
Tor	4.3.5	derived functor of $\otimes$
$\mathrm{Tot}(-)$	11.7.1	total complex of a double complex. Also used for a simplicial set formed from a cosimplicial simplicial set
$\mathrm{Tot}^{\pi}(-)$	11.7.1	product total complex formed from a double complex
$U(-)$	10.5.1	universal enveloping algebra of a Lie algebra
$\mathbb{Z}$		integers
$\mathbb{Z}\,(-)$	8.5.4	simplicial set formed by taking the free abelian group in each degree
$\hat{\mathbb{Z}}_p$	12.2.3	p-adics
$Z_n(-)$	4.1.1	cycles of a chain complex
$Z^r(-)$	11.2	cycles in E^r-term of a spectral sequence
α	4.5	natural transformation from the identity arising from a pair of adjoint functors
β	4.5	natural transformation to the identity arising from a pair of adjoint functors
$\beta^{(r)}$	11.8.1	rth Bockstein homomorphism when the prime is understood
$\beta_p^{(r)}$	11.8.1	rth mod p Bockstein homomorphism
$\Gamma(V)$		divided polynomial algebra on a vector space or free module
$\Gamma[S]$	10.1.7	divided polynomial algebra on the vector space or free module generated by a set
$\gamma_j(-)$	10.1.7	a basis element in a divided polynomial algebra
∂	4.1.5	connecting homomorphism in a long exact homology sequence
∂	5.2.1	boundary map in the singular chain complex of a space
∂_j	5.2.1	homomorphism of singular homology groups defined by composition with inclusion of a face

δ	4.4.3	coboundary in a cochain complex or connecting homomorphism in a long exact cohomology sequence. In a cochain complex obtained by dualizing a chain complex, it is the Hom-dual of ∂, up to sign
δ^j	5.2	inclusion of a face into the standard n-simplex
Δ^*	8.1.6	category which can be used to define simplicial sets
Δ^n	5.2	the standard n-simplex
$\Delta[n]$	8.1.3	simplicial set whose realization is the standard n-simplex
ϵ		coaugmentation of a coalgebra or the "coaugmentation" map in a cotriple
η		augmentation of an algebra or the "augmentation" map in a triple
η	14.6.3	Hopf map from S^3 to S^2 or its suspension
η_n	14.6.3	map from S^{n+1} to S^n obtained by the suspending Hopf map
μ		multiplication of an H-space or Hopf algebra or the action of an H-space. Also used for the "multiplication" in a triple.
ι_n		generator of $H_n(S^n)$
θ_X	14.3.1	natural transformation used in constructing Steenrod operations
π'		projection onto the first factor of a product
π''		projection onto the second factor of a product
$\pi_n(-)$	3.1.3	homotopy group
$\pi_n^S(-)$	3.1.3	stable homotopy group
ψ		comultiplication of an co-H-space or Hopf algebra. Also used for the "comulitiplication" in a cotriple
σ_j	8.1.3	map from the standard $(n+1)$-simplex to the standard n-simplex which collapses a face
$\Omega(-)$	7.1.4	loop space
$\Omega'(-)$	7.9.1	Moore loop space
$\tau(-)$	11.9.4	transgression homomorphism
τ_k	14.5.1	generator of the dual of the Steenrod algebra
ξ_k	14.5.1	generator of the dual of the Steenrod algebra
χ		canonical conjugation of a Hopf algebra
$2^{(-)}$		set of subsets of a set
$\underline{C}(-,-)$	1.1.1	set of morphisms in a category. Equivalent to $\mathrm{Hom}_{\underline{C}}(-,-)$
$(-)^c$		complement of a set
$\overline{(-)}$		closure of a set
$\overset{\circ}{(-)}$		interior of a set. Equivalent to $\mathrm{Int}\,(-)$
$\partial(-)$		boundary of a set

Y^X	2.3.8	space of continuous functions or of continuous basepoint preserving functions if the spaces are pointed. Equivalent respectively to either $\mathrm{Map}\,(X,Y)$ or $\mathrm{Map}_*(X,Y)$. Can also mean function space of simplicial sets
$\cong$	1.1.2	isomorphism in a category, including homeomorphism between topological spaces
$\approx$	3.1.2	homotopy equivalence between pointed spaces
$\simeq$	3.1.1	homotopic for maps of spaces or chain complexes or simplicial sets
$\simeq\big(\mathrm{rel}\,(-)\big)$	3.1.1	homotopic relative to a subset
$\vee$	3.1.1	one-point union of pointed spaces
$\wedge$	3.1.1	reduced smash product of pointed spaces
$\coprod$	1.2	coproduct of objects in a category
$\prod$	1.2	product of objects in a category
$\perp$	1.2	map induced using the universal property of a coproduct
$\top$	1.2	map induced using the universal property of a product
$\otimes$	4.3.5	tensor product
$\square$	10.3.1	cotensor product
$(-)_\#$	3.1.1	map on homotopy induced by left composition
$\cup$		union
$\cup$	5.6.1	cup product or relative cup product
$\cap$		intersection
$\cap$	5.7.2	cap product or relative cap product
$X\cup_f Y$	1.7	attaching construction. The space obtained by gluing X to Y via f.
$\dashv$	4.5	adjoint functors
$\lvert-\rvert$		degree of an element in a graded set
$\lvert-\rvert$	8.1.3	realization of a simplicial set
$\alpha*\beta$	3.1.4	product of paths
$(-)*(-)$	7.7.1	reduced join of pointed spaces
$*$		basepoint of a pointed space or the map sending every element to the basepoint. Also used for the basepoint of a simplicial set
$G*_A H$	3.3.1	amalgamated free product
$[\alpha]\cdot[\beta]$	7.2.6	action of the fundamental groupoid
$(-)^\#$	3.1.1	map on homotopy induced by right composition
$(-)_*$	4.1.2	map induced on homology
$(-)^*$		map induced on cohomology
$(-)^*$		dual space
$(-)^\bullet$	12.2	cosimplicial object obtained from a triple
$(\hat{-})$		completion with respect to a filtration
$(-)_{(p)}$	12.2.2	localization of a space or module
$(-)^{(n)}$	2.7.1	n-skeleton of a CW-complex or simplicial set
$(-)^{(n)}$		n-fold reduced smash product of a pointed space

$\gamma_!$	7.1.3	
$f^!$	9.1.2	pullback of a bundle
$E^r \Rightarrow \mathcal{F}_{(-)}$	14.4.4	spectral sequence abuts to filtered object
$\langle -, - \rangle$	4.4.3	Kroncker product
$\langle -, - \rangle$	7.8.1	Samelson product
$[-, -]$	7.8.1	Whitehead product. Also used for the operation in a Lie algebra
$[-, -]$	3.1.3	pointed homotopy classes of pointed maps rel basepoint
$\{-, -\}$	3.1.3	stable homotopy classes of maps
$\{\gamma, \beta, \alpha\}$	14.7	Toda bracket
X/G	3.2.3	space of cosets
X/A	7.1.6	cofibre of an inclusion
$\underline{n}$		degree n self-map of a sphere
$\underline{R}$	4.4.3	chain complex concentrated in degree 0
$R(V)$		polynomial algebra on an R-module
$R[S]$		polynomial algebra on the R-module generated by a set
$R[G]$		group ring of the group G over the ring R
$R_\infty(-)$	12.2	Bousfield-Kan contruction of the ring R on a pointed simplical set

Part 1

PREREQUISITES

CHAPTER 1

Prerequisites from Category Theory

1 Definitions

Definition 1.1.1 A *category* $\underline{C}$ consists of:

E1) a class Obj $\underline{C}$ known as the *objects* of $\underline{C}$ (which need not form a set);

E2) for each pair X, Y of objects of Obj $\underline{C}$ a set, denoted $\underline{C}(X, Y)$ or $\mathrm{Hom}_{\underline{C}}(X, Y)$, called the *morphisms* in the category $\underline{C}$ from X to Y;

E3) for each triple X, Y, Z of objects of Obj $\underline{C}$, a set function $\circ : \underline{C}(X, Y) \times \underline{C}(Y, Z) \to \underline{C}(X, Z)$ called *composition* (where $\circ(f, g)$ is usually written $g \circ f$ or simply gf);

E4) for each object X of Obj $\underline{C}$, an element $1_X \in \underline{C}(X, X)$ called the *identity morphism* of X;

such that:

A1) $h \circ (g \circ f) = (h \circ g) \circ f$ for all $f \in \underline{C}(W, X)$, $g \in \underline{C}(X, Y)$, and $h \in \underline{C}(Y, Z)$;

A2) $1_Y \circ f = f$ and $f \circ 1_X = f$ for all $f \in \underline{C}(X, Y)$.

When the category is understood we often write $f : X \to Y$ to indicate that f is a morphism from X to Y. We will sometimes use the "Milnor-Moore convention" under which the name of an object may be used to denote the identity morphism on that object when it is clear from the context that a morphism is required. If there exist morphisms $f : X \to Y$ and $g : Y \to X$ such that $gf = 1_X$ and $fg = 1_Y$ then X and Y are called *isomorphic* and f and g are called *isomorphisms*. We write $X \cong Y$ for "X is isomorphic to Y".

Definition 1.1.2 Let $\underline{C}$ and $\underline{D}$ be categories. A (*covariant*) *functor* $F : \underline{C} \to \underline{D}$ consists of:

E1) an object $F(X)$ of Obj $\underline{D}$ for each object X of Obj $\underline{C}$;

E2) a morphism $F(g) \in \underline{D}(F(X), F(Y))$ for each morphism $g \in \underline{C}(X, Y)$;

such that:

A1) $F(h \circ g) = F(h) \circ F(g)$ for all $g \in \underline{C}(X, Y)$, $h \in \underline{C}(Y, Z)$;

A2) $F(1_X) = 1_{F(X)}$ for all objects X of Obj $\underline{C}$.

Similarly, a *contravariant functor* is defined by reversing the objects $F(X)$ and $F(Y)$ in E2 and reversing the order of composition in A1.

Definition 1.1.3 Let $\underline{C}$ and $\underline{D}$ be categories and let F, G be functors from $\underline{C}$ to $\underline{D}$. A *natural transformation* $\eta : F \to G$ consists of a morphism $\eta_X \in \underline{D}(F(X), G(X))$ for each object X of Obj $\underline{X}$ such that

3

$$
\begin{array}{ccc}
F(X) & \xrightarrow{\;\eta_X\;} & G(X) \\
\Big\downarrow{\scriptstyle F(f)} & & \Big\downarrow{\scriptstyle G(f)} \\
F(Y) & \xrightarrow{\;\eta_Y\;} & G(Y)
\end{array}
$$

commutes for all $f \in \underline{C}(X, Y)$. A natural transformation η such that η_X is an isomorphism for every object X is called a *natural equivalence*.

While strictly speaking one should say "the category of groups and group homomorphisms" one often hears phrases like "the category of groups" when it is clear what the intended morphisms are.

Example 1.1.4 Let S be a partially ordered set. From S we can construct a category $\underline{C_S}$ by setting $\mathrm{Obj}\,\underline{C_S} = S$ and

$$
\underline{C_S}(x, y) = \begin{cases} \text{set with one element} & \text{if } x \le y; \\ \emptyset & \text{otherwise} \end{cases}
$$

for $x, y \in S$. Since the relevant sets are either empty or singletons there is no choice as to the definition of $\circ$ or the element $1_x \in \underline{C_S}(x, x)$, and the axioms for partially ordered sets guarantee that the axioms for a category are satisfied. $\qquad\square$

A category (such as that in the example) in which the objects form a set is called a *small category*.

2 Universal Constructions

If $\{X_j\}_{j \in J}$ is a set of objects in $\underline{C}$ we say that an object X of $\mathrm{Obj}\,\underline{C}$ together with a set of morphisms $\{\pi_j : X \to X_j\}$ forms the *product* of $\{X_j\}_{j \in J}$, written $X = \prod_{j \in J} X_j$, if given any object Y of $\mathrm{Obj}\,\underline{C}$ and morphisms $\{p_j : Y \to X_j\}$ there exists a unique morphism $\phi : Y \to X$ such that $\pi_j \phi = p_j$ for all j. This is an example of an object defined by a *universal property* or *universal construction*. For arbitrary objects in a category there need not be any object satisfying the definition of product, but it is easy to see that an object defined by a universal property is unique in the sense that any two objects satisfying the condition are isomorphic.

Another important universal construction is the "pullback". Let $f : A \to X$ and $g : B \to X$ be morphisms in some category $\underline{C}$. An object P of $\underline{C}$ together with morphisms $p : P \to A$ and $q : P \to B$ is called the *pullback* of f and g if $fp = gq$ and if given any object Q together with morphisms $\alpha : Q \to A$, $\beta : Q \to B$ such that $f\alpha = g\beta$, there exists a unique morphism $\phi : Q \to P$ such that $p\phi = \alpha$ and $q\phi = \beta$.

The dual concepts, obtained by reversing all of the arrows, are called "coproduct" and "pushout" respectively. It is customary to call these universal constructions as well, although it would be consistent with the rest of the terminology to refer to them as co-universal constructions.

Notational Remark The coproduct symbol is "$\coprod$". Given $f : A \to Y$ and $g : B \to Y$ the universal property of coproduct yields an induced map $A \coprod B \to Y$ which we denote by $f \perp g$. It is the composite $A \coprod B \xrightarrow{\;f \coprod g\;} Y \coprod Y \xrightarrow{\;\nabla\;} Y$ where

$\nabla = 1_Y \perp 1_Y$ is the "fold" map. The dual concept could be written $f \top g : Y \to A \times B$ but we will usually write $(f, g) : Y \to A \times B$ instead.

The above concepts can be generalized as follows.

A *diagram* in a category $\underline{X}$ consists of a functor $D : \underline{C_S} \to \underline{X}$ where $\underline{C_S}$ is the category coming from some partially ordered set S. An object L of Obj $\underline{X}$ together with morphisms $\{\lambda_s : L \to D(s)\}_{s \in S}$ is called a *limit* of the diagram D if $\lambda_t = D(s \prec t) \circ \lambda_s$ for every morphism $s \prec t$ in $\underline{C_S}$, and if given any object Y of Obj $\underline{X}$ together with morphisms $\{y_s : Y \to D(s)\}_{s \in S}$ satisfying $y_t = D(s \prec t) \circ y_s$ for every morphism $s \prec t$ in $\underline{C_S}$, there exists unique $\phi : Y \to L$ such that $\lambda_s \phi = y_s$ for all $s \in S$.

Dually a *colimit* of the diagram D is an object L of Obj $\underline{X}$ together with morphisms $\{\lambda_s : D(s) \to L\}_{s \in S}$ such that $\lambda_s = \lambda_t \circ D(s \prec t)$ for every morphism $s \prec t$ in $\underline{C_S}$, and given any object Y of Obj $\underline{X}$ together with morphisms $\{y_s : D(s) \to Y\}_{s \in S}$ satisfying $y_s = y_t \circ D(s \prec t)$ for every morphism $s \prec t$ in $\underline{C_S}$, there exists unique $\phi : L \to Y$ such that $\phi \lambda_s = y_s$ for all $s \in S$.

Since it is unique up to isomorphism we usually refer to "the" limit or colimit of a diagram rather than "a" limit or colimit. In many cases when it is clear what the morphisms λ_s are, we drop them and refer to L as being the limit or colimit.

A pullback is the limit of a diagram coming from the partially ordered set $\{a, b, c\}$ with $a \prec c$ and $b \prec c$. Aside from the examples given above, another important special case of a limit occurs when the set S is the negative integers. In this case we say that L is the *inverse limit* of the objects X^n where $X^n = D(-n)$, written $L = \varprojlim_n X^n$. Dually, if S is the positive integers the colimit L is called the *direct limit* of the objects $X_n = D(n)$ and written $L = \varinjlim_n X_n$.

Exercise 1.2.1 Express the kernel of a group homomorphism as a pullback.

3 Basic Notions from Categorical Theory

Let $\underline{C}$ be a category. A morphism $f : X \to Y$ in $\underline{C}$ which has the property that for any $\alpha, \beta : W \to X$, $f\alpha = f\beta$ only when $\alpha = \beta$ is called a *monomorphism* in $\underline{C}$. A morphism $f : X \to Y$ in $\underline{C}$ which has the property that for any $\alpha, \beta : Y \to \overline{Z}$, $\alpha f = \beta f$ only when $\alpha = \beta$ is called an *epimorphism*. An object I of $\underline{C}$ is called an *initial object* if there is a unique morphism $I \to X$ for each object X of $\underline{C}$. An object T of $\underline{C}$ is called a *terminal object* if there is a unique morphism $X \to T$ for each object X of $\underline{C}$. An object of $\underline{C}$ which is both an initial object and a terminal object is called a *zero object*. It is clear that up to isomorphism, a category can have at most one zero object. If $\underline{C}$ has a zero object 0, then for any pair of objects X, Y of Obj $\underline{C}$ there is a unique composition $X \to 0 \to Y$, known as the *zero morphism* from X to Y and written $0 : X \to Y$.

Definition 1.3.1 A category $\underline{C}$ is called an *additive category* if:
1) it has a zero object;
2) the product exists for every pair of objects of Obj $\underline{C}$;
3) for every pair of objects X, Y of Obj $\underline{C}$, $\underline{C}(X, Y)$ has an abelian group structure under which for all objects W, X, Y of Obj $\underline{C}$ composition $\underline{C}(W, X) \times \underline{C}(X, Y) \to \underline{C}(W, Y)$ is bilinear.

Suppose that $\underline{C}$ has a zero object. A map $j : A \to X$ is called the *kernel* of a map $f : X \to Y$ if $fj = 0$ and given any $j' : B \to X$ such that $fj' = 0$ there exists unique $\alpha : B \to A$ such that $j\alpha = j'$. (That is, j is universal with respect to the property $fj = 0$.) A map $k : Y \to C$ is called the *cokernel* of a map $f : X \to Y$ if $kf = 0$ and given any $k' : Y \to D$ such that $k'f = 0$ there exists unique $\beta : C \to D$ such that $\beta k = k'$.

Definition 1.3.2 A category $\underline{C}$ is called an *abelian category* if:

1) it is additive;
2) every morphism has a kernel and a cokernel;
3) every monomorphism is the kernel of its cokernel;
4) every epimorphism is the cokernel of its kernel;
5) every morphism can be factored as the composition of an epimorphism and a monomorphism.

Exercise 1.3.3 Give an example of a category which has both an initial object and a terminal object, but no zero object.

Exercise 1.3.4 Show that the inclusion $\mathbb{Z} \hookrightarrow \mathbb{Q}$, although not a bijection, is both a monomorphism and an epimorphism in the category of rings.

Prerequisites from Point Set Topology

1 Definitions and Elementary Constructions

Definition 2.1.1 A *topological space* $(X, \mathcal{T})$ consists of a set X together with a subset $\mathcal{T} \subset 2^X$ (called the *open* sets) having the properties:

1) $X \in \mathcal{T}$ and $\emptyset \in \mathcal{T}$;
2) $U_j \in \mathcal{T}$ for all $j \in J$ implies $\bigcup_{j \in J} U_j \in \mathcal{T}$;
3) $U, V \in \mathcal{T}$ implies $U \cap V \in \mathcal{T}$.

In the notation, $\mathcal{T}$ is usually omitted and we refer to "the topological space X". The topology in which every subset of X is open is called the *discrete topology on X*. An open set containing the point $x \in X$ is sometimes called an *open neighbourhood* of x. $(X, \mathcal{T})$ is called *metrizable* if there exists a metric d on X such that $\mathcal{T}$ equals the set of open subsets in (X, d).

If $\mathcal{T}$ and $\mathcal{T}'$ are topologies on X such that $\mathcal{T} \subset \mathcal{T}'$ (as subsets of 2^X) then we say that $\mathcal{T}$ is *weaker* than $\mathcal{T}'$ (or $\mathcal{T}'$ is *stronger* than $\mathcal{T}$). A function $f : X \to Y$ between topological spaces is called *continuous* if $f^{-1}(U)$ is open in X for all open subsets U of Y. If there exist continuous functions $f : X \to Y$ and $g : Y \to X$ such that $gf = 1_X$ and $fg = 1_Y$ then X and Y are called *homeomorphic* and f and g are called *homeomorphisms*. We write $X \cong Y$ for X homeomorphic to Y. We write $\mathcal{T}op$ for the category of topological spaces and continuous functions.

Let A be a subset of X. A is called *closed* if A^c is open, where A^c denotes the complement, $X - A$, of A in X. The *closure* of A, written $\overline{A}$, is the smallest closed subset of X which contains A. Thus $\overline{A} = \bigcap_{\{B \mid B \supset A, B \text{ closed}\}} B$. The *interior* of A, written $\mathring{A}$ or $\operatorname{Int} A$, is the largest open subset of X contained in A. Thus $\mathring{A} = \bigcup_{\{B \mid B \subset A, B \text{ open}\}} B = \left(\overline{A^c}\right)^c$. The *boundary* of A, written ∂A, is defined by $\partial A = \overline{A} \cap \overline{A^c}$. A is called *dense* if $\overline{A} = X$.

Let X be a set and suppose given, for each $j \in J$, a set function $f_j : X \to Y_j$ from X to some topological space Y_j. The *topology on X induced from the functions f_j* is the weakest topology on X such that f_j is continuous for all j. Explicitly, this topology is the intersection (within 2^X) of all topologies on X under which f_j is continuous for all $j \in J$. Dually, suppose given a set X and a collection $g_j : W_j \to X$ of set functions from topological spaces W_j to X. The *topology on X induced from the functions g_j* is defined as the strongest topology on X such that g_j is continuous for all j. Explicitly, a set $U \subset X$ is open in this topology if and only if $g_j^{-1}(U)$ is open in W_j for all j.

If X is a topological space and $A \subset X$ then the *subspace topology* on A is the topology induced by the single function $\{i\}$ where $i : A \hookrightarrow X$ is the inclusion map. Dually, suppose X is a topological space, let $\sim$ be an equivalence relation on X, and let $Y = X/\sim$ denote the set of equivalence classes. Then the *quotient topology* on Y is the topology induced by the single function $\{q\}$ where $q : X \twoheadrightarrow Y$ is the quotient map.

Given a collection of spaces $\{X_j\}_{j \in J}$, the *product topology* on $\prod_{j \in J} X_j$ is the topology induced by the projection maps $\pi_i : \prod_{j \in J} X_j \to X_i$. With this topology, $\prod_{j \in J} X_j$ is the product of the spaces X_j in the category of topological spaces.

A collection $\mathcal{U}$ of open subsets of X is called a *base* for the topology on X if it is closed under intersections ($U, U' \in \mathcal{U} \Rightarrow U \cap U' \in \mathcal{U}$) and every open set of X is a union of sets in $\mathcal{U}$. A collection $\mathcal{U}$ of open subsets of X is called a *subbase* for the topology on X if every open set V of X can be written as a union $V = \bigcup_{i \in I} W_i$ where each W_i is the intersection of finitely many sets in $\mathcal{U}$. Given any subset $\mathcal{S} \subset 2^X$, the topology generated by $\mathcal{S}$ is the weakest topology containing $\mathcal{S}$; the set $\mathcal{S}$ will be a subbase for this topology.

2 Separation

Let X be a topological space. X is called T_0 if for all $x, y \in X$ such that $x \neq y$ either there exists open U such that $x \in U$, $y \notin U$ or else there exists open U such that $y \in U$, $x \notin U$. X is called T_1 if for all $x, y \in X$ such that $x \neq y$ there exists open U such that $x \in U$, $y \notin U$ and there exists open V such that $y \in V$, $x \notin V$. X is called T_2 or *Hausdorff* if for all $x, y \in X$ such that $x \neq y$ there exist open U, V such that $x \in U$, $y \in V$ and $U \cap V = \emptyset$. X is called T_3 or *regular* if X is T_1 and for all $x \in X$ and closed $F \subset X$ with $x \notin F$ there exist open U, V such that $x \in U$, $F \subset V$ and $U \cap V = \emptyset$. X is called $T_{3\frac{1}{2}}$ or *completely regular* if X is T_1 and for all $x \in X$ and closed $F \subset X$ with $x \notin F$ there exists continuous $f : X \to [0,1]$ such that $f(x) = 0$ and $f(F) = 1$. X is called T_4 or *normal* if X is T_1 and for all closed $F, G \subset X$ with $F \cap G = \emptyset$ there exist open U, V such that $F \subset U$, $G \subset V$, and $U \cap V = \emptyset$.

It is easy to see that

$$\text{metrizable} \Rightarrow \text{normal} \Rightarrow \text{completely regular} \Rightarrow \text{regular} \Rightarrow \text{Hausdorff} \Rightarrow T_1 \Rightarrow T_0.$$

There are counterexamples to each of the reverse implications, although some (particularly the example of a space which is regular but not completely regular) are rather messy. Some useful equivalent formulations are given by the following proposition.

Proposition 2.2.1
   a) *X is T_1 if and only if the points of X are closed subsets of X.*
   b) *X is Hausdorff if and only if $\Delta(X)$ is a closed subset of $X \times X$, where $\Delta(X) = \{(x, x) \mid x \in X\}$ denotes the diagonal of $X \times X$.*
   c) *X is regular if and only if X is Hausdorff and given $x \in U$, where U is open, there exists open V such that $x \in V \subset \overline{V} \subset U$.*
   d) *X is normal if and only if X is Hausdorff and given $F \subset U$, where F is closed and U is open, there exists open V such that $F \subset V \subset \overline{V} \subset U$.* $\qquad\square$

The following lemma shows why there is no need for a concept of "completely normal" analogous to "completely regular".

Theorem 2.2.2 (Urysohn's Lemma) *Let X be normal and let $F, G \subset X$ be closed subsets such that $F \cap G = \emptyset$. Then exists continuous $f : X \to [0,1]$ such that $f(F) = 0$ and $f(G) = 1$.* $\qquad\square$

3 Compactness

Definition 2.3.1 A topological space X is called *compact* if given any collection of open subsets $\{U_j\}_{j \in J}$ such that $\bigcup_{j \in J} U_j = X$, there exists a finite subcollection $U_{j_1}, U_{j_2}, \ldots, U_{j_n}$ such that $U_{j_1} \cup U_{j_2} \cup \ldots \cup U_{j_n} = X$.

Some consequences derived from the definition are given in the following proposition.

Proposition 2.3.2

a) $f : X \to Y$ *continuous, X compact* $\Rightarrow f(X)$ *compact.*

b) X *Hausdorff, A compact subset of X* $\Rightarrow A$ *closed in X.*

c) $f : X \to Y$ *continuous bijection, X compact, Y Hausdorff* $\Rightarrow f$ *is a homeomorphism.*

d) X *compact Hausdorff* $\Rightarrow X$ *normal.* $\qquad\qquad\square$

Another important property of compact sets is given by the following theorem whose proof makes use of the axiom of choice in the case where the index set is infinite.

Theorem 2.3.3 (Tychonoff's Theorem) *If X_j is compact for all $j \in J$ then $\prod_{j \in J} X_j$ is compact.* $\qquad\qquad\square$

Definition 2.3.4 A topological space X is called *locally compact* if every point has a neighbourhood whose closure is compact.

Theorem 2.3.5 (One-point Compactification) *Let X be a locally compact Hausdorff space. Then there exists a compact Hausdorff space X_∞ and an inclusion $i : X \hookrightarrow X_\infty$ such that $X_\infty - X$ is a single point.* $\qquad\qquad\square$

In metric spaces there are other characterizations and consequences of compactness. Let (X, d) be a metric space. Recall the following definitions. Given an open cover $\mathcal{U} = \{U_j\}_{j \in J}$ of X, $\epsilon > 0$ is called a *Lebesgue number* for $\mathcal{U}$ if every subset of X having diameter less than ϵ is contained in U_j for some $j \in J$. X is called *sequentially compact* if every sequence in X has a convergent subsequence. X is called *complete* if every Cauchy sequence in X converges. X is called *totally bounded* if for every $\epsilon > 0$ there exists a finite subset S_ϵ of X such that the balls of radius ϵ centred at the points of S_ϵ cover X.

Theorem 2.3.6 (Cantor Intersection Theorem) *Let (F_n) be a monotonically decreasing (nested) sequence of nonempty closed subsets of a complete metric space such that $\lim_{n \to \infty} \operatorname{diam}(F_n) = 0$. Then $\bigcap_n F_n$ consists of exactly one point.*

$$\square$$

Theorem 2.3.7 *Let (X, d) be a metric space. Then*

a) X *is compact* $\Longleftrightarrow X$ *is sequentially compact* $\Longleftrightarrow X$ *is complete and totally bounded.*

b) X *compact* $\Rightarrow$ *every open cover of X has a Lebesgue number.* $\qquad\square$

For topological spaces X and Y, the set of all continuous functions from X to Y is denoted Y^X, or $\operatorname{Map}(X, Y)$. We make it into a topological space by means of the "compact-open" topology, described as follows. Given subsets $A \subset X$ and $B \subset Y$, let $\langle A, B \rangle = \{f \in Y^X \mid f(A) \subset B\}$. The *compact-open* topology on Y^X is the topology generated by the subsets $\{\langle K, U\rangle \mid K \subset X \text{ is compact and } U \subset Y \text{ is}$ open$\}$.

Theorem 2.3.8 (Exponential Law) *If X is locally compact Hausdorff then* $(Y^X)^Z \cong Y^{Z \times X}$ *for all Y and Z.* $\qquad\square$

4 Connectedness

A continuous function from $[0,1]$ to X is called a *path* in X.

Definition 2.4.1 Let X be a topological space.

a) X is called *connected* if there do not exist open subsets U, V of X such that $U \cup V = X$ and $U \cap V = \emptyset$.

b) X is called *path connected* if for all $x, y \in X$ there exists a path ω such that $\omega(0) = x$ and $\omega(1) = y$.

c) A maximal connected subspace is called a *connected component*. A maximal path connected subspace is called a *connected path component* or simply a *path component*.

d) X is called *locally connected* if the connected components of every open subset of X are open. It is called *locally path connected* if the path components of every open subset of X are open.

Some elementary properties of these concepts are given in the following proposition.

Proposition 2.4.2 *Let X be a topological space.*

a) *X path connected $\Rightarrow$ X connected.*

b) *X locally path connected $\Rightarrow$ X locally connected.*

c) *X connected and locally path connected $\Rightarrow$ X path connected.*

d) *$f : X \to Y$ continuous, X connected $\Rightarrow$ $f(X)$ connected.*

e) *$f : X \to Y$ continuous, X path connected $\Rightarrow$ $f(X)$ path connected.*

f) *Each point of X is in a unique connected component of X, so X is the disjoint union of its connected components.*

g) *Each point of X is in a unique path component of X, so X is the disjoint union of its path components.* $\qquad\square$

The standard example of space which is connected but not path connected is the subset $\{(0, y) \mid y \in \mathbb{R}\} \cup \{(x, \sin(1/x)) \mid 0 < x \le 1\}$ of $\mathbb{R}^2$. Note that X connected does not imply X locally connected and similarly X path connected does not imply X locally path connected. An example of a space which is path connected but not locally path connected is the comb space

$$\{(x, 0) \mid 0 \le x \le 1\} \cup \{(0, y) \mid 0 \le y \le 1\} \cup \{(1/n, y) \mid 0 \le y \le 1\}_{n=1}^{\infty}.$$

5 Coverings and Paracompactness

Let X be a topological space. A cover $\{U_i\}_{i \in I}$ of X is called a *refinement* of the cover $\{V_j\}_{j \in J}$ if for all $i \in I$ there exists $j(i) \in J$ such that $U_i \subset V_{j(i)}$. A collection $\{V_j\}_{j \in J}$ of subsets of X is called *locally finite* if every point $x \in X$ has an open neighbourhood U_x such that $U_x \cap V_j = \emptyset$ for all but finitely many j.

Locally finite collections of subsets have the following convenient property.

Lemma 2.5.1 *Let X be a topological space and let $\{W_j\}_{j \in J}$ be a locally finite set of arbitrary (not necessarily open) subsets of X. Then $\bigcup_{j \in J} \overline{W_j} = \overline{\bigcup_{j \in J} W_j}$.* $\qquad\square$

Definition 2.5.2 A topological space X is called *paracompact* if every open cover of X has a locally finite open refinement.

As can be seen from the following theorem, most common spaces are paracompact.

Theorem 2.5.3 *Let X be a topological space.*

a) *X metrizable $\Rightarrow$ X paracompact.*
b) *X regular and $X = \bigcup_{n=1}^{\infty} K_n$ where K_n is compact $\Rightarrow$ X paracompact.*
c) *X paracompact Hausdorff $\Rightarrow$ X normal.*

See Munkres [Mu, pp. 255–259] for parts (a) and (c) and Dugundji [D, p. 174] for part (b). $\qquad\qquad\square$

Looking ahead, all CW-complexes are paracompact according to Theorem 2.7.2d.

The main application of paracompactness is the construction of "partitions of unity" and "numerable covers".

Definition 2.5.4 Let X be a topological space and let $\{U_j\}_{j\in J}$ be an open cover of X. A *partition of unity* relative to the cover $\{U_j\}_{j\in J}$ consists of a set of functions $f_j : X \to [0,1]$ such that:

1) $\overline{f_j^{-1}\big((0,1]\big)} \subset U_j$ for all $j \in J$;
2) $\{\overline{f_j^{-1}\big((0,1]\big)}\}_{j\in J}$ is locally finite;
3) $\sum_{j\in J} f_j(x) = 1$ for all $x \in X$.

Note that condition (2) implies that for each $x \in X$, $f_j(x) = 0$ for all but finitely many j, and so the sum in condition (3) makes sense.

Definition 2.5.5 A *numerable cover* of a topological space X consists of a locally finite open cover $\{V_j\}_{j\in J}$ such that for all j there exists a function $f_j : X \to [0,1]$ with $V_j = f^{-1}\big((0,1]\big)$.

Clearly if $\{f_j\}_{j\in J}$ forms a partition of unity relative to some cover $\{U_j\}_{j\in J}$ then $f^{-1}\big((0,1]\big)$ forms a numerable cover which is a refinement of $\{U_j\}_{j\in J}$.

Theorem 2.5.6 *Let X be Hausdorff. Then X is paracompact $\Longleftrightarrow$ for every open cover $\mathcal{U}$ of X there exists a partition of unity relative to $\mathcal{U}$ $\Longleftrightarrow$ every open cover of X has a numerable refinement.*

See Munkres [Mu, pp. 224–225] for hints leading to the proof. $\qquad\square$

Partitions of unity are useful in proving global properties from local ones. That is, they often can be applied to show that if every point of a space X has an open neighbourhood which has some property P, then X has property P.

6 Coherent and Compactly Generated Topologies

Let X be a set and let $\{A_j\}_{j\in J}$ be a collection of subsets of X such that $X = \bigcup_{j\in J} A_j$. Suppose that each A_j has been given a topology such that A_j has the subspace topology induced from $A_{j'}$ whenever $A_j \subset A_{j'}$. Then the topology on X induced from the inclusion maps $A_j \hookrightarrow X$ (in the sense defined in Section 2.1) is called the *coherent topology on X from the subspaces A_j.* Explicitly, a subset $W \subset X$ is closed in this topology if and only if $W \cap A_j$ is closed in A_j for all $j \in J$.

Remark 1 The definition makes sense without the conditions that $X = \bigcup_{j \in J} A_j$ and that A_j has the subspace topology induced from $A_{j'}$ whenever $A_j \subset A_{j'}$, but I know of no examples where this construction is useful without those conditions.

Remark 2 This topology is sometimes called the "weak" topology on X coming from the subspaces A_j, however this use of the term "weak" is not related to the term "weaker" in the phrase "$\mathcal{T}$ is weaker than $\mathcal{T}'$" as defined in Section 2.1.

In many cases, one begins with a topological space X together with a covering $\mathcal{A}$ of X and uses the sets in $\mathcal{A}$ to define a new topology on X. An important example of this is where one takes $\mathcal{A}$ to be the set of all compact subspaces of a Hausdorff space X.

Definition 2.6.1 Let X be a Hausdorff topological space. Then the *compactly generated* topology on X is the coherent topology on X coming from the set of all compact subsets of X.

Given a Hausdorff space X, we will denote the topological space consisting of the set X with its compactly generated topology by $X_\mathcal{K}$. If $X = X_\mathcal{K}$ then we say that X is compactly generated.

Proposition 2.6.2 *Let X be a Hausdorff space. Then*

a) $1_X : X_\mathcal{K} \to X$ *is continuous.*
b) $(X_\mathcal{K})_\mathcal{K} = X_\mathcal{K}$.

Proof The Hausdorff hypothesis plus Proposition 2.3.2b implies (a) and in fact shows that the restriction of 1_X to any compact subset is a homeomorphism to its image. Therefore the set of compact subspaces of $X_\mathcal{K}$ is the same as that of X and so part (b) follows. $\qquad\square$

Proposition 2.6.3

a) X *locally compact Hausdorff* $\Rightarrow$ X *compactly generated.*
b) X *metrizable* $\Rightarrow$ X *compactly generated.* $\qquad\square$

Proposition 2.6.4 (Exponential Law) $\left(\left((Y^X)_\mathcal{K} \right)^Z \right)_\mathcal{K} \cong \left(Y^{(Z \times X)_\mathcal{K}} \right)_\mathcal{K}$ *for all compactly generated spaces X, Y, and Z.*

See Steenrod [St1]. $\qquad\square$

Exercise 2.6.5 Compare the topology on $\mathbb{R}^\infty$ which results from the metric $(x, y) = \sqrt{\sum_{k=1}^{\infty} (x_k - y_k)^2 / 2^k}$ with the coherent topology from the subspaces $\mathbb{R}^n$. In which direction is the identity map continuous?

7 CW-complexes

Let X, Y be topological spaces, let A be a subspace of X, and let $f : A \to Y$ be continuous. The "attaching construction" builds a new space, written $X \cup_f Y$, formed by joining X to Y by gluing points in A to their images as follows. $X \cup_f Y$ is defined as $(X \coprod Y)/\sim$ with the quotient topology, where $a \sim f(a)$ for all $a \in A$.

For $m \geq 1$, set $D^m = \{x \in \mathbb{R}^m \mid \|x\| \leq 1\}$, $S^{m-1} = \partial(D^m)$, and let D^0 be the space consisting of a single point. Sometimes instead of the spaces D^m it will be more convenient to use the homeomorphic spaces $[0,1]^m$ which have the advantage that they satisfy $[0,1]^k \times [0,1]^n = [0,1]^{k+n}$. If $g_j : \partial D^m \to X$ are continuous maps for $j \in J$ and $Y = \left(\coprod_{j \in J} D^m \right) \bigcup_{(\perp g_j)} X$ then we say that Y is obtained from X by attaching m-cells using the attaching maps $\{g_j\}$. We could define a finite CW-complex as anything which can be obtained by the following procedure. Begin with any finite set of points, $X^{(0)}$, with the discrete topology. Choose an integer n and inductively for $k = 1, \ldots, n$ choose maps $g_j : \partial D^k \to X^{(k-1)}$ for all j in some finite (possibly empty) set J_k. Then attach k-cells to $X^{(k-1)}$ using $\{g_j\}_{j \in J_k}$ and label the resulting space $X^{(k)}$. While this describes finite CW-complexes in an intuitive way, to deal with infinite CW-complexes it is more convenient to make the following definition.

Definition 2.7.1 A CW-*structure* on a Hausdorff space X consists of a set of maps $f_j : D^{m(j)} \to X$ for all j in some set J such that, setting $e_j = f_j \left(\text{Int} \left(D^{m(j)} \right) \right)$:

1) $X = \coprod_{j \in J} e_j$ (disjoint union as sets);
2)   (i) $f_j|_{\text{Int}\left(D^{m(j)} \right)} : \text{Int} \left(D^{m(j)} \right) \to e_j$ is a homeomorphism;

      (ii) for $m(j) > 0$, $f_j \left(\partial \left(D^{m(j)} \right) \right)$ is contained in the union of finitely many cells e_i, each satisfying $m(i) < m(j)$;
3) $A \subset X$ is closed in X if and only if $A \cap \overline{e_j}$ is closed in $\overline{e_j}$ for all $j \in J$.

The map f_j is called the *characteristic map* for the cell; e_j is called an *open cell* in X of dimension $m(j)$; $\overline{e_j}$ is called a *closed cell*. Condition (3) of the definition says that X has the coherent topology from its closed cells. Note that an open cell in X is not necessarily open as a subset of X.

A CW-*complex* consists of a space together with a CW-structure on that space, although occasionally we will be sloppy and use the term to mean a space for which a CW-structure exists.

Set $X^{(n)} = \bigcup_{k \leq n} \{$open k-cells of $X\}$, called the *n-skeleton* of X. If X and Y are CW-complexes, a continuous map $f : X \to Y$ is called *cellular* if $f \left(X^{(n)} \right) \subset Y^{(n)}$ for all n.

One can also define a "relative CW-complex" (X, A) in which $X = A \coprod_{j \in J} e_j$ and the cells are attached as above, where A rather than $\emptyset$ is the (-1)-skeleton.

The basic properties of CW-complexes are given by the following proposition.

Theorem 2.7.2

a) $\overline{e_j} = f_j \left(D^{m(j)} \right)$.
b) *K compact subset of X $\Rightarrow$ $K \cap e_j = \emptyset$ for all but finitely many j.*
c) *X CW $\Rightarrow$ X compactly generated.*
d) *X CW $\Rightarrow$ X paracompact and normal.*
e) *A connected CW-complex is path connected.* $\qquad\qquad\qquad$ $\square$

For countable CW-complexes, part (d) follows from Theorem 2.5.3b; the general case is a theorem of Miyazaki [Miy].

Let $X = \bigcup_{i \in I} e_i$ and $Y = \bigcup_{j \in J} e_j$ be CW-complexes. It is easy to see that maps $D^{m(i)+m(j)} \cong D^{m(i)} \times D^{m(j)} \xrightarrow{f_i \times f_j} X \times Y$ satisfy the first two conditions in the definition of a CW structure. Unless at least one of X, Y is finite, $X \times Y$ need not be compactly generated, so the third condition will fail. However the third condition holds for $(X \times Y)_\mathcal{K}$, which in fact satisfies the definition of the product of X and Y in the category of CW-complexes and cellular maps.

The next two theorems show why CW-complexes play a central role in homotopy theory. For most types of questions in homotopy theory they allow one to restrict attention to the case of CW-complexes and cellular maps.

Theorem 7.5.1 (Cellular Approximation Theorem) *Let $f : X \to Y$. Then there exists $g : X \to Y$ such that $g \simeq f$ and g is cellular.*

Theorem 7.5.8 (CW Approximation Theorem) *Given a topological space Y there exists a CW-complex X and a map $f : X \to Y$ such that $f_\# : \pi_n(X) \to \pi_n(Y)$ is an isomorphism for all n.*

Theorem 2.7.3 (Milnor) *If X is a CW-complex and K is compact, then X^K is a CW-complex.*

See Milnor [Mi5]. $\square$

Exercise 2.7.4 Is the subset $\bigcup_{n=1}^{\infty} \{(x, y) \mid x^2 - x/n + y^2 = 0\}$ of $\mathbb{R}^2$ a CW-complex?

CHAPTER 3

The Fundamental Group

1 Homotopy

Let $I = [0, 1]$.

Definition 3.1.1 Let X, Y be topological spaces and let A be a subspace of X. Let f, $g : X \to Y$ be continuous maps such that $f|_A = g|_A$. Then we say f is *homotopic* to g relative to A, written $f \simeq g$ (rel A), if there exists continuous $H : X \times I \to Y$ (called a *homotopy*) such that $H(x, 0) = f(x)$ for all $x \in X$, $H(x, 1) = g(x)$ for all $x \in X$, and $H(a, t) = f(a) = g(a)$ for all $a \in A$ and $t \in I$.

The following notations are in common use: $H : f \simeq g$ (rel A) means that H is a homotopy rel A from f to g; $H_t : X \to Y$ is the function defined by $H_t(x) = H(x, t)$.

A *topological pair* (X, A) consists of a topological space X together with a subspace A of X. A map $f : (X, A) \to (Y, B)$ of topological pairs consists of a continuous function $f : X \to Y$ such that $f(A) \subset B$. A homotopy between maps f, g of pairs consists of a continuous function $H : X \times I \to Y$ such that $H(x, 0) = f(x)$ for all $x \in X$, $H(x, 1) = g(x)$ for all $x \in X$, and $H(a, t) \in B$ for all $a \in A$ and $t \in I$. In the special case where A consists of a single point x_0 of X we say that (X, x_0) is a *pointed topological space*. For pointed topological spaces X and Y, the space of pointed maps from X to Y is denoted Y^X or $\mathrm{Map}_*(X, Y)$ and topologized as a subspace of $\mathrm{Map}\,(X, Y)$ with the compact-open topology. In cases where the choice of basepoint x_0 is not important we might omit it in the notation and refer to the pointed space X. The *wedge* of pointed spaces X, Y, written $X \vee Y$, is the space $\{(x, y) \in X \times Y \mid x = x_0 \text{ or } y = y_0\}$ with basepoint (x_0, y_0), where x_0 and y_0 are the basepoints of X and Y. The *smash product* of pointed spaces X, Y, written $X \wedge Y$, is defined as $(X \times Y)/(X \vee Y)$. The pointed version of exponential law is

Theorem 3.1.2 (Exponential Law) *Let X, Y, and Z be pointed spaces with X locally compact Hausdorff. Then $(Y^X)^Z \cong Y^{Z \wedge X}$.* $\qquad\square$

Let (X, x_0), (Y, y_0) be pointed spaces. If there exist pointed maps $f : (X, x_0) \to (Y, y_0)$ and $g : (Y, y_0) \to (X, x_0)$ such that $gf \simeq 1_X$ (rel $\{x_0\}$) and $fg \simeq 1_Y$ (rel $\{y_0\}$) then X and Y are called *homotopy equivalent* or said to have the same *homotopy type*, and f and g are called *homotopy equivalences*. We write $(X, x_0) \approx (Y, y_0)$ for "(X, x_0) is homotopy equivalent to (Y, y_0)". If $(X, x_0) \approx (\{x_0\}, x_0)$ then (X, x_0) is called *contractible*.

Proposition 3.1.3

a) *Let $f, f' : (X, A) \to (Y, B)$ and $g, g' : (Y, B) \to (Z, C)$ be maps with $f \simeq f'$ (rel A) and $g \simeq g'$ (rel B). Then $g \circ f \simeq g' \circ f'$ (rel A).*

b) *Let X, Y be topological spaces, let A be a subspace of X, and let $\alpha : A \to Y$ be continuous. Then $\simeq$ (rel A) is an equivalence relation on*

$$\{f : X \to Y \mid f \text{ is continuous and } f|_A = \alpha\}.$$

15

c) *If $X \subset \mathbb{R}^n$ is convex, then $(X, x_0) \approx (\{x_0\}, x_0)$ for any $x_0 \in X$.*   □

The set of equivalence classes of pointed maps from (X, x_0) to (Y, y_0) under $\simeq$ (rel $\{x_0\}$) is denoted $[(X, x_0), (Y, y_0)]$. If $f : (Y, y_0) \to (Z, z_0)$ is a map of pointed spaces then Proposition 3.1.3 implies that left composition with f induces a well defined map denoted $f_\# : [(X, x_0), (Y, y_0)] \to [(X, x_0), (Z, z_0)]$. Similarly if $g : (W, w_0) \to (X, x_0)$ then right composition with g induces a well defined map denoted $g^\# : [(X, x_0), (Y, y_0)] \to [(W, w_0), (Y, y_0)]$.

Let $\pi_n(Y, y_0)$ denote $[(S^n, \epsilon_1), (Y, y_0)]$, using $\epsilon_1 = (1, 0, \dots, 0)$ as the basepoint for S^n. For $n > 0$, $\pi_n(Y, y_0)$ has a natural group structure and is called the *nth homotopy group* of (Y, y_0). In the special case $n = 1$, $\pi_1(Y, y_0)$ is called the *fundamental group* of (Y, y_0). We shall now describe this group structure in the case $n = 1$, leaving the general case to Chapter 7.

In describing the group structure on $\pi_1(Y, y_0)$, in place of S^1 we will use the homeomorphic space $I/\sim$ (where $0 \sim 1$) with the point $0 \sim 1$ as basepoint. Thus an element of $\pi_1(Y, y_0)$ will be represented by a closed path in Y beginning and ending at y_0. Let $[f]$, $[g]$ belong to $\pi_1(Y, y_0)$. Intuitively the product of $[f]$ and $[g]$ in $\pi_1(Y, y_0)$ is represented by the path which does first f and then g, reparameterized to be a function on I. More precisely, we define the product of $[f]$ and $[g]$ to be the path represented by $f * g$ where for paths α, β with $\beta(0) = \alpha(1)$ we set

$$(\alpha * \beta)(t) = \begin{cases} \alpha(2t) & \text{if } 0 \leq t \leq 1/2; \\ \beta(2t - 1) & \text{if } 1/2 \leq t \leq 1. \end{cases}$$

Given a path f, define the path f^{-1} by $f^{-1}(t) = f(1 - t)$ and let c_{y_0} denote the constant path given by $c_{y_0}(t) = y_0$ for all $t \in I$. A map homotopic to the constant map is called *null homotopic*.

Proposition 3.1.4

a) $([f], [g]) \mapsto [f * g]$ *is a well defined associative operation on $\pi_1(Y, y_0)$.*

b) $[c_{y_0}][f] = [f][c_{y_0}] = [f]$ *for all $[f] \in \pi_1(Y, y_0)$.*

c) *For $[f] \in \pi_1(Y, y_0)$, $[f][f^{-1}] = [f^{-1}][f] = [c_{y_0}]$.*
   Therefore $\pi_1(Y, y_0)$ forms a group under the operation $*$.

d) *If $f : (Y, y_0) \to (Z, z_0)$ is a map of pointed spaces then $f_\# : \pi_1(Y, y_0) \to \pi_1(Z, z_0)$ is a group homomorphism. If $f \simeq g$ (rel $\{y_0\}$) then $f_\# = g_\#$.*

e) *Let α be a path in Y from y_0 to y_1. Then $[f] \mapsto [\alpha * f * \alpha^{-1}]$ is an isomorphism from $\pi_1(Y, y_0)$ to $\pi_1(Y, y_1)$.*   □

Note that none of the equalities in parts (a)–(c) holds for the paths themselves, but only after passing to homotopy classes.

If Y is path connected then according to part (e), $\pi_1(Y, y_0)$ is independent of the choice of y_0, (up to isomorphism,) and so we sometimes write simply $\pi_1(Y)$ in this case. If Y is path connected and $\pi_1(Y) = \{1\}$ (where $\{1\}$ denotes the trivial group) then Y is called *simply connected*. Part (d) implies that a homotopy equivalence induces an isomorphism of fundamental groups. In particular, Y contractible implies Y is simply connected, although the converse is far from true. As we shall see, fundamental groups need not be abelian.

There are two standard methods of computing homotopy groups: coming down from above by means of "covering spaces" and coming up from within by Van Kampen's Theorem. We shall explore these methods in the next two sections.

The *fundamental groupoid* of a space Y is the category whose objects are the points of Y, where a morphism from b to b' is homotopy class (rel $\{0, 1\}$) of paths from b to b', and composition is given by $[\beta] \circ [\alpha] = [\alpha * \beta]$.

Remark A *groupoid* is a category in which every morphism is an isomorphism.

2 Covering Spaces

Definition 3.2.1 A map $p : E \to X$ is called a *covering projection* if every point $x \in X$ has an open neighbourhood U_x such that $p^{-1}(U_x)$ is a disjoint union of open sets each of which is homeomorphic to U_x. E is called the *covering space* and X the *base space* of the covering projection.

If $p : E \to X$ is a covering projection, an open subset $U \subset X$ such that $p^{-1}(U)$ is a disjoint union of open sets each homeomorphic to U is called *evenly covered* and if $p^{-1}(U) = \coprod_{j \in J} S_j$ with $S_j \cong U$ then S_j is called a *sheet* over U. For $x \in X$, $p^{-1}(x)$ is called the *fibre* of p over x.

Proposition 3.2.2 *A covering space of a locally path connected space is locally path connected.* $\qquad\square$

The standard example of a covering projection is $\exp : \mathbb{R} \to S^1$ given by $\exp(t) = e^{2\pi i t}$ where S^1 is regarded as the unit circle in $\mathbb{C}$. This is a special case of a covering space coming from the action of a topological group on a space.

A *topological group* G is a topological space with a group structure in which the group operations $* : G \times G \to G$ and inverse $: G \to G$ are continuous, where $G \times G$ is given the product topology. Any group becomes a topological group when given the discrete topology (every set open). A *(left) action* of a topological group G on a topological space X consists of a continuous function $\phi : G \times X \to X$ (where $G \times X$ is given the product topology) such that (using $g \cdot x$ to denote $\phi(g, x)$) the following are satisfied:

   1) $g_1 \cdot (g_2 \cdot x) = (g_1 g_2) \cdot x$ for all $g_1,\, g_2 \in G$, $x \in X$;
   2) $e \cdot x = x$ for all $x \in X$, where e denotes the identity of G.

A space together with an action of G on that space is called a *G-space*. A *G-map* from a G-space X to a G-space Y is a map $f : X \to Y$ satisfying $f(g \cdot x) = g \cdot \big(f(x)\big)$ for all $g \in G$ and $x \in X$. Given an action of G on X, X/G denotes the topological space formed by the set of equivalence classes of X under $x \sim g \cdot x$, with the quotient topology.

Proposition 3.2.3 *Suppose a group G acts on a space X such that for all $x \in X$ there exists an open neighbourhood V_x such that $V_x \cap (g \cdot V_x) = \emptyset$ for all $g \neq e$ in G. Then the quotient map $p : X \to X/G$ is a covering projection.*

$\qquad\square$

The action $\mathbb{Z} \times \mathbb{R} \to \mathbb{R}$ given by $t \cdot x = t + x$ yields the covering projection $\mathbb{R} \to \mathbb{R}/\mathbb{Z} \cong S^1$ mentioned above. Another important example is the action $(\mathbb{Z}/2\mathbb{Z}) \times S^n \to S^n$ where the nontrivial element of $\mathbb{Z}/2\mathbb{Z} = \{1, -1\}$ acts by $(-1) \cdot x = -x$. This yields the covering projection $S^n \to \mathbb{R}P^n$ where $\mathbb{R}P^n = S^n/(\mathbb{Z}/2\mathbb{Z})$. The space $\mathbb{R}P^n$ is called *real projective n-space*.

A key feature of covering spaces is the ability to uniquely "lift" maps from X to E. Given $p : E \to X$, and $f : Y \to X$, a map $f' : Y \to E$ is called a *lift* of f if $pf' = f$.

Theorem 3.2.4 (Covering Homotopy Theorem) *Let $p : E \to X$ be a covering projection, let $f : Y \to X$, and suppose $f' : Y \to E$ is a lift of f. Let $H : Y \times I \to X$ be a homotopy such that $H_0 = f$. Then H lifts uniquely to a homotopy $H' : Y \times I \to E$ such that $H'_0 = f'$.*

Outline of Proof

1) The lift is unique if it exists since if H', H'' are lifts then for each $y \in Y$ the sets $\{t \in I \mid H'(y,t) = H''(y,t)\}$ and $\{t \in I \mid H'(y,t) \neq H''(y,t)\}$ are open (using that p is a covering projection) and will produce a disconnection of I if the latter is not empty.
2) For all $y \in Y$ there exists an open neighbourhood V_y of y and a partition $0 = t_0 < t_1 < \ldots < t_n = 1$ (depending upon y) such that $H(V_y \times [t_i, t_{i+1}])$ is contained in an evenly covered set for all $i = 0, \ldots, n - 1$. (This uses compactness of I.)
3) For all y, there exists continuous $H'_y : V_y \times I \to E$ lifting $H|_{V_y \times I}$ and extending $H'_y|_{V_y \times 0} = f'|_{V_y}$.
4) The various liftings $H_{y'}$ from (3) combine to produce a well defined map of sets $H' : Y \times I \to E$ (using the uniqueness shown in step 1).
5) H' is continuous. $\qquad\qquad\qquad\qquad\qquad\qquad\qquad\qquad\qquad\qquad\square$

The main point appears in step (3), where having constructed the lift of $H|_{V_y \times [0, t_i]}$ one uses the sheet over $H(V_y \times [t_i, t_{i+1}])$ containing $H'(V_y \times t_i)$ to extend to $V_y \times [t_i, t_{i+1}]$.

This theorem implies that a covering projection is a fibration. (See Chapter 7.) The special case where Y is a single point says that every path ω in X lifts uniquely to a path ω' in E once $\omega'(0)$ is specified. Note however that the lift of a closed path (one satisfying $\omega(0) = \omega(1)$) need not be closed.

The Covering Homotopy Theorem has the following Corollary.

Corollary 3.2.5 *Let $p : (E, e_0) \to (X, x_0)$ be a covering projection.*

a) *Let $\sigma, \tau : I \to X$ be paths in X beginning at x_0 and let σ', τ' be the lifts of σ and τ beginning at e_0. Suppose $\sigma(1) = \tau(1)$ and $\sigma \simeq \tau$ (rel $\{0,1\}$). Then $\sigma'(1) = \tau'(1)$ and $\sigma' \simeq \tau'$ (rel $\{0,1\}$).*
b) *$p_\# : \pi_1(E, e_0) \to \pi_1(X, x_0)$ is a monomorphism.* $\qquad\qquad\qquad\square$

Suppose now that Y is connected. Notice that since $Y \times I \approx Y$, the existence of the lift f' implies that $H_\#\big(\pi_1(Y \times I)\big) \subset p_\#\big(\pi_1(E)\big)$. More generally (except for the addition of an extra locally path connectedness hypothesis) we have the following theorem.

Theorem 3.2.6 (Lifting Theorem) *Let $p : E \to X$ be a covering projection and let $f : W \to X$. Suppose W is connected and locally path connected. Then f lifts to $f' : W \to E$ if and only if $f_\#\big(\pi_1(W)\big) \subset p_\#\big(\pi_1(E)\big)$. If it exists, the lift is uniquely determined once $f'(w)$ is specified for any $w \in W$.* $\qquad\square$

Universal Covering Spaces

Let $p : E \to X$ and $p' : E' \to X$ be covering projections. A *morphism from E to E' of covering spaces over X* consists of a map $\phi : E \to E'$ such that $p'\phi = p$. An *equivalence of covering spaces* is a morphism of covering spaces which is also a homeomorphism.

Definition 3.2.7 A basepoint preserving covering projection $\tilde{p} : \tilde{X} \to X$ between pointed spaces is called the *universal covering projection* of X (and $\tilde{X}$ is called the *universal covering space* of X) if for any basepoint preserving covering projection $p : E \to X$ there exists a unique basepoint preserving morphism of covering projections from $\tilde{p}$ to p.

The standard categorical argument shows that if X has a universal covering projection then it is unique up to equivalence. From the Lifting Theorem we get the following corollary.

Corollary 3.2.8 *A simply connected locally path connected covering space is a universal covering space.* $\square$

A space X is called *semilocally simply connected* if each point $x \in X$ has an open neighbourhood U_x such that the group homomorphism $i_\# : \pi_1(U_x, x) \to \pi_1(X, x)$ induced by the inclusion map $i : U_x \hookrightarrow X$ is trivial.

Theorem 3.2.9 *Every connected, locally path connected, semilocally simply connected space has a universal covering space.* $\square$

Computing Fundamental Groups from Covering Spaces

A self-equivalence of a covering projection is called a *covering transformation*. Thus a covering transformation of $p : E \to X$ is a self-homeomorphism $\phi : E \to E$ such that $p\phi = p$. The covering transformations of p form a group under composition.

Theorem 3.2.10 *Let $p : E \to X$ be a covering projection such that E is simply connected and locally path connected (thus a universal covering space). Then $\pi_1(X) \cong$ the group of covering transformations of p.* $\square$

Note The hypotheses imply that X is path connected so $\pi_1(X)$ is independent of the choice of basepoint.

Corollary 3.2.11 *Suppose a group G acts on a simply connected and locally path connected space X in such a way that for all $x \in X$ there exists an open neighbourhood V_x satisfying $V_x \cap (g \cdot V_x) = \emptyset$ for all $g \neq e$ in G. Then $\pi_1(X/G) \cong G$.* $\square$

In particular, writing $S^1 = \mathbb{R}/\mathbb{Z}$ yields $\pi_1(S^1) = \mathbb{Z}$. In Section 3.3 we will use Van Kampen's Theorem to show that S^n is simply connected for $n > 1$. It will then follow from Corollary 3.2.11 that $\pi_1(\mathbb{R}P^n) = \mathbb{Z}/2\mathbb{Z}$ for $n > 1$.

Exercise 3.2.12 Show that for $n \geq 2$, any map from $\mathbb{R}P^n$ to S^1 is null homotopic.

Exercise 3.2.13

  a) Let $f : S^1 \to S^1$ be a basepoint preserving map such that $f(-x) = -f(x)$. Show that f represents an odd multiple of the generator of $\pi_1(S^1)$.
  b) Show that for $n \geq 2$, there does not exist a continuous map $f : S^n \to S^1$ satisfying $f(-x) = -f(x)$.

Exercise 3.2.14 Show that $\prod_{j=1}^{\infty} S^1$ (with the product topology) does not have a universal covering space.

Theorem 3.2.15 ("Galois Theory" of Covering Spaces) *Let $p : E \to X$ be a covering projection such that E is simply connected and locally path connected (thus a universal covering space). Then for every subgroup $H \subset \pi_1(X)$ there exists a covering projection $p_H : E_H \to X$, unique up to equivalence of covering spaces, such that $(p_H)_\#\big(\pi_1(E_H)\big) = H$.* $\square$

In the above situation, the map $E \to E_H$ is also a covering projection; this follows from Proposition 3.2.3.

3 Van Kampen's Theorem

Let A, G, H be groups and let $\alpha : A \to G$, $\beta : A \to H$ be group homomorphisms. The pushout, $G *_A H$, of α and β is the *amalgamated free product* of G and H over A defined as follows. The elements of $G *_A H$ are "words" $w_1 w_2 \cdots w_n$, where for each j either $w_j \in G$ or $w_j \in H$, modulo the equivalence relation generated by $\alpha(a) \sim \beta(a)$ for all $a \in A$. Words are multiplied by juxtaposition. In the special case where $A = \{1\}$, this yields the free product $G * H$ of G and H.

Theorem 3.3.1 (Van Kampen's Theorem) *Let U, V be path connected open subsets of X such that $U \cup V = X$ and $U \cap V$ is nonempty and path connected. Then $\pi_1(X) = \pi_1(U) *_{\pi_1(U \cap V)} \pi_1(V)$.*

The proof relies upon the compactness of I. See Massey [Mas, Chapter 4].
  $\square$

From the theorem we can immediately calculate that $\pi_1(S^n) = \{1\}$ for $n > 1$ and $\pi_1\big(S^1 \vee S^1\big) = \mathbb{Z} * \mathbb{Z}$.

CHAPTER 4

Homological Algebra

1 Chain Complexes

Definition 4.1.1 A *chain complex* (C, d) in the category of abelian groups consists of an abelian group C_n for each integer n together with a homomorphism $d_n : C_n \to C_{n-1}$ such that $d_{n-1} \circ d_n = 0$. The maps d_n are called *boundary operators* or *differentials*.

In the notation, often d is omitted and we refer to the "chain complex C". The subgroup $\operatorname{Ker} d_n$ of C_n is denoted $Z_n(C)$; its elements are called *cycles*. The subgroup $\operatorname{Im} d_{n+1}$ of C_n is denoted $B_n(C)$; its elements are called *boundaries*. $d_n d_{n+1} = 0$ implies $B_n(C) \subset Z_n(C)$. The quotient group $Z_n(C)/B_n(C)$ is denoted $H_n(C)$ and called nth *homology group* of C; its elements are called *homology classes*. If x, y belong to C_n such that $x - y$ belongs to $B_n(C)$ then x and y are called *homologous*.

Definition 4.1.2 Let C and D be chain complexes. A *chain map* $f : C \to D$ consists of a group homomorphism f_n for all n such that

$$
\begin{array}{ccc}
C_n & \xrightarrow{\ d_n\ } & C_{n-1} \\
\downarrow{\scriptstyle f_n} & & \downarrow{\scriptstyle f_{n-1}} \\
D_n & \xrightarrow{\ d_n\ } & D_{n-1}
\end{array}
$$

commutes.

The subscripts are usually omitted; thus the condition that f be a chain map is usually written $fd = df$. If $f : C \to D$ is a chain map, then for all n the association $[x] \mapsto [f(x)]$ induces a well defined homomorphism denoted $f_* : H_n(C) \to H_n(D)$.

The *suspension* of the chain complex C, written SC, is the chain complex with $(SC)_n = C_{n-1}$ and differential d_n on $(SC)_n$ defined to be (-1) times the differential d_{n-1} on C_{n-1}. (The sign convention is in keeping with a more general philosophy discussed at the end of Section 4.4.)

If $H_n(C) = 0$ for $n \neq 0$ then C is called *acyclic*. A composition of homomorphisms of abelian groups $X \xrightarrow{f} Y \xrightarrow{g} Z$ is called *exact at* Y if $\operatorname{Ker} g = \operatorname{Im} f$. Thus $H_n(C)$ can be thought of as the deviation from exactness of the composition $d_n d_{n+1}$. A sequence $X_n \longrightarrow X_{n-1} \longrightarrow \ldots \longrightarrow X_0$ is called *exact* if it is exact at X_i for all $i = 1, \ldots, n - 1$. A 5-term exact sequence of the form $0 \longrightarrow A \longrightarrow B \longrightarrow C \longrightarrow 0$ is called a *short exact sequence*.

Proposition 4.1.3

a) $0 \longrightarrow A \overset{f}{\longrightarrow} B \overset{g}{\longrightarrow} C \longrightarrow 0$ exact $\Rightarrow f$ is injective, g is surjective, and $C \cong B/f(A)$.

b) $0 \longrightarrow A \overset{f}{\longrightarrow} B \longrightarrow 0$ exact $\Rightarrow f$ is an isomorphism.

c) $0 \longrightarrow A \longrightarrow 0$ exact $\Rightarrow A = 0$. $\qquad\qquad\square$

A map $i : A \to B$ is called a *split monomorphism* if there exists $r : B \to A$ (called a *retraction*) such that $ri = 1_A$. A map $p : A \to B$ is called a *split epimorphism* if there exists $s : B \to A$ (called a *section* or *cross-section*) such that $ps = 1_B$. It is trivial to show that a split monomorphism is indeed a monomorphism and a split epimorphism is indeed an epimorphism.

Proposition 4.1.4 *Let* $0 \longrightarrow A \overset{i}{\longrightarrow} B \overset{p}{\longrightarrow} C \longrightarrow 0$ *be a short exact sequence. Then the following are equivalent:*

1) i *is a split monomorphism;*
2) p *is a split epimorphism;*
3) $B \cong A \oplus C.$ $\qquad\qquad\square$

A short exact sequence is said to *split* if the conditions of the previous proposition hold.

Lemma 4.1.5 (Snake Lemma) *Let*

$$
\begin{array}{ccccccccc}
0 & \longrightarrow & A' & \overset{i'}{\longrightarrow} & A & \overset{i''}{\longrightarrow} & A'' & \longrightarrow & 0 \\
& & \downarrow{\scriptstyle f'} & & \downarrow{\scriptstyle f} & & \downarrow{\scriptstyle f''} & & \\
0 & \longrightarrow & B' & \overset{j'}{\longrightarrow} & B & \overset{j''}{\longrightarrow} & B'' & \longrightarrow & 0
\end{array}
$$

be a commutative diagram with exact rows. Then there is an induced exact sequence

$$
0 \longrightarrow \operatorname{Ker} f' \overset{i'_*}{\longrightarrow} \operatorname{Ker} f \overset{i''_*}{\longrightarrow} \operatorname{Ker} f'' \overset{\partial}{\longrightarrow} \operatorname{CoKer} f' \overset{j'_*}{\longrightarrow} \operatorname{CoKer} f \overset{j''_*}{\longrightarrow} \operatorname{CoKer} f'' \longrightarrow 0.
$$

$\qquad\qquad\square$

The map ∂, called the *connecting homomorphism*, is defined as follows. Given $x \in \operatorname{Ker} f''$ pick $y \in A$ such that $i''(y) = x$ (using that exactness implies that i'' is surjective). Commutativity plus exactness plus $f''(x) = 0$ imply that there exists $z \in B'$ such that $j'z = fy$. Set $\partial x = [z] \in \operatorname{CoKer} f'$ and check that this is well defined.

Lemma 4.1.6 (5-Lemma) *Let*

$$
\begin{array}{ccccccccc}
A & \longrightarrow & B & \longrightarrow & C & \longrightarrow & D & \longrightarrow & E \\
\downarrow{\scriptstyle f} & & \downarrow{\scriptstyle g} & & \downarrow{\scriptstyle h} & & \downarrow{\scriptstyle i} & & \downarrow{\scriptstyle j} \\
A' & \longrightarrow & B' & \longrightarrow & C' & \longrightarrow & D' & \longrightarrow & E'
\end{array}
$$

be a commutative diagram with exact rows. Suppose that g and i are isomorphisms, f an epimorphism, and j a monomorphism. Then h is an isomorphism. $\qquad\square$

Lemma 4.1.7 *Let* $0 \longrightarrow P \overset{f}{\longrightarrow} Q \overset{g}{\longrightarrow} R \longrightarrow 0$ *be a short exact sequence of chain complexes. Then there is an induced natural (long) exact sequence of homology groups*

$$\ldots \longrightarrow H_n(P) \overset{f_*}{\longrightarrow} H_n(Q) \overset{g_*}{\longrightarrow} H_n(R) \overset{\partial}{\longrightarrow}$$
$$H_{n-1}(P) \overset{f_*}{\longrightarrow} H_{n-1}(Q) \overset{g_*}{\longrightarrow} H_{n-1}(R) \longrightarrow \ldots .$$

$\square$

Remark "Short exact" for a sequence of chain complexes means that the corresponding sequence of abelian groups is exact for all n. "Natural" means that a commuting diagram consisting of two short exact rows and joined by vertical maps yields a commutating diagram with the long exact sequences as rows and induced vertical maps.

Lemma 4.1.8 (Algebraic Mayer-Vietoris Sequence) *Let*

$$
\begin{array}{ccccccccccc}
\cdots \longrightarrow & A_n & \overset{i}{\longrightarrow} & B_n & \overset{j}{\longrightarrow} & C_n & \overset{\partial}{\longrightarrow} & A_{n-1} & \overset{i}{\longrightarrow} & B_{n-1} & \overset{j}{\longrightarrow} & C_{n-1} & \longrightarrow \cdots \\
& \downarrow{\alpha} & & \downarrow{\beta} & & \downarrow{\gamma} & & \downarrow{\alpha} & & \downarrow{\beta} & & \downarrow{\gamma} & \\
\cdots \longrightarrow & A'_n & \overset{i'}{\longrightarrow} & B'_n & \overset{j'}{\longrightarrow} & C'_n & \overset{\partial}{\longrightarrow} & A'_{n-1} & \overset{i'}{\longrightarrow} & B'_{n-1} & \overset{j'}{\longrightarrow} & C'_{n-1} & \longrightarrow \cdots
\end{array}
$$

be a commutative diagram with exact rows. Suppose that $\gamma : C_n \to C'_n$ *is an isomorphism for all n. Then there is a an induced long exact sequence*

$$\ldots \longrightarrow A_n \longrightarrow B_n \oplus A'_n \longrightarrow B'_n \overset{\Delta}{\longrightarrow} A_{n-1} \longrightarrow B_{n-1} \oplus A'_{n-1} \longrightarrow B'_{n-1} \longrightarrow \ldots .$$

$\square$

The first map in this long exact sequence is defined by $a \mapsto (ia, \alpha a)$ and second is defined by $(b, a') \mapsto \beta b - i'a'$. The connecting homomorphism is given by $\Delta = \partial \gamma^{-1} j'$.

Definition 4.1.9 Let C and D be chain complexes and let $f, g : C \to D$ be chain maps. A *chain homotopy* s from f to g consists of a homomorphism $s_n : C_n \to D_{n+1}$ for all n such that $d_{n+1}s_n + s_{n-1}d_n = f_n - g_n$ for all n.

If there exists a chain homotopy s from f to g then we say f is *chain homotopic* to g and write $f \simeq g$ or $s : f \simeq g$. Some motivation for this definition is given at the beginning of Section 4.4 although later theorems such as Theorem 5.5.2 are required to see why this is an analogue of the geometric definition of homotopy. If there exist chain maps $f : C \to D$ and $g : D \to C$ such that $gf \simeq 1_C$ and $fg \simeq 1_D$ then C and D are called *chain homotopy equivalent* and f and g are called *chain homotopy equivalences*; we write $C \approx D$.

Proposition 4.1.10

   a) *Chain homotopy is an equivalence relation on the chain maps from C to D.*
   b) $f \simeq g \Rightarrow f_* = g_* : H_*(C) \to H_*(D)$. $\square$

Given a chain map $f : A \to X$, it is often useful to be able to replace X by a chain homotopy equivalent complex X' such that the corresponding map $A \to X'$ is an injection, thus allowing the formation of a quotient complex and resulting long exact homology sequence. The construction of X' presented below is motivated by its geometric counterpart, Theorem 7.1.16.

Theorem 4.1.11 *Let $f : A \to X$ be a chain map. Then there exists a factorization $f = \phi i$ where $i : A \to X'$ is an injection and $\phi : X' \to X$ is a chain homotopy equivalence.*

Proof Set $X'_n = A_n \oplus X_n \oplus A_{n-1}$. Define $d : X'_n \to X'_{n-1}$ by $d(a, x, b) = (da - b, dx + fb, -db)$ and check that $d^2 = 0$. Define chain maps $j : X \to X'$ and $\phi : X' \to X$ by $j(x) = (0, x, 0)$ and $\phi(a, x, b) = fa + x$. Then $\phi j = 1_X$ and a chain homotopy $s : j\phi \simeq 1_{X'}$ is given by $s(a, x, b) = (0, 0, a)$. The inclusion $i : A \to X'$ given by $i(a) = (a, 0, 0)$ is a chain map and satisfies $f = \phi i$. $\qquad\square$

The chain complex X' is called the *algebraic mapping cylinder* of f and the quotient complex X'/A is called the *algebraic mapping cone* of f by analogy with their geometric counterparts. Notice that the composite $X \xrightarrow{\ j\ } X' \twoheadrightarrow X'/A$ is also an injection and that the quotient $(X'/A)/X$ is isomorphic to SA, which is analogous to the geometric situation discussed after Proposition 7.1.17.

2 Direct Limits

Let CC denote the category of chain complexes and chain maps.

Proposition 4.2.1 *Let J be a partially ordered set and let $D : J \to \underline{C}$ be a diagram in an abelian category $\underline{C}$. Then $L = \varinjlim_{J} D$ exists and is given by $L =$*

$\coprod_{j \in J} D(j)/\sim$ *where $\big(D(i \prec j)\big)(x) \sim \big(D(i \prec k)\big)(x)$ for all i, j, $k \in J$ and $x \in D(i)$. The map $\lambda_j : D(j) \to L$ is the composite of the inclusion of $D(j)$ into the coproduct followed by projection to L.* $\qquad\square$

Definition 4.2.2 A partially ordered set J is called a *directed set* if for all $i, j \in J$ there exists k such that $i \prec k$ and $j \prec k$. If J is the category coming from a directed set (see Example 1.1.4), a diagram $D : J \to \underline{C}$ is called a *direct system* in $\underline{C}$.

Lemma 4.2.3 *Let $D : J \to \underline{C}$ be a direct system and let $x \in D(j)$ be such that $\lambda_j(x) = 0$ where $\lambda_j : D(j) \to \varinjlim_{J} D$ is the canonical map. Then there exists $k \succ j$ such that $\big(D(j \prec k)\big)(x) = 0$.* $\qquad\square$

The colimit of a direct system is more commonly called its *direct limit*.

Theorem 4.2.4 (Homology Commutes with Direct Limits) *Let $D : J \to CC$ be a direct system. Then $H_*(\varinjlim_{J} D) = \varinjlim_{J} H_*(D)$ where $H_*(D)$ denotes the direct system of abelian groups obtained by applying homology to each $D(j)$ for $j \in J$.*

Proof Let $C = \varinjlim_{J} D$ and let $\lambda_j : D(j) \to C$ be the canonical map. Let $\theta : \varinjlim_{J} H_*(D) \to H_*(C)$ be the map induced by the universal property of colimit. Given $[x] \in H_*(C)$ (where x belongs to C) choose a representative $x_j \in D(j)$ such that $\lambda_j(x_j) = x$. Since $\lambda_j : D(j) \to C$ is a chain map and x is a cycle, $\lambda_j(dx_j) = dx = 0$ in C, so by Lemma 4.2.3 there exists $k \succ j$ such that $D(j \prec k)(dx_j) = 0$. Setting $x_k = D(j \prec k)(x_j)$, we see that x_k is a cycle such that $\lambda_k(x_k) = x$ and so $\theta\big((\lambda_k)_*([x_k])\big) = [x]$. Therefore θ is surjective.

Suppose now that y belongs to $\varinjlim_J H_*(D)$ such that $\theta(y) = 0$. Choose a representative $[x_j] \in H_*\big(D(j)\big)$ such that $(\lambda_j)_*([x_j]) = y$. Then $[\lambda_j(x_j)] = \theta(y) = 0$ in $H_*(C)$ so there exists $v \in C$ such that $dv = \lambda_j(x_j)$. Write $v = \lambda_k(v_k)$ for some $v_k \in D(k)$ and find m such that $j \prec m$ and $k \prec m$. Setting $x_m = D(j \prec m)(x_j)$ and $v_m = D(k \prec m)(v_k)$, we have $\lambda_m(dv_m - x_m) = 0$ so by Lemma 4.2.3 there exists $r \succ m$ such that $dv_r - x_r = 0$, where $x_r = D(m \prec r)(x_m)$ and $v_r = D(m \prec r)(v_m)$. Thus $[x_r] = 0$ and so $y = (\lambda_r)_*([x_r]) = 0$. Therefore θ is injective. $\qquad\square$

From the explicit construction of Proposition 4.2.1 it is easy to see that if $D : J \to CC$ is a direct system of chain complexes in which $D(i \prec j)$ is a monomorphism for all $i, j \in J$ then $\varinjlim_J D = \bigcup_{j \in J} D(j)$, so Theorem 4.2.4 shows how to compute the homology of such a union.

The dual concept to direct system, a functor from a category obtained from a partially ordered set having the property that for all $i, j \in J$ there exists k such that $k \prec i$ and $k \prec j$, is called an *inverse system*. Equivalently an inverse system could be regarded as a contravariant functor from the category obtained from a directed set. The limit of an inverse system is called its *inverse limit*. Homology does not commute with inverse limits although the two processes are related by the algebraic analogue of Theorem 13.1.3b.

3 Derived Functors

Let R be a ring. If we assume that all the objects are (left) R-modules and if we replace "homomorphism of abelian groups" by "R-module homomorphism" throughout the preceding two sections then all of the concepts and conclusions carry over intact.

Given a set S, the module $R^S = \bigoplus_{s \in S} R$ is called the *free R-module* on the set S. It has the property that for every R-module M, an R-module homomorphism from R^S to M is uniquely determined by its values on the set S, where S is included into R^S by regarding s as the element whose sth component is 1 and whose other components are 0.

Proposition 4.3.1 *Let R be a ring and let P be an R-module. The following are equivalent:*

   a) *there exists an R-module Q such that $P \oplus Q$ is a free R-module;*

   b) *given a surjective homomorphism $g : B \twoheadrightarrow A$ and a homomorphism $f : P \to A$ there exists a "lift" $f' : P \to B$ (not necessarily unique) such that $gf' = f$;*

   c) *any short exact sequence $0 \to M \to N \to P \to 0$ of R-modules ending in P splits.* $\qquad\square$

An R-module satisfying the conditions of Proposition 4.3.1 is called a *projective R-module*.

Definition 4.3.2 Let M be an R-module. A nonnegatively graded chain complex P_* of R-modules is called a *projective resolution* of M if P_n is a projective

R-module for all n and if

$$H_n(P_*) = \begin{cases} M & \text{if } n = 0; \\ 0 & \text{if } n \neq 0. \end{cases}$$

A projective resolution P_* exists for every R-module M; in fact it is always possible to construct a resolution in which P_n is free for all n.

Theorem 4.3.3 *Let P_* and Q_* be nonnegatively graded chain complexes of R-modules such that P_n is projective for all n and Q_* is acyclic. Then*

  a) *Given any homomorphism $f : H_0(P_*) \to H_0(Q_*)$ there exists a chain map $\phi : P_* \to Q_*$ such that $\phi_* = f$ on $H_0(\ )$.*
  b) *If ϕ, $\psi : P_* \to Q_*$ are two chain maps inducing the same homomorphism on $H_0(\ )$ then $\phi \simeq \psi$.* $\square$

Proposition 4.3.4 *Let C be a nonnegatively graded chain complex of R-modules. Then it is possible to choose projective resolutions $P_*(C_k)$ and chain maps $P_*(C_k) \to P_*(C_{k-1})$ such that:*

  1) *$P_q(C_*)$ forms a chain complex for each fixed q;*
  2) *$P_q(C_k) \to P_q(C_{k-1})$ is the composition of a split epimorphism and a split monomorphism for each q and k;*
  3) *$H_k\big(P_q(C_*)\big)$ forms a projective resolution of $H_k(C)$ for each k.* $\square$

Let M be an R-module and let $F : R\text{-Modules} \to S\text{-Modules}$ be an additive functor. (See Definition 1.3.1.) Pick a projective resolution P_* of M. Then $F(P_*)$ is a chain complex of S-modules. If F preserves exactness (image under F of an exact sequence is exact) then $F(P_*)$ will be acyclic with $H_0\big(F(P_*)\big) = F(M)$, however this will not be true for arbitrary F. We let $(L_nF)(M) = H_n\big(F(P_*)\big)$ and call L_nF the nth *left derived functor* of F. It is easy to see that L_nF is indeed a functor from R-Modules to S-Modules and it follows from Theorem 4.3.3 that $(L_nF)(M)$ is well defined in the sense of being independent (up to isomorphism) of the choice of the projective resolution P_*. Intuitively L_nF measures the deviation of F from preserving exactness.

A functor which preserves exactness is called *exact*. F is exact if and only if F preserves kernels and cokernels. A functor which preserves kernels is called *left exact* and a functor which preservers cokernels is called *right exact*. Left derived functors are most useful in the case where F is right exact, since this is sufficient to conclude that $L_0F = F$.

All of these concepts can be dualized by turning the arrows around. Thus, for example, a *cochain complex* C consists of an R-module C^n for every n together with R-module homomorphisms $d^n : C^{n-1} \to C^n$ satisfying $d^n \circ d^{n-1} = 0$. It is customary to use upper indices for cohomology. A cochain complex becomes a chain complex under the assignment $C_n = C^{-n}$. Dualizing the other concepts gives the definitions of *injective module*, *injective resolution*, and *right derived functor*.

The two most important examples of derived functors are Tor and Ext which are described next. These are the derived functors of the tensor product and Hom functors respectively.

Let R be a ring, let M be a right R-module, and let N be a left R-module. The *tensor product* of M and N over R, written $M \otimes_R N$, is (co-)universal with the property that there is an R-bilinear form from (M, N) to $M \otimes_R N$. Explicitly, $M \otimes_R N$ is the free abelian group on the set $M \times N$ modulo the equivalence relation generated by $(m, n_1 + n_2) \sim (m, n_1) + (m, n_2)$, $(m_1 + m_2, n) \sim (m_1, n) + (m_2, n)$, and $(mr, n) \sim (m, rn)$ for all m, m_1, $m_2 \in M$, n, n_1, $n_2 \in N$, and $r \in R$. The equivalence class of (m, n) is denoted $m \otimes n$. For fixed M, the association $N \mapsto M \otimes_R N$ forms a functor from R-modules to abelian groups; its nth left derived functor evaluated at N is written $\mathrm{Tor}_n^R(M, N)$. If R is omitted then it is conventionally assumed to be $\mathbb{Z}$. $N \mapsto M \otimes_R N$ is right exact, so $\mathrm{Tor}_0^R(M, N) = M \otimes_R N$. A module Q having the property that tensoring with Q preserves exactness (equivalently, $\mathrm{Tor}_n^R(Q, N)=0$ for all $n > 0$ and all N) is called *flat*.

In general $M \otimes_R N$ has no natural R-module structure. However if M has in addition the structure of a left R-module, then this can be used to put a left R-module structure on $M \otimes_R N$ which in turn induces a left R-module structure on $\mathrm{Tor}_n^R(M, N)$. In the same way a right R-module structure on N induces a right R-module structure on $\mathrm{Tor}_n^R(M, N)$. In particular, if R is a commutative ring then $\mathrm{Tor}_n^R(M, N)$ becomes an R-module.

Given a fixed left R-module M, $N \mapsto \mathrm{Hom}_R(M, N)$ forms a functor from (left) R-modules to abelian groups; its nth right derived functor evaluated at M is written $\mathrm{Ext}_R^n(M, N)$, an abelian group in general with natural R-module structure in special cases. The functor is left exact and so $\mathrm{Ext}_R^0(M, N) = \mathrm{Hom}_R(M, N)$.

In the definition of Tor one could reverse the roles of M and N, tensoring a projective resolution of M with N rather than vice versa, however the resulting derived functors turn out again to be $\mathrm{Tor}_n^R(M, N)$. (See Example 11.7.3.) Similarly $\mathrm{Ext}_R^n(M, N)$ can be computed either from a projective resolution of M or from an injective resolution of N.

By convention $\mathrm{Tor}^R(M, N) = \mathrm{Tor}_1^R(M, N)$ and $\mathrm{Ext}_R(M, N) = \mathrm{Ext}_R^1(M, N)$. As the names suggest, Tor has something to do with torsion and Ext has something to do with extensions. In fact,

Proposition 4.3.5 *Let R be a commutative ring. Then*

a) $\mathrm{Tor}^R(M, R/rR) \cong \{x \in M \mid rx = 0\}$ *(the "r-torsion" of M).*

b) $\mathrm{Ext}_R(M, N) \cong \{$*short exact sequences* $0 \to N \to X \to M \to 0\}$ *(the R-module extensions of M by N).* $\qquad\square$

Theorem 4.3.6 *Let R be a PID. Then*

a) *A submodule of a free R-module is free.*

b) *A submodule of a flat R-module is flat.*

c) *If $n > 1$, $\mathrm{Tor}_n^R(M, N) = 0$ and $\mathrm{Ext}_R^n(M, N) = 0$ for all R-modules M and N.* $\qquad\square$

Proposition 4.3.7

a) $R \otimes_R N \cong \mathrm{Hom}_R(R, N) \cong N$ *for all R-modules N.*

b) $\mathrm{Tor}^R(R, N) = \mathrm{Ext}_R(R, N) = 0$ *for all R-modules N.*

c) *For any abelian group G, $0 \to \mathrm{Tor}(G, \mathbb{Z}/n\mathbb{Z}) \to G \xrightarrow{n} G \to G \otimes \mathbb{Z}/n\mathbb{Z} \to 0$ and $0 \to \mathrm{Hom}(\mathbb{Z}/n\mathbb{Z}, G) \to G \xrightarrow{n} G \to \mathrm{Ext}(\mathbb{Z}/n\mathbb{Z}, G) \to 0$ are exact, where the map named n is multiplication by n. In particular,*

    i) $\mathrm{Tor}(\mathbb{Q}, \mathbb{Z}/n\mathbb{Z}) = \mathbb{Q} \otimes \mathbb{Z}/n\mathbb{Z} = \mathrm{Ext}(\mathbb{Z}/n\mathbb{Z}, \mathbb{Q}) = \mathrm{Hom}(\mathbb{Z}/n\mathbb{Z}, \mathbb{Q}) = 0;$

    ii) $\mathrm{Tor}(\mathbb{Z}/m\mathbb{Z}, \mathbb{Z}/n\mathbb{Z}) \cong \mathbb{Z}/m\mathbb{Z} \otimes \mathbb{Z}/n\mathbb{Z} \cong \mathrm{Ext}(\mathbb{Z}/n\mathbb{Z}, \mathbb{Z}/m\mathbb{Z}) \cong \mathrm{Hom}(\mathbb{Z}/n\mathbb{Z}, \mathbb{Z}/m\mathbb{Z}) \cong \mathbb{Z}/d\mathbb{Z}$ *where d is the greatest common divisor of m and n.* $\qquad\square$

Exercise 4.3.8 Show that $\mathbb{Q}$, although not projective, is a flat $\mathbb{Z}$-module.

Given a group G, make $\mathbb{Z}$ into a module over the group ring $\mathbb{Z}[G]$ by letting G act trivially on $\mathbb{Z}$. The algebraic group homology of G is defined by $H_*^{\mathrm{alg}}(G) = \mathrm{Tor}_*^{\mathbb{Z}[G]}(\mathbb{Z}, \mathbb{Z})$. Similarly set $H_{\mathrm{alg}}^*(G) = \mathrm{Ext}_{\mathbb{Z}[G]}^*(\mathbb{Z}, \mathbb{Z})$.

Remark This is usually called simply the "homology of the group" and written $H_*(G)$, however we shall reserve the notation $H_*(G)$ for the homology of the underlying space of a topological group G. In Chapter 10 a space BH is associated to each topological group H and $H_*^{\mathrm{alg}}(G)$ equals the homology of the space BG, where G is regarded as a topological group with the discrete topology.

Consider the case $G = \mathbb{Z}/n\mathbb{Z}$. In the group ring $\mathbb{Z}[\mathbb{Z}/n\mathbb{Z}]$, let T denote a generator of $\mathbb{Z}/n\mathbb{Z}$ and let $N = 1 + T + T^2 + \ldots + T^{n-1}$. Then $N(T-1) = (T-1)N = 0$. Therefore letting W_k be the free $\mathbb{Z}[\mathbb{Z}/n\mathbb{Z}]$-module on a single generator e_k with $d : W_k \to W_{k-1}$ given by

$$de_k = \begin{cases} Ne_{k-1} & \text{if } k \text{ is even;} \\ (T-1)e_{k-1} & \text{if } k \text{ is odd} \end{cases}$$

gives a projective resolution of $\mathbb{Z}$ over $\mathbb{Z}[\mathbb{Z}/n\mathbb{Z}]$. Computing with this projective resolution shows

$$H_k^{\mathrm{alg}}(\mathbb{Z}/n\mathbb{Z}) = \begin{cases} \mathbb{Z} & \text{if } k = 0; \\ \mathbb{Z}/n\mathbb{Z} & \text{if } k \text{ is odd;} \\ 0 & \text{if } k > 0 \text{ is even.} \end{cases}$$

4 Tensor Products, Duals, and Universal Coefficient Theorems

If M and N are graded R-modules we make $M \otimes_R N$ into a graded abelian group by setting $(M \otimes_R N)_n = \bigoplus_{i+j=n} M_i \otimes_R N_j$. If C and D are chain complexes then setting $d(x \otimes y) = dx \otimes y + (-1)^{|x|} x \otimes dy$, where x is homogeneous of degree $|x|$, yields a homomorphism d on $C \otimes D$ satisfying $d^2 = 0$ which is used to make $C \otimes D$ into a chain complex.

Let I denote the chain complex with R-module basis $c \in I_1$, $b \in I_0$, $a \in I_0$ and differential given by $dc = b - a$. For any chain complex X, there are chain maps $i_0, i_1 : X \to I \otimes X$ given by $i_0(x) = a \otimes x$ and $i_1(x) = b \otimes x$ for $x \in X$. A chain map $H : I \otimes X \to Y$ provides a chain homotopy $s : Hi_1 \simeq Hi_0$ given by $s(x) = H(c \otimes x)$ and conversely a chain homotopy $s : f \simeq g$ determines a chain map $H : I \otimes X \to Y$ such that $Hi_1 = f$ and $Hi_0 = g$. This fact is part of the motivation for the definition of chain homotopy.

Theorem 4.4.1 (Künneth Theorem) *Let C be a chain complex of right R-modules and let D be a chain complex of left R-modules. Suppose that C_n and the boundaries $B_n(C)$ are flat R-modules for all n. Then for all n there is a natural short exact sequence of abelian groups*

$$0 \to \big(H_*(C) \otimes_R H_*(D)\big)_n \to H_n(C \otimes_R D) \to \mathrm{Tor}^R\big(H_*(C), H_*(D)\big)_{n-1} \to 0$$

which is split (although not split naturally).   □

Remark 1 If R is a PID then the hypothesis that $B_n(C)$ be flat is redundant.

Remark 2 In almost all of the cases of interest in homotopy theory, R is a PID and C is a complex of free R-modules so the hypotheses are satisfied.

Remark 3 Even if C or D has additional structure which induces an R-module structure on the sequence, the sequence need not split as R-modules unless the hypotheses on $B_n(C)$ are strengthened from flat to projective.

For any left R-module G, let $\underline{G}$ denote the chain complex

$$\underline{G}_n = \begin{cases} G & \text{if } n = 0; \\ 0 & \text{if } n \neq 0 \end{cases}$$

with (perforce) the zero differential. Then for any chain complex C of right R-modules, $(C \otimes_R \underline{G})_n = C_n \otimes_R G$ for all n. The homology of this chain complex is called the *homology of C with coefficients in G*, written $H_*(C; \underline{G})$. As a special case of the Künneth Theorem we get

Theorem 4.4.2 (Universal Coefficient Theorem I) *Let C be a chain complex of right R-modules and let G be a left R-module. Suppose that C_n and $B_n(C)$ are flat R-modules for all n. Then for all n there is a natural short exact sequence of abelian groups*

$$0 \to H_n(C) \otimes G \to H_n(C; G) \to \mathrm{Tor}^R\big(H_{n-1}(C), G\big) \to 0$$

which is split (although not split naturally). $\qquad\square$

If M and N are graded R-modules, define a graded abelian group $\mathrm{Hom}^R(M, N)$ by setting $\big(\mathrm{Hom}^R(M, N)\big)_n$ to be the set of all families $(f_j)_{j \in \mathbb{Z}}$ of degree n maps $f_j : M_j \to N_{j+n}$. Note that since the Hom-groups are formed dimensionwise, the graded group $\mathrm{Hom}^R(M, N)$ does **not** become isomorphic to the ungraded group of R-module homomorphisms from M to N when the grading is ignored, since it will be a direct sum rather than a direct product.

If C and D are chain complexes then a differential is defined on $\mathrm{Hom}^R(C, D)$ by requiring that $d\big(f(v)\big) = (df)(v) + (-1)^{|f|} f(dv)$. Thus we set $(df)(v) = d\big(f(v)\big) - (-1)^{|f|} f(dv)$. In particular, given a chain complex (C, ∂) of R-modules we can form the dual complex $\big(\mathrm{Hom}(C, \underline{R}), \delta\big)$ where $\underline{R}$ is the chain complex consisting of R in degree 0 and 0 elsewhere. We usually regard $\mathrm{Hom}(C, \underline{R})$ as a cochain complex by the standard convention $\mathrm{Hom}(C, \underline{R})^n = \mathrm{Hom}(C, \underline{R})_{-n}$. Since $\underline{R}$ is concentrated in degree 0, one of the terms in the formula for δ vanishes and we get

$$\delta^n = (-1)^n (\partial_n)^* : \mathrm{Hom}_R(C_{n-1}, R) \to \mathrm{Hom}_R(C_n, R).$$

A dual version of Theorem 4.4.2 is:

Theorem 4.4.3 (Universal Coefficient Theorem II) *Let C be a chain complex of R-modules. Suppose that C_n and $B_n(C)$ are projective R-modules for all n. Then for all n there is a natural short exact sequence of abelian groups*

$$0 \to \mathrm{Ext}^R\big(H_{n-1}(C), R\big) \to H^n\big(\mathrm{Hom}_R(C, \underline{R})\big) \to \mathrm{Hom}_R\big(H_n(C), R\big) \to 0$$

which is split (although not split naturally). $\qquad\square$

Let $D = \mathrm{Hom}_R(C, \underline{R})$. For $f \in D^n$ and $c \in C_n$ the element $f(c) \in R$ is often written $\langle f, c \rangle$. From the formula $\langle \delta f, c \rangle = (-1)^{|f|+1} \langle f, \partial c \rangle$ defining δ it is easy to see that if f is a cocycle and c is a cycle then $\langle [f], [c] \rangle$ is independent of the choice of representatives for the cohomology class $[f]$ and the homology class $[c]$. Thus we get a well defined pairing $\langle \, , \, \rangle : H^n(D) \otimes H_n(C) \to R$ called the *Kronecker product* or *Kronecker pairing* .

Remark Various books have various conventions regarding the sign in the definition $\delta^n = (\pm)(\partial_n)^*$. The guiding philosophy should be that when symbols of degree p and q are interchanged one should introduce a sign $(-1)^{pq}$. For example, regarding d as having degree -1, the formula for $d(a \otimes b)$ satisfies this. Whitehead's book [W] uses the convention $\delta^n = (-1)^{n-1}(\partial_n)^*$ so that $(\delta f)(c) = (-1)^{|f|} f(\partial c)$, which, superficially at least, seems to fit the philosophy better. However this results in an awkward sign in the formula for the boundary of a cap product (Proposition 5.7.2) and also results in getting only commutativity up to sign in one of the natural diagrams occurring in relation to Poincaré duality which strictly commutes under our convention. (See discussion following Theorem 6.3.1.) The sign used here is the one that appears in Milnor [MS], Dold [D], and MacLane[Mac] and it is in fact consistent with the philosophy, as seen earlier. An equivalent explanation for it can be found by thinking of duality as a map of chain complexes $\phi : \mathrm{Hom}(C, \underline{R}) \otimes C \to \underline{R}$. (To define the tensor product of a cochain complex and a chain complex regard the cochain complex as a chain complex by $(\ )^n = (\ )_{-n}$ and apply the definition of the tensor product of chain complexes; thus $f \otimes c$ has degree 0 if and only if $|f| = |c|$.) Then ϕ becomes a chain map under our definition of δ but not under the Whitehead convention. There are also other sign conventions in the literature (e.g., Spanier [Sp] and Greenberg-Harper [GH] and Eilenberg-Steenrod [EilS]) but they lead to less aesthetic formulas in several places.

5 Adjoint Functors

Let $\underline{C}$ and $\underline{D}$ be categories and let $F : \underline{C} \to \underline{D}$ and $G : \underline{D} \to \underline{C}$ be functors. If there is a natural bijection of sets $\underline{D}(FX, Y) \to \underline{C}(X, GY)$ for every object X of $\mathrm{Obj}\,\underline{C}$ and Y of $\mathrm{Obj}\,\underline{D}$ then F and G are called *adjoint functors*, written $F \dashv G$. (Natural means that it commutes with morphisms in both variables.) F is called the *left adjoint* or *coadjoint* and G is called the *right adjoint* or *adjoint*. Let $I_{\underline{C}}$ and $I_{\underline{D}}$ denote the identity functors on $\underline{C}$ and $\underline{D}$. If $F \dashv G$ then there are induced natural transformations $\alpha : I_{\underline{C}} \to GF$ and $\beta : FG \to I_{\underline{D}}$ where 1_{FX} corresponds to α_X and β_Y to 1_{GY} under the bijections giving the adjointness.

Proposition 4.5.1 *Let* $F : \underline{C} \to \underline{D}$ *and* $G : \underline{D} \to \underline{C}$ *be adjoint functors.*

a) $G(\beta_Y) \circ \alpha_{GY} = 1_{GY}$ *for all objects* Y *of* $\mathrm{Obj}\,\underline{D}$.

b) $\beta_{FX} \circ F(\alpha_X) = 1_{FX}$ *for all objects* X *of* $\mathrm{Obj}\,\underline{C}$.

c) *The adjoint of a map* $p : X \to GY$ *in* $\underline{C}$ *is* $\beta_Y \circ F(p)$.

d) *The adjoint of a map* $q : FX \to Y$ *in* $\underline{D}$ *is* $G(q) \circ \alpha_X$. $\square$

6 Relative Derived Functors

Let $\mathcal{E}$ be a class of epimorphisms in a category $\underline{C}$. An object P of $\underline{C}$ is called *projective relative to* $\mathcal{E}$ if for every epimorphism $g : B \to A$ of $\mathcal{E}$ and homomorphism $f : P \to A$ there exists a "lift" $f' : P \to B$ (not necessarily unique) such that $gf' = f$. The *closure* $C(\mathcal{E})$ of a class $\mathcal{E}$ of epimorphisms consists of those epimorphisms $g : B \to A$ such that for every projective P relative to $\mathcal{E}$ and every map $f : P \to A$ there exists a lift $f' : P \to B$ such that $gf' = f$. The category $\underline{C}$ is said to have *enough projectives* relative to class $\mathcal{E}$ if for every object M of $\underline{C}$ there exists an epimorphism $\epsilon : P \to M$ of $\mathcal{E}$ with P projective relative to $\mathcal{E}$.

If $\mathcal{E}$ is a closed class of epimorphisms such that $\underline{C}$ has enough projectives relative to $\mathcal{E}$ then we can form projective resolutions relative to $\mathcal{E}$ for every object of $\underline{C}$, define derived functors $L_q^{\mathcal{E}}(T)$ relative to $\mathcal{E}$ for any additive functor $T : \underline{C} \to \underline{A}$

to an abelian category $\underline{A}$, and all the standard properties will carry over to this generalization.

Proposition 4.6.1 *Let* $F : \underline{C} \to \underline{D}$ *and* $G : \underline{D} \to \underline{C}$ *be functors between abelian categories* $\underline{C}$, $\underline{D}$ *such that* $F \dashv G$. *Suppose that* $G : \underline{D}(A, B) \to \underline{C}(GA, GB)$ *is a set injection for every pair of objects* A, B *of* $\underline{D}$. *Let* $\mathcal{E}$ *be a closed class of epimorphisms in* $\underline{C}$ *such that* $\underline{C}$ *has enough projectives relative to* $\mathcal{E}$. *Then* $G^{-1}(\mathcal{E})$ *is a closed class of epimorphisms in* $\underline{D}$ *such that* $\underline{D}$ *has enough projectives relative to* $G^{-1}(\mathcal{E})$, *and the projectives of* $\underline{D}$ *relative to* $G^{-1}(\mathcal{E})$ *are the direct summands of* FP, *where* P *is a projective in* $\underline{C}$ *relative to* $\mathcal{E}$. $\qquad\square$

Note that the injectivity condition of G on morphisms guarantees that e is an epimorphism whenever $G(e)$ is.

Similarly we can define injectives relative to a class of monomorphisms, relative injective resolutions, and use an adjoint pair $F \dashv G$ in which F is an injection on morphisms to transport these from one category to another.

7 Derived Functors of $\varprojlim$

Let $\mathcal{AB}$ denote the category of abelian groups and let $\mathcal{I}nv$ be the category of inverse systems of abelian groups indexed over the nonpositive integers. Then $\varprojlim$ is a functor from $\mathcal{I}nv$ to $\mathcal{AB}$. Unlike $\varinjlim$, this functor fails to preserve exactness.

Let $\underline{G}$ be the inverse system $\ldots \longrightarrow G^{n+1} \xrightarrow{\phi^n} G^n \longrightarrow \ldots \xrightarrow{\phi^0} G^0$ where we use the convention $G^n = G_{-n}$. Define $\phi : \prod_k G^k \to \prod_k G^k$ to be the map whose projection onto G^n is the composite, $\prod_k G^k \twoheadrightarrow G^{n+1} \xrightarrow{\phi^n} G^n$. Then

$$\varprojlim(\underline{G}) = \mathrm{Ker}\left((1 - \phi) : \prod_k G^k \to \prod_k G^k \right).$$

Set $\varprojlim^1(\underline{G}) = \mathrm{CoKer}\,(1 - \phi)$. From the definition it is easy to see that if ϕ^n is onto for all n then $1 - \phi$ is onto. In other words,

Proposition 4.7.1 *Let* $\underline{G} = \{G^n\}$ *be the inverse system in which* $\phi^n : G^{n+1} \to G^n$ *is onto for all* n. *Then* $\varprojlim^1(\underline{G}) = 0$. $\qquad\square$

The Snake Lemma implies

Proposition 4.7.2 *Let* $0 \to \underline{G}' \to \underline{G} \to \underline{G}'' \to 0$ *be a short exact sequence of inverse systems. Then there is an induced exact sequence*

$$0 \to \varprojlim(\underline{G}') \to \varprojlim(\underline{G}) \to \varprojlim(\underline{G}'') \to \varprojlim^1(\underline{G}') \to \varprojlim^1(\underline{G}) \to \varprojlim^1(\underline{G}'') \to 0.$$

$\qquad\square$

Proposition 4.7.3 *Let* $I = \{I^n\}$ *be an injective in* $\mathcal{I}nv$. *Then* $\phi^n : I^{n+1} \to I^n$ *is surjective for all* n. *In particular,* $\varprojlim^1(\underline{I}) = 0$.

Proof Construct an injection from I to a system J whose structure maps are surjective. Since I is injective, there is a retraction $J \to I$ and this implies that the structure maps in I are surjective too. $\qquad\square$

The definition of derived functors combined with Proposition 4.7.2 and Proposition 4.7.3 yield the following proposition which justifies the notation.

Proposition 4.7.4 $\varprojlim^1$ *is the first right derived functor of* $\varprojlim$. $\qquad\square$

CHAPTER 5

Homology of Spaces

1 Eilenberg-Steenrod Homology Axioms

Definition 5.1.1 (Eilenberg-Steenrod) Let $\mathcal{A}$ be a class of topological pairs such that:

1) (X, A) in $\mathcal{A} \Rightarrow (X, X)$, $(X, \emptyset)$, (A, A), $(A, \emptyset)$, and $(X \times I, A \times I)$ are in $\mathcal{A}$;
2) $(*, \emptyset)$ is in $\mathcal{A}$ (where $*$ denotes a space with one point).

A *homology theory* on $\mathcal{A}$ consists of:

E1) an abelian group $H_n(X, A)$ for each pair (X, A) in $\mathcal{A}$ and each integer n;
E2) a homomorphism $f_* : H_n(X, A) \to H_n(Y, B)$ for each map of pairs $f : (X, A) \to (Y, B)$;
E3) a homomorphism $\partial : H_n(X, A) \to H_{n-1}(A)$ for each integer n (where $H_n(A)$ is an abbreviation for $H_n(A, \emptyset)$),

such that:

A1) $1_* = 1$;
A2) $(gf)_* = g_* f_*$;
A3) ∂ is natural. That is, given $f : (X, A) \to (Y, B)$, the diagram

$$
\begin{array}{ccc}
H_n(X, A) & \xrightarrow{\ f_*\ } & H_n(Y, B) \\
\Big\downarrow{\partial} & & \Big\downarrow{\partial} \\
H_{n-1}(A) & \xrightarrow{\ f|_{A_*}\ } & H_{n-1}(B)
\end{array}
$$

commutes;
A4) Exactness:

$$
\cdots \longrightarrow H_n(A) \longrightarrow H_n(X) \longrightarrow H_n(X, A) \xrightarrow{\ \partial\ }
$$
$$
H_{n-1}(A) \longrightarrow H_{n-1}(X) \longrightarrow H_{n-1}(X, A) \longrightarrow \cdots
$$

is exact for every pair (X, A) in $\mathcal{A}$, where $H_*(A) \to H_*(X)$ and $H_*(X) \to H_*(X, A)$ are induced by the inclusion maps $(A, \emptyset) \to (X, \emptyset)$ and $(X, \emptyset) \to (X, A)$;
A5) Homotopy: $f \simeq g \Rightarrow f_* = g_*$.
A6) Excision: If (X, A) is in $\mathcal{A}$ and U is an open subset of X such that $\overline{U} \subset \mathring{A}$ and $(X - U, A - U)$ is in $\mathcal{A}$ then the inclusion map $(X - U, A - U) \to (X, A)$ induces an isomorphism $H_n(X - U, A - U) \xrightarrow{\cong} H_n(X, A)$ for all n;
A7) Dimension: $H_n(*) = \begin{cases} \mathbb{Z} & \text{if } n = 0; \\ 0 & \text{if } n \neq 0. \end{cases}$

Many homology theories also satisfy the following "Compactness Axiom".

A8) For each $\alpha \in H_n(X, A)$ there exists a pair of compact subspaces (X_0, A_0) in $\mathcal{A}$ such that $\alpha \in \operatorname{Im} j_*$, where $j : (X_0, A_0) \to (X, A)$ is the inclusion map.

A theory satisfying A1–A6 is called a *generalized homology theory*. (See Chapter 13.)

Theorem 13.1.7 (Equivalence of Theories) Let Y and Y' be unreduced homology theories satisfying the Milnor wedge axiom. (See Chapter 13.) Let $\eta : Y \to Y'$ be a natural transformation of homology theories. Suppose that $\eta_* : Y_*(*) \to Y'_*(*)$ is an isomorphism. Then $\eta_X : Y_*(X) \to Y'_*(X)$ is an isomorphism for all CW-complexes X.

2 Singular Homology Theory

A set of points $\{a_0, a_1, \dots, a_n\} \in \mathbb{R}^N$ is called *geometrically independent* if the set

$$\{a_1 - a_0, a_2 - a_0, \dots, a_n - a_0\}$$

is linearly independent. The convex hull of $\{a_0, a_1, \dots, a_n\}$ is called the *n-simplex* spanned by $a_0, a_1, \dots, a_n$, and $a_0, a_1, \dots, a_n$ are called its *vertices*. If $\{a_0, a_1, \dots, a_n\}$ is geometrically independent, every point x in the n-simplex spanned by $a_0, a_1, \dots, a_n$ can be written uniquely as $x = \sum_{i=0}^n t_i a_i$ where $t_i \geq 0$ for all i and $\sum_{i=0}^n t_i = 1$; the t_i's are called the *barycentric coordinates* of x with respect to $a_0, a_1, \dots, a_n$. The *barycentre* of the n-simplex is the point all of whose barycentric coordinates are $1/(n+1)$. Let $\epsilon_0 = (0, 0, \dots, 0)$, $\epsilon_1 = (1, 0, \dots, 0)$, $\epsilon_2 = (0, 1, 0, \dots, 0)$, $\epsilon_n = (0, 0, \dots, 0, 1)$ in $\mathbb{R}^n$. The n-simplex spanned by $\epsilon_0, \epsilon_1, \dots, \epsilon_n$ is called the *standard n-simplex*, written Δ^n. If $A \subset \mathbb{R}^m$ is convex, a function $f : A \to \mathbb{R}^k$ is called *affine* if $f(ta + (1-t)b) = tf(a) + (1-t)f(b)$ for all $a, b \in \mathbb{R}^m$ and $t \in I$. Thus an affine function on an n-simplex is determined by its values on the vertices.

Given a topological space X, a continuous function $T : \Delta^n \to X$ is called a *singular n-simplex* of X. Let $S_n(X)$ be the free abelian group on $\{$singular n-simplices of $X\}$. For $0 \leq j \leq n$, let $\delta^j : \Delta^{n-1} \to \Delta^n$ denote the inclusion of Δ^{n-1} as the jth face of Δ^n. That is, δ^j is the affine map defined on vertices by

$$\delta^j(\epsilon_i) = \begin{cases} \epsilon_i & \text{if } i < j; \\ \epsilon_{i+1} & \text{if } i \geq j. \end{cases}$$

Let $\partial_j : S_n(X) \to S_{n-1}(X)$ be the homomorphism defined on generators by $\partial_j(T) = T \circ \delta^j$ and let $\partial = \sum_{j=0}^n (-1)^j \partial_j : S_n(X) \to S_{n-1}(X)$.

Proposition 5.2.1 *The maps ∂_0, ∂_1, $\dots$, ∂_n satisfy the "simplicial identity" $\partial_i \partial_j = \partial_{j-1} \partial_i$ for $i < j$ and thus $\partial \partial = 0$.* $\qquad\square$

The chain complex $\big(S_*(X), \partial\big)$ is called the *singular chain complex* of X and its homology is called the *singular homology* of X, written $H_*(X)$. A map of spaces $f : X \to Y$ yields a chain map $f_* : S_*(X) \to S_*(Y)$ which satisfies $1_* = 1$ and $(gf)_* = g_* f_*$. Thus $S_*(\ )$ forms a functor from the category of topological spaces to the category of chain complexes.

Given a topological pair (X, A), define $H_n(X, A)$ to be the nth homology group of the quotient chain complex $S_*(X)/S_*(A)$.

Verification that Singular Homology is a Homology Theory

Singular homology is defined on all topological pairs. From the definitions and theorems on chain complexes it is trivial that singular homology satisfies conditions E1, E2, E3, A1, A2, A3, and A4 of the Eilenberg-Steenrod axioms. Axiom A7 can also be checked directly from the definitions. Using compactness of Δ^n and the fact that a finite union of compact sets is compact it is easily verified that singular homology satisfies the compactness axiom A8 as well.

The proofs that singular homology satisfies the remaining two axioms use an inductive method of constructing chain homotopies called "acyclic models". The idea behind acyclic models is as follows. Suppose we are attempting to construct a chain homotopy s between chain maps $f, g : C \to D$ and that we have already constructed s_k for $k < n$ in such a way that $ds_k + s_{k-1}d = f - g$ for $k < n$. If C_n is a free abelian group then to define s_n we need, for each generator T of C, a solution z to the equation $dz = (f - g - s_{n-1}d)(T)$. It is easy to check that the necessary condition for existence of a solution, $d(f - g - s_{n-1}d)(T) = 0$, is satisfied but unless $H_n(C) = 0$ this is not sufficient to guarantee a solution. In the method of acyclic models one solves such equations in spaces with trivial homology (the acyclic models) and then uses naturality to transport the solutions elsewhere.

Theorem 5.2.2 (Homotopy Axiom) *Singular homology satisfies the homotopy axiom A5.*

Outline of Proof Let $i_X, j_X : X \to X \times I$ be the inclusions $i_X(x) = (x, 0)$, $j_X(x) = (x, 1)$. Since $H : f \simeq g$ implies $f = Hi_X$ and $g = Hj_X$, it suffices to show that $(i_X)_* = (j_X)_*$.

1) If $C \subset \mathbb{R}^N$ is convex then

$$H_k(C) = H_k(*) = \begin{cases} \mathbb{Z} & \text{if } k = 0; \\ 0 & \text{if } k \neq 0. \end{cases}$$

In particular $H_*(\Delta^n) = H_*(*)$.

2) For $k < 0$, define $D_X : S_k(X) \to S_{k+1}(X \times I)$ to be the zero map for all spaces X. Assume by induction that there exists an integer $n \geq 0$ such that for all $k < n$ and for all spaces X, homomorphisms $D_X : S_k(X) \to S_{k+1}(X \times I)$ have been constructed satisfying:
   (i) $\partial D_X + D_X \partial = (j_X)_* - (i_X)_*$;
   (ii) for all $f : X \to Y$

$$
\begin{array}{ccc}
S_k(X) & \xrightarrow{\;D_X\;} & S_{k+1}(X \times I) \\
\downarrow{\scriptstyle f_*} & & \downarrow{\scriptstyle (f \times I)_*} \\
S_k(Y) & \xrightarrow{\;D_Y\;} & S_{k+1}(Y \times I)
\end{array}
$$

   commutes.

3) Given a generator $T : \Delta^n \to X$ of $S_n(X)$, since $T = T_*(1_{\Delta^n})$, once $D_{\Delta^n}(1_{\Delta^n})$ has been defined there is only one possibility for $D_X(T)$ which is consistent with diagram 2(ii) applied to $T : \Delta^n \to X$. If $n = 0$, noting that $\Delta^0 = *$ and $\Delta^1 = I$, define $D_{\Delta^0}(1_{\Delta^0}) = 1_I \in S_1(\Delta^0 \times I)$. To define $D_{\Delta^n}(1_{\Delta^n})$ when $n > 0$, use the induction hypotheses to check that

$$\partial\big(j_{\Delta^n *}(1_{\Delta^n}) - i_{\Delta^n *}(1_{\Delta^n}) - D_{\Delta^n}\partial(1_{\Delta^n})\big) = 0.$$

Therefore $H_n(\Delta^n \times I) = 0$ implies that there exists $z \in S_{n+1}(\Delta^n)$ such that

$$\partial z = j_{\Delta^n *}(1_{\Delta^n}) - i_{\Delta^n *}(1_{\Delta^n}) - D_{\Delta^n}\partial(1_{\Delta^n}).$$

Pick any such z and set $D_{\Delta^n}(1_{\Delta^n}) = z$.

4) Check that the inductive hypotheses of 2 are satisfied for $k = n$.

5) Given a topological pair (X, A) use 2(ii) to show that D_A and D_X induce a chain homotopy $D_{X,A} : S_n(X, A) \to S_{n+1}(X, A)$. $\qquad\square$

A key technical point is that the induction must be done simultaneously for all spaces. Note that even though the construction of D_X is not canonical since it involves choices, it is natural (in the technical sense: commutes with maps of spaces) since the same $z \in S_{n+1}(\Delta^n)$ is used to define D_X on $S_n(X)$ for all X. Notice also that the success of the method depends on the fact that every element of $S_*(X)$ is in the image of some map from one of the acyclic models Δ^n. Examination of the above proof shows that the only properties of $(i_X)_* - (j_X)_*$ used in the construction of the natural chain homotopy $D_X : (i_X)_* - (j_X)_* \simeq 0$ were that $(i_X)_* - (j_X)_*$ is a natural chain map such that $(i_X)_* - (j_X)_* = 0 : H_0(X) \to H_0(X \times I)$.

Summarizing:

Theorem 5.2.3 (Method of Acyclic Models) *Let $\phi_X : S_*(X) \to S_*\big(F(X)\big)$ be a natural chain map such that $(\phi_X)_0 = 0 : H_0(X) \to H_0\big(F(X)\big)$, where F is a functor from $\mathcal{T}op$ to $\mathcal{T}op$. Then there exists a natural chain homotopy $D_X : \phi_X \simeq 0$.* $\qquad\square$

In preparation for the proof of excision we introduce the "barycentric subdivision operator". Its construction is very similar to the acyclic model construction of the chain homotopy in the preceding proof.

Every point in Δ^{n+1} can be written uniquely as $(1-t)z + t\epsilon_{n+1}$ where $z \in \Delta^n$, $t \in I$, and $\epsilon_{n+1} = (0, 0, \dots, 0, 1)$. Given a convex subset C of $\mathbb{R}^N$ and a point $c \in C$, for all n define $s_c : S_n(C) \to S_{n+1}(C)$ as follows. For a generator $G \in S_n(C)$ set $s_c(G)$ to be the generator of $S_{n+1}(C)$ given by $s_c(G)\big((1-t)z + t\epsilon_{n+1}\big) = (1-t)G(z) + tc$. (This chain homotopy can be used to do step (1) of the preceding proof.) We define the *barycentric subdivision operator*, $\mathrm{sd}_X : S_n(X) \to S_n(X)$ for all spaces X by first defining $\mathrm{sd}_{\Delta^n}(1_{\Delta^n})$ and then transporting the definition elsewhere by naturality. Explicitly, set $\mathrm{sd}_X : S_0(X) \to S_0(X)$ to be the identity map and suppose by induction that sd_X has been defined on $S_k(X)$ for all $k < n$ and for all spaces X in such a way that $\partial\,\mathrm{sd}_X = \mathrm{sd}_X\,\partial$ and $\mathrm{sd}_Y\,f_* = f_*\,\mathrm{sd}_X$ for all $f : X \to Y$. Set $\mathrm{sd}_{\Delta^n}(1_{\Delta^n}) = (-1)^n s_{\hat\sigma}\big(\mathrm{sd}(\partial(1_{\Delta^n}))\big)$ where $\hat\sigma$ is the barycentre of Δ^n. For an arbitrary generator $T : \Delta^n \to X$ of $S_n(X)$ define $\mathrm{sd}_X(T) = T_*\big(\mathrm{sd}_{\Delta^n}(1_{\Delta^n})\big)$.

Theorem 5.2.4

a) sd_X *is a natural chain map.*

b) *Every n-simplex of $\mathrm{sd}_{\Delta^n}(1_{\Delta^n})$ is spanned by vertices $\hat\sigma_0, \hat\sigma_1, \dots, \hat\sigma_n$ where $\sigma_0, \sigma_1, \dots, \sigma_n$ are an increasing collection of subfaces of Δ^n (each a proper face of the next) and $\hat\sigma_j$ is the barycentre of σ_j.*

c) *Let $\mathcal{W}$ be a collection of subsets of X whose interiors cover X. Let $T : \Delta^n \to X$ be a generator of $S_n(X)$. Then there exists N such that $\mathrm{sd}^N(T) = \sum m_i T_i$ with each $\mathrm{Im}\,T_i$ contained in some set in $\mathcal{W}$.*

d) *For each m there exists a natural chain homotopy $D_X : 1_{X*} \simeq (\mathrm{sd}_X)^m$.*

Part (c) is proved by choosing a Lebesgue number for the covering $\{T^{-1}W\}_{W \in \mathcal{W}}$ of Δ^n and using that for all $\epsilon > 0$ there exists N such that the diameter of each simplex of $\mathrm{sd}_{\Delta^n}^N (1_{\Delta^n})$ is less than ϵ. The proof of (d) is by acyclic models, making use of the fact that $\mathrm{sd} - 1_*$ is a natural chain map which is 0 on $H_0(\ )$. $\qquad \square$

Let $\mathcal{W}$ be a collection of subsets of X whose interiors cover X. Set $S_n^{\mathcal{W}} = $ free abelian group on $\{T : \Delta^n \to X \mid \mathrm{Im}\, T \subset W \text{ for some } W \in \mathcal{W}\}$. It is easy to check that $\partial \left(S_n^{\mathcal{W}}(X) \right) \subset S_{n-1}^{\mathcal{W}}(X)$.

Theorem 5.2.5 *Let $\mathcal{W}$ be a collection of subsets of X whose interiors cover X. Then the inclusion $S_*^{\mathcal{W}}(X) \hookrightarrow S_*(X)$ induces a chain homotopy equivalence.* $\qquad \square$

Theorem 5.2.6 (Excision Axiom) *Let A be a subspace of X and suppose U is a subspace of A such that $\overline{U} \subset \mathring{A}$. Then the inclusion $j : (X-U, A-U) \to (X, A)$ induces an isomorphism on singular homology.* $\qquad \square$

This is actually slightly stronger than the excision axiom A6 in that U is not required to be open. The proof makes use of Theorem 5.2.4 applied to the collection $\mathcal{W} = \{X - U, A\}$, noting that the hypothesis $\overline{U} \subset \mathring{A}$ guarantees that the interiors of the sets in this collection cover X.

Consequences of the Axioms

Theorem 5.2.7 (Mayer-Vietoris) *Let X_1, X_2 be subspaces of X such that $X = X_1 \cup X_2$ and let $A = X_1 \cap X_2$. Suppose that the inclusion $(X_1, A) \hookrightarrow (X, X_2)$ induces on isomorphism on homology. Then there is a long exact homology sequence*

$$\ldots \longrightarrow H_n(A) \xrightarrow{(i_{1*}, i_{2*})} H_n(X_1) \oplus H_n(X_2) \xrightarrow{j_{1*} - j_{2*}} H_n(X) \xrightarrow{\Delta}$$
$$H_{n-1}(A) \longrightarrow H_{n-1}(X_1) \oplus H_{n-1}(X_2) \longrightarrow H_{n-1}(X) \longrightarrow \ldots$$

with the maps other than Δ induced by inclusions. $\qquad \square$

Notice that by excision the hypothesis that $(X_1, A) \hookrightarrow (X, X_2)$ induce an isomorphism on homology is automatically satisfied if X_1 and X_2 are open. In fact, if we were to strengthen the hypotheses slightly by making this assumption the resulting statement becomes equivalent to A5 (given the axioms A1–A4 and A6) and is sometimes used in place of A5 in the list of Eilenberg-Steenrod axioms.

Proposition 5.2.8 (Homology Sequence of a Triple) *Suppose $A \subset B \subset X$. Then there is a long exact homology sequence*

$$\ldots \longrightarrow H_n(B, A) \longrightarrow H_n(X, A) \longrightarrow H_n(X, B) \xrightarrow{\partial}$$
$$H_{n-1}(B, A) \longrightarrow H_{n-1}(X, A) \longrightarrow H_{n-1}(X, B) \longrightarrow \ldots$$

with the maps other than ∂ induced by inclusions. If j_* denotes the canonical map $H_*(B) \to H_*(B, A)$ then $\partial = j_* \circ \tilde{\partial}$ where $\tilde{\partial}$ is the connecting map in the long exact homology sequence of the pair (X, B). $\qquad \square$

If there exist maps $i : A \to X$ and $r : X \to A$ such that $ri = 1_A$ then the homology of X contains the homology of A as a direct summand. Thus every nonempty space contains $H_*(*)$ as a direct summand. It is often convenient to ignore this summand, partly because it makes the statement or proofs of some theorems cleaner (avoiding the need to single out the case $n = 0$ for special treatment).

Let $\underline{\mathbb{Z}}$ denote the chain complex

$$\underline{\mathbb{Z}}_n = \begin{cases} \mathbb{Z} & \text{if } n = 0; \\ 0 & \text{if } n \neq 0 \end{cases}$$

with (perforce) the zero differential. Given nonempty X, define $\epsilon : S_*(X) \to \underline{\mathbb{Z}}$ by $\epsilon_0(\sum m_i T_i) = \sum m_i$ and (perforce) $\epsilon_n = 0$ for $n > 0$. ϵ is called the *augmentation map*. It is easy to see that $\epsilon : S_*(X) \to \underline{\mathbb{Z}}$ is a chain map and in the case $X = *$ is a chain homotopy equivalence. The kernel of $\epsilon : S_*(X) \to \underline{\mathbb{Z}}$ is called the *reduced singular chain complex* of X, written $\tilde{S}_*(X)$, and its homology is called the *reduced homology* of X, written $\tilde{H}_*(X)$.

Proposition 5.2.9 $H_*(X) \cong \tilde{H}_*(X) \oplus H_*(*)$. $\qquad\qquad\qquad\qquad\qquad\square$

3 Homology Calculations and Elementary Applications

Proposition 5.3.1

a) $H_q(S^n) = \begin{cases} \mathbb{Z} & \text{if } q = 0 \text{ or } n; \\ 0 & \text{otherwise.} \end{cases}$

b) $H_q(S^1 \times S^1) = \begin{cases} \mathbb{Z} & \text{if } q = 0 \text{ or } 2; \\ \mathbb{Z} \oplus \mathbb{Z} & \text{if } q = 1; \\ 0 & \text{otherwise.} \end{cases}$

c) $H_q(\mathbb{R}P^2) = \begin{cases} \mathbb{Z} & \text{if } q = 0; \\ \mathbb{Z}/2\mathbb{Z} & \text{if } q = 1; \\ 0 & \text{otherwise.} \end{cases}$

d) $H_q(\text{Klein bottle}) = \begin{cases} \mathbb{Z} & \text{if } q = 0; \\ \mathbb{Z} \oplus (\mathbb{Z}/2\mathbb{Z}) & \text{if } q = 1; \\ 0 & \text{otherwise.} \end{cases} \qquad\square$

Proposition 5.3.2

a) *Let $r : S^n \to S^n$ be the reflection $r(x_0, x_1, \ldots, x_n) = (-x_0, x_1, \ldots, x_n)$. Then r_* is multiplication by -1 on $\tilde{H}_n(S^n)$.*

b) *Let $a : S^n \to S^n$ be the antipodal map $a(x) = -x$. Then a_* is multiplication by $(-1)^{n+1}$ on $\tilde{H}_n(S^n)$.* $\qquad\qquad\qquad\qquad\square$

These calculations have the following consequences.

Proposition 5.3.3 *There does not exist a map $r : D^n \to S^{n-1}$ such that the composite $S^{n-1} \hookrightarrow D^n \xrightarrow{r} S^{n-1}$ is homotopic to $1_{S^{n-1}}$.* $\qquad\qquad\square$

Theorem 5.3.4 (Brouwer Fixed Point Theorem) *Let $g : D^n \to D^n$ be continuous. Then there exists $x \in D^n$ such that $g(x) = x$.* $\qquad\qquad\square$

Theorem 5.3.5 *Let TS^n denote the tangent space of S^n. There exists a continuous function $v : S^n \to TS^n$ such that $v(x) \neq 0$ for all $x \in S^n$ if and only if n is odd.* $\qquad\qquad\square$

Theorem 5.3.6 (Jordan Curve Theorem) *Suppose $n > 0$. Let C be a subset of S^n which is homeomorphic to S^{n-1}. Then $S^n - C$ has precisely two path components and C is their common boundary.*

In the proof of this theorem, one first uses Mayer-Vietoris sequences, induction on k, and the Cantor Intersection Theorem to show that if B be a subset of S^n which is homeomorphic to the cube I^k then $S^n - B$ is acyclic. It follows by Mayer-Vietoris and induction on k that if A is a subset of S^n which is homeomorphic to S^k then

$$\tilde{H}_i(S^n - A) = \begin{cases} \mathbb{Z} & \text{if } i = n - k - 1; \\ 0 & \text{otherwise.} \end{cases}$$

In particular, $\tilde{H}_i(S^n - A) = \mathbb{Z}$, so $S^n - C$ has two components. The fact that C is the common boundary of the two components is proved using point set topology. See [Mas, Corollary 6.4] for details. $\qquad\square$

Exercise 5.3.7 Compute the map induced on $H_1(S^1)$ by the map $x \mapsto x^2$ from S^1 to S^1.

Exercise 5.3.8 Give an example of two CW-complexes which have the same homology groups but are not homotopy equivalent.

Exercise 5.3.9 Show that every self-map of $\mathbb{R}P^{2n}$ has a fixed point.

4 Homology of CW-complexes; Cellular Homology

Let X be a CW-complex. By axiom A8, every element of $H_*(X)$ lies in $\operatorname{Im} j_*$ for some $j : K \to X$ with K compact. Since K compact implies $K \subset X^{(n)}$ for some n, $S_*(X) = \bigcup_n S_* \left(X^{(n)} \right)$ which by Section 4.2 is the same as $\varinjlim_n S_* \left(X^{(n)} \right)$. Therefore we get

Theorem 5.4.1 *Let X be a CW-complex. Then $H_*(X) = \varinjlim_n H_* \left(X^{(n)} \right)$.*

$\qquad\square$

For a CW-complex X define a chain complex $\left(C_*(X), \partial_C \right)$ as follows. Set $C_n(X) = H_n \left(X^{(n)}, X^{(n-1)} \right)$. Define $\partial_C : C_n(X) \to C_{n-1}(X)$ to be the connecting homomorphism coming from the long exact homology sequence of the triple $\left(X^{(n)}, X^{(n-1)}, X^{(n-2)} \right)$. Equivalently $\partial_C = (p_{\partial\langle n-1\rangle})_* \partial\langle n\rangle$ where

$$p_{\partial\langle n-1\rangle} : \left(X^{(n-1)}, \emptyset \right) \to \left(X^{(n-1)}, X^{(n-2)} \right)$$

and $\partial\langle n\rangle$ is the connecting homomorphism from the long exact homology sequence of the pair $\left(X^{(n)}, X^{(n-1)} \right)$. Therefore $\partial_C \partial_C = (p_{\partial\langle n-2\rangle})_* \partial\langle n-1\rangle (p_{\partial\langle n-1\rangle})_* \partial\langle n\rangle = 0$ since $\partial\langle n-1\rangle (p_{\partial\langle n-1\rangle})_* = 0$. Thus $\left(C_*(X), \partial_C \right)$ forms a chain complex called the *cellular chain complex* of X. Its homology is called the *cellular homology* of X.

Theorem 5.4.2 (Cellular Homology Equals Singular Homology) *Let X be a CW-complex. Then $H\left(C_*(X) \right) \cong H_*(X)$.*

Proof Let $J_n = \{n\text{-cells of } X\}$. In each n-cell e_j^n of X select a point x_j. Using the homotopy equivalence $X^{(n-1)} \approx X^{(n-1)} \cup \bigcup_{j \in J_n} (e_j^n - x_j) = X^{(n)} - \{x_j\}_{j \in J_n}$ and excising $X^{(n-1)}$ gives

$$H_q\left(X^{(n)}, X^{(n-1)}\right) \cong H_q\left(X^{(n)}, X^{(n)} - \{x_j\}_{j \in J_n}\right)$$

$$\cong H_q\left(\bigcup_{j \in J_n} e_j^n, \bigcup_{j \in J_n}(e_j^n - x_j)\right)$$

$$\cong \bigoplus_{j \in J_n} H_q(e_j^n, e_j^n - x_j)$$

$$\cong \begin{cases} \text{free abelian group on } \{n\text{-cells of } X\} & \text{if } q = n; \\ 0 & \text{if } q \neq n. \end{cases}$$

Using long exact sequences and induction gives

$$H_q\left(X^{(n)}\right) = \begin{cases} H_q(X) & \text{if } q < n; \\ 0 & \text{if } q > n. \end{cases}$$

Since $H_n\left(X^{(n-1)}, X^{(n-2)}\right) = H_n\left(X^{(n+1)}, X^{(n)}\right) = 0$ there is a diagram

$$
\begin{array}{ccccccc}
 & & 0 & & & & \\
 & & \downarrow & & & & \\
H_{n+1}(X^{(n+1)}, X^{(n)}) & \xrightarrow{\Delta} & H_n(X^{(n)}, X^{(n-2)}) & \longrightarrow & H_n(X^{(n+1)}, X^{(n-1)}) & \longrightarrow & 0 \\
 & & \downarrow{\scriptstyle j_*} & & & & \\
 & & H_n(X^{(n)}, X^{(n-1)}) & & & & \\
 & & \downarrow{\scriptstyle \partial_C} & & & & \\
 & & H_{n-1}(X^{(n-1)}, X^{(n-2)}) & & & &
\end{array}
$$

where the column is the long exact homology sequence of the triple $\left(X^{(n)}, X^{(n-1)}, X^{(n-2)}\right)$ and the row is the long exact homology sequence of the triple $\left(X^{(n+1)}, X^{(n)}, X^{(n-2)}\right)$. By Proposition 5.2.8, $j_* \circ \Delta = \partial_C$. The diagram shows that

$$H_n\left(C_*(X)\right) = \operatorname{Ker} \partial_C / \operatorname{Im} \partial_C \cong H_n\left(X^{(n)}, X^{(n-2)}\right) / \operatorname{Im} \Delta$$

$$\cong H_n\left(X^{(n+1)}, X^{(n-2)}\right)$$

and so $H_n\left(C_*(X)\right) \cong H_n\left(X^{(n+1)}, X^{(n-2)}\right) \cong H_n\left(X^{(n+1)}\right) \cong H_n(X).$ $\qquad \square$

Remark This argument is a special case of that used to establish the Serre spectral sequence (Theorem 11.9.2) applied to the case of $1_X : X \to X$.

Proposition 5.4.3

a) $H_q\left(\mathbb{R}P^n\right) = \begin{cases} \mathbb{Z} & \text{if } q = 0 \text{ or if } q = n \text{ and } n \text{ odd;} \\ \mathbb{Z}/2\mathbb{Z} & \text{if } q < n \text{ and } q \text{ odd;} \\ 0 & \text{otherwise.} \end{cases}$

b) $H_q\left(\mathbb{C}P^n\right) = \begin{cases} \mathbb{Z} & \text{if } q \leq 2n \text{ and } q \text{ even;} \\ 0 & \text{otherwise.} \end{cases}$ $\qquad\qquad\square$

Note $\mathbb{C}P^n$ denotes n-dimensional complex projective space, defined as S^{2n+1} modulo the action $z \cdot (z_1, z_2, \dots, z_n) = (zz_1, zz_2, \dots, zz_n)$ of S^1, where S^1 and S^{2n+1} are thought of as the unit balls in $\mathbb{C}$ and $\mathbb{C}^{n+1}$ respectively.

5 Acyclic Models

As we noted in the discussion preceding Theorem 5.2.3, the method of acyclic models depends on the fact that every element of $S_*(X)$ is in the image of some map from a chain complex having trivial homology, which suggests that there could be other functors from topological spaces to chain complexes (besides $S_*(X)$) for which the theorem is valid. Furthermore, the domain and range of the natural transformation ϕ_X need not be the same, although the conditions required of the domain functor are different from those of the range. For example,

Theorem 5.2.3' *Let F, G be functors from topological spaces to nonnegatively graded chain complexes. Suppose that $F_n(X)$ is a free abelian group for all n and X and suppose that every element of $F(X)$ is in $\operatorname{Im} F(j)$ for some map $j : C \to X$ where $G(C)$ is acyclic. Let $\phi_X, \psi_X : F(X) \to G(X)$ be a natural chain maps such that $(\phi_X)_0 = (\psi_X)_0 : H_0\big(F(X)\big) \to H_0\big(G(X)\big)$. Then there exists a natural chain homotopy $D_X : \phi_X \simeq \psi_X$.* $\qquad\square$

In this form the theorem is reminiscent of Theorem 4.3.3(b), combined with naturality. There is an analogue of Theorem 4.3.3(a) as well.

Theorem 5.5.1 *Let F, G be functors from topological spaces to nonnegatively graded chain complexes. Suppose that $F_n(X)$ is a free abelian group for all n and X and suppose that every element of $F(X)$ is in $\operatorname{Im} F(j)$ for some map $j : C \to X$ where $G(C)$ is acyclic. Let $\phi_0 : F_0 \to G_0$ be a natural transformation. Then $(\phi_0)_X$ extends to a natural chain map $\phi_X : F(X) \to G(X)$. (Furthermore, any two extensions are naturally chain homotopic by Theorem 5.2.3'.) In particular, if both F and G satisfy the freeness and acyclicity conditions then any natural isomorphism $H_0\big(F(X)\big) \xrightarrow{\cong} H_0\big(G(X)\big)$ extends to a natural chain homotopy equivalence $F(X) \to G(X)$.* $\qquad\square$

Let $\mathcal{T}op$ denote the category of topological spaces and CC the category of chain complexes. The assignment $(X, Y) \mapsto S_*(X) \otimes S_*(Y)$ is a functor from $\mathcal{T}op \times \mathcal{T}op$ to CC which yields a free chain complex for every pair of spaces X, Y. Furthermore, every generator of $S_p(X) \otimes S_q(Y)$ is the image under some map of the element $1_{\Delta^p} \otimes 1_{\Delta^q}$ in the acyclic complex $S_*(\Delta^p) \otimes S_*(\Delta^q)$. Similarly, let $D_n : \Delta^n \to \Delta^n \times \Delta^n$ be the diagonal map. Then the assignment $(X, Y) \mapsto S_*(X \times Y)$ yields a free chain complex for every pair of spaces X, Y such that every generator of $S_n(X \times Y)$ is the image of D_n in the acyclic complex $S_*(\Delta^n \times \Delta^n)$. Since there is a natural isomorphism $H_0(X) \otimes H_0(Y) \xrightarrow{\cong} H_0(X \times Y)$, the extension of Theorem 5.5.1 to the case of functors on $\mathcal{T}op \times \mathcal{T}op$ shows

Theorem 5.5.2 (Eilenberg-Zilber) *There is a natural chain homotopy equivalence* $S_*(X) \otimes S_*(Y) \approx S_*(X \times Y)$. $\qquad\qquad\square$

Together with the Künneth formula, this determines the homology of a product in terms of the homology of the factors.

For many purposes it is sufficient merely to know of the existence of natural chain homotopy equivalences between $S_*(X) \otimes S_*(Y)$ and $S_*(X \times Y)$ but at times it is convenient to have explicit formulas for a pair of inverse natural equivalences. The best known example of such a pair are the Eilenberg-Zilber map (actually invented by Eilenberg and MacLane) $EZ : S_*(X) \otimes S_*(Y) \to S_*(X \times Y)$ and the Alexander-Whitney map $AW : S_*(X \times Y) \to S_*(X) \otimes S_*(Y)$, which are described next.

A natural chain map on $S_*(X \times Y) \to S_*(X) \otimes S_*(Y)$ is determined by its values on the elements $D_n \in S_n(\Delta^n \times \Delta^n)$. For $i \leq n$ let $\sigma_{\{0,1,\dots,i\}} : \Delta^i \to \Delta^n$ be the affine map taking k to k for $0 \leq k \leq i$ (inclusion of the "front" i-face) and let $\tau_{\{0,1,\dots,i\}} : \Delta^i \to \Delta^n$ be the affine map taking k to $n - i + k$ for $0 \leq k \leq i$ (inclusion of the "back" i-face). AW is the map determined by setting $AW(D_n) = \sum_{p+q=n}(-1)^{pq}\sigma_{\{0,1,\dots,p\}} \otimes \tau_{\{0,1,\dots,q\}}$.

For $0 \leq j \leq n - 1$, let $\sigma_j : \Delta^n \to \Delta^{n-1}$ denote the affine map defined on vertices by

$$\sigma_j(\epsilon_i) = \begin{cases} \epsilon_i & \text{if } i \leq j; \\ \epsilon_{i-1} & \text{if } i > j, \end{cases}$$

called the jth *degeneracy* map. A permutation α in the symmetric group S_{p+q} is called a p,q-*shuffle* if $\alpha(i) < \alpha(j)$ whenever either $1 \leq i < j \leq p$ or $p \leq i < j \leq p+q$. Given a p,q-shuffle α, define $\sigma_\alpha : \Delta^{p+q} \to \Delta^p \times \Delta^q$ to be the affine map whose components are the composition of degeneracies

$$\sigma_\alpha = \big(\sigma_{\alpha(p+q)} \circ \sigma_{\alpha(p+q-1)} \circ \dots \circ \sigma_{\alpha(p+1)}, \sigma_{\alpha(p)} \circ \sigma_{\alpha(p-1)} \circ \dots \circ \sigma_{\alpha(1)}\big).$$

EZ is the map determined by setting

$$EZ(1_{\Delta_p} \otimes 1_{\Delta_q}) = \sum_{\alpha \in \{p,q\text{-shuffles}\}} (-1)^{\mathrm{sgn}(\alpha)}\sigma_\alpha.$$

One advantage of the pair EZ, AW is that EZ is associative and AW is coassociative in the sense that

$$EZ \circ \big(EZ \otimes 1_{S_*(Z)}\big) = EZ \circ \big(1_{S_*(X)} \otimes EZ\big) : S_*(X) \otimes S_*(Y) \otimes S_*(Z) \to S_*(X \times Y \times Z)$$

and

$$\big(1_{S_*(X)} \otimes AW\big) \circ AW = \big(AW \otimes 1_{S_*(Z)}\big) \circ AW : S_*(X \times Y \times Z) \to S_*(X) \otimes S_*(Y) \otimes S_*(Z)$$

for any spaces X, Y, and Z; for an arbitrary pair of natural equivalences, we could claim only "homotopic" rather than "equals". Furthermore, the two are compatible in the sense that for any spaces W, X, Y, and Z

$$S_*(W \times X) \otimes S_*(Y \times Z) \xrightarrow{\quad EZ \quad} S_*(W \times X \times Y \times Z)$$

$$\downarrow AW \otimes AW \qquad\qquad\qquad\qquad \downarrow (W \times \tau_{X,Y} \times Z)_*$$

$$S_*(W) \times S_*(X) \otimes X_*(Y) \otimes S_*(Z) \qquad\qquad S_*(W \times Y \times X \times Z)$$

$$\downarrow S_*(W) \otimes T_{X,Y} \otimes S_*(Z) \qquad\qquad\qquad \downarrow AW$$

$$S_*(W) \otimes S_*(Y) \otimes S_*(X) \otimes S_*(Z) \xrightarrow{\quad EZ \otimes EZ \quad} S_*(W \times Y) \otimes S_*(X \times Z)$$

commutes, where $\tau_{X,Y}$ interchanges X and Y, and $T_{X,Y}(f \otimes g) = (-1)^{|f||g|} g \otimes f$. These equalities have the consequence that AW can be used to turn $S_*(X)$ into an associative graded coalgebra as discussed in the next section, and if X is a topological monoid then the combination of EZ and AW induces an associative coassociative graded Hopf algebra structure on $S_*(X)$. (See Chapter 10 for the definition and properties of graded coalgebras and graded Hopf algebras.)

6 Singular Cohomology

For a topological space X and a ring R we define the singular cochain complex of X with coefficients in R, written $S^*(X; R)$, to be $\mathrm{Hom}_R\big(S_*(X), \underline{R}\big)$ (with differential as in Section 4.4). Its cohomology is called the *singular cohomology* of X with coefficients in R, written $H^*(X; R)$. The universal coefficient theorem gives an algebraic formula for $H^*(X; R)$ in terms of $H_*(X; R)$, making it clear that the groups $H^*(X; R)$ contain no more information than $H_*(X; R)$. Since the singular cohomology groups contain no information not already available from the singular homology groups one might be tempted to ask, "Why consider $H^*(X; R)$?" Perhaps the best reason for introducing cohomology is that the cohomology groups have additional structure; $H^*(X; R)$ can be given a natural ring structure described below.

Given R-modules M, N there is a canonical map

$$\mu : \mathrm{Hom}(A, R) \otimes_R \mathrm{Hom}(B, R) \to \mathrm{Hom}(A \otimes_R B, R).$$

(This map is an isomorphism if either A or B is a finitely generated free R-module, but not in general.) Let $\mathrm{diag} : X \to X \times X$ be the diagonal map. The composition

$$S^*(X; R) \otimes S^*(X; R) \xrightarrow{\mu} \mathrm{Hom}_R\big(S_*(X) \otimes S_*(X), R\big) \xrightarrow{AW^*} S^*(X \times X; R) \xrightarrow{\mathrm{diag}^*} S^*(X; R)$$

turns $S^*(X; R)$ into an associative algebra over R. This multiplication on $S^*(X; R)$ is called the *cup product* and the image of $f \otimes g$ under it is often written $f \cup g$, or simply fg. Explicitly, $\langle f \cup g, T \rangle = (-1)^{|f||g|} \langle f, T \circ \sigma_{\{0,\ldots,|f|\}} \rangle \langle g, T \circ \tau_{\{0,\ldots,|g|\}} \rangle$ for a generator $T \in S_{|f|+|g|}(X; R)$.

Proposition 5.6.1

a) $\delta(f \cup g) = \delta f \cup g + (-1)^{|f|} f \cup \delta g$ for all $f, g \in S^*(X; R)$.

b) *For cohomology classes $[f]$, $[g] \in H^*(X; R)$, $\delta(f \cup g) = 0$ and its cohomology class is independent of the choice of representatives f and g. Therefore the cup product on the singular cohomology complexes induces a well defined cup product on $H^*(X; R)$ turning $H^*(X; R)$ into an associative algebra over R.*

c) *For every continuous map $q : X \to Y$ the induced maps $q^* : S^*(Y; R) \to S^*(X; R)$ and $q^* : H^*(Y; R) \to H^*(X; R)$ are algebra homomorphisms.*

$\square$

Note See the discussion following Theorem 4.4.3 for the definition of δ including the sign convention used.

Instead of looking at cohomology we could have considered the composite

$$S_*(X; R) \xrightarrow{\text{diag}_*} S_*(X \times X; R) \xrightarrow{AW} S_*(X; R) \otimes S_*(X; R).$$

This map gives $S_*(X; R)$ and $H_*(X; R)$ the structure of coalgebras and any information derived from the cohomology multiplication could instead have been obtained from this coalgebra. Therefore perhaps the reason given for considering cohomology is unconvincing. However algebras are intuitively easier to work with than coalgebras and there are contexts and generalizations where cohomology seems to arise more naturally than homology and does not always lend itself readily to dualization, so it remains an object worth studying. There is however a noticeable trend towards using the homology coalgebra structure directly for calculations which, thirty years ago, might have been expressed in terms of cohomology.

Of course, the map AW could equally well have been replaced by some other (necessarily chain homotopic) natural equivalence and the same cup product would be obtained at the cohomology level.

A graded ring A is called *(graded) commutative* if $ab = -1^{|a||b|}ba$ for all a, $b \in A$. At one time this concept was called "graded commutative" but nowadays it is usually called "commutative".

Theorem 5.6.2 *For all X, $H^*(X; R)$ is commutative.*

Proof By acyclic models, the natural chain self-map $T_X : S_*(X) \otimes S_*(X) \to S_*(X) \otimes S_*(X)$ given by $T(x \otimes y) = (-1)^{|x||y|} y \otimes x$ is naturally chain homotopic to the identity. $\square$

Remark Note the cup product on $S^*(X)$ is not commutative; it only becomes commutative after passing to cohomology.

7 Relative Cohomology

Let $j : A \hookrightarrow X$ be an inclusion of topological spaces. The notation $S^*(X, A; R)$ is used stand for $\text{Hom}_R\big(S_*(X, A), \underline{R}\big)$. Since $S_*(X, A; R)$ is a free R-module we get

Proposition 5.7.1 $0 \longrightarrow S^*(X, A; R) \longrightarrow S^*(X) \xrightarrow{j^*} S^*(A) \longrightarrow 0$ *is exact.*

$\square$

The Eilenberg-Steenrod axioms can be dualized in the obvious way, and it is clear that singular cohomology satisfies the dual axioms (with exactness following from the preceding proposition). Since $S^*(X, A; R) = \text{Ker } j^*$ is an ideal in $S^*(X)$, the cup product on $S^*(X; R)$ induces pairings $S^*(X; R) \otimes S^*(X, A; R) \to S^*(X, A; R)$ and $S^*(X, A; R) \otimes S^*(X; R) \to S^*(X, A; R)$, called *relative cup products*, with the image of $f \otimes g$ under a relative cup product again written as $f \cup g$. These maps pass to homology inducing well defined pairings $H^*(X; R) \otimes H^*(X, A; R) \to H^*(X, A; R)$ and $H^*(X, A; R) \otimes H^*(X; R) \to H^*(X, A; R)$ also called *relative cup products* and denoted the same way.

Given $g \in S^q(X; R)$ and $x \in S_{p+q}(X; R)$ we can dualize to define the *cap product* $g \cap x \in S_p(X; R)$ by letting $\langle f, g \cap x \rangle$ equal $\langle f \cup g, x \rangle$. For a generator $T : \Delta_{p+q} \to$

X of $S_{p+q}(X)$, $g \cap T$ is given explicitly by $g \cap T = (-1)^{pq} \langle g, T \circ \tau_{\{0,\dots,q\}} \rangle T \circ \sigma_{\{0,\dots,p\}}$.
(Verification that this element has the desired property justifies the definition of
cap product by demonstrating that the element of the double dual described in the
definition does in fact come from an element of $S_p(X; R)$.)

Proposition 5.7.2 $\partial(g \cap x) = \delta g \cap x + (-1)^{|g|}(g \cap \partial x)$ *for* $g \in S^*(X; R)$ *and*
$x \in S_*(X; R)$. $\qquad\qquad\square$

It follows that the cap product induces a well defined map on cohomology

$$\cap : H^q(X; R) \otimes H_{p+q}(X; R) \to H_p(X; R).$$

There are also two versions of a relative cap product: $S^q(X; R) \otimes S_{p+q}(X, A; R) \to$
$S_p(X, A; R)$ and $S^q(X, A; R) \otimes S_{p+q}(X, A; R) \to S_p(X; R)$. These also produce well
defined maps at the homology level.

Proposition 5.7.3 *Let* $\phi : (X, A) \to (Y, B)$ *be a map of topological pairs. Let*
g *belong to* $S^*(Y, B)$ *and let* x *belong to* $S_{p+q}(X, A)$. *Then* $\phi_*(\phi^* g \cap x) = g \cap \phi_* x$
in $S_*(Y)$. $\qquad\qquad\square$

CHAPTER 6

Manifolds

1 Definitions

Definition 6.1.1 A (*topological*) *n-dimensional manifold* (without boundary) is a paracompact Hausdorff space in which every point has an open neighbourhood (called a *coordinate neighbourhood* or *chart*) which is homeomorphic to $\mathbb{R}^n$.

Definition 6.1.2 A *topological n-dimensional manifold with boundary* is a paracompact Hausdorff space in which every point has an open neighbourhood which is either homeomorphic to $\mathbb{R}^n$ or homeomorphic to the upper half plane $\{(x_1, x_2, \dots, x_n) \in \mathbb{R}^n \mid x_n \geq 0\}$. If M is a manifold with boundary, the *boundary* of M consists of those points in M having open neighbourhoods homeomorphic to the upper half plane.

2 Orientation on Manifolds

Excision yields

Proposition 6.2.1 *Let M be an n-dimensional manifold. Then for all $x \in M$,*

$$H_q(M, M - x) = \left\{ \begin{array}{ll} \mathbb{Z} & \text{if } q = n; \\ 0 & \text{if } q \neq 0. \end{array} \right.$$

□

Definition 6.2.2 A choice of one of the two generators for $H_n(M, M - x) \cong \mathbb{Z}$ is called a *local orientation* for M at x.

For $A \subset B \subset M$, let $j_A^B : (M, M - B) \to (M, M - A)$ be the corresponding map of pairs. We write j_A for j_A^M.

Theorem 6.2.3 *Let M be an n-dimensional manifold. For each $x \in M$ let α_x be a local orientation for M at x. Suppose that these generators are compatible in the sense that for all x there exists an open neighbourhood U_x of x and there exists $\alpha_{U_x} \in H_n(M, M - U_x)$ such that $j_y^{U_x}(\alpha_{U_x}) = \alpha_y$ for all $y \in U_x$. Then given compact $K \subset M$, there exists unique $\alpha_K \in H_n(M, M - K)$ such that $(j_y^K)_*(\alpha_K) = \alpha_y$ for all $y \in K$.*

Existence is trivial if K is contained in a Euclidean chart. It is proved in general by writing K as a union of finitely many compact subsets each of which is contained in a Euclidean neighbourhood and then applying Mayer-Vietoris using the following Lemma which also implies uniqueness.

Lemma 6.2.4 *Let M be an n-dimensional manifold and let K be a compact subset of M. Then*

    a) *$H_q(M - K) = 0$ for $q > n$;*
    b) *If ζ is an element of $H_n(M, M - K)$ such that $(j_x^K)_*(\zeta) = 0$ for all $x \in K$, then $\zeta = 0$.*

47

Outline of Proof of Lemma

1) Check the case $M = \mathbb{R}^n$, K a compact convex subset.
2) Use Mayer-Vietoris to show that (for arbitrary M) if $K = K_1 \cup K_1$ where the lemma holds for K_1, K_2, and $K_1 \cap K_2$, then the lemma holds for K.
3) Use induction and steps (1) and (2) to show that the lemma holds when $M = \mathbb{R}^n$ and K is a finite union of compact convex subsets.
4) Show that the lemma holds when $M = \mathbb{R}^n$ and K is an arbitrary compact subset.
5) Use step (4) to check that the lemma holds for $K \subset U \subset M$ where U is a Euclidean chart of M.
6) Do the general case by writing K as a finite union of compact subsets each of which is contained in a Euclidean chart. $\qquad\qquad\square$

The heart of the proof is in step (4), where for each $z \in H_{q-1}(\mathbb{R}^n - K)$ one chooses a compact set $L_z \subset \mathbb{R}^n - K$ such that z lies in image $H_{q-1}(L_z) \to H_{q-1}(\mathbb{R}^n - K)$ (axiom A8), then finds a finite union of compact convex sets A_z such that $K \subset A_z \subset \mathbb{R}^n - L_z$, and applies step (3) to A_z.

Definition 6.2.5 Let M be an n-dimensional manifold. Suppose that there exists a family of elements $\zeta = (\zeta_K)_{\{K \text{ compact subset of } M\}}$ with $\zeta_K \in H_n(M, M-K)$ such that $(j_x^K)_*(\zeta_K)$ is a local orientation for M at x for all $x \in K$. Suppose further that the family is compatible in the sense that if x belongs to $K_1 \cap K_2$ then $(j_x^{K_1})_*(\zeta_{K_1}) = (j_x^{K_2})_*(\zeta_{K_2})$. Then M is called *orientable* and ζ is called a (global) *orientation* for M.

If M is connected and orientable then it will have precisely two orientations, $\{\zeta_K\}$ and $\{-\zeta_K\}$ and the local orientation at any point uniquely determines one of the two orientations. For an arbitrary ring R we can define *orientation with coefficients* in R by replacing $H_*(\ )$ by $H_*(\ ; R)$ in the preceding discussion. In this case, whenever M is connected and orientable with coefficients in R, it will have one orientation corresponding to every unit of the ring R. In particular, every manifold is uniquely $(\mathbb{Z}/2\mathbb{Z})$-orientable. For this reason, $(\mathbb{Z}/2\mathbb{Z})$-orientability is sometimes useful when dealing with nonorientable manifolds.

Theorem 6.2.6

a) *Let X be a connected nonorientable manifold. Then there is a 2-fold covering space $p : E \to X$ such that E is a connected orientable manifold.*
b) *If M is a simply connected manifold then M is orientable.* $\qquad\square$

Exercise 6.2.7 Show that there is no orientation-reversing self-homotopy-equivalence of $\mathbb{C}P^{2n}$.

3 Poincaré Duality

Let M be an oriented n-dimensional manifold and let $(\zeta_K)_{\{K \text{ compact subset of } M\}}$ be its chosen orientation. Alternatively, if M is not orientable, let (ζ_K) be a $(\mathbb{Z}/2\mathbb{Z})$-orientation of M and use homology and cohomology with $(\mathbb{Z}/2\mathbb{Z})$-coefficients throughout. For each compact $K \subset M$, define $D_K : H^i(M, M - K) \to H_{n-i}(M)$ by $D_K(z) = z \cap \zeta_K$. If $K \subset L \subset M$ with K, L compact then by Theorem 6.2.3, $(j_K^L)_*(\zeta_L) = \zeta_K$. It follows that $D_K = D_L(j_K^L)^* : H^i(M, M - K) \to H_{n-i}(M)$. Therefore the various maps D_K induce (by universal property) a unique map

$$D : \varinjlim_{\substack{K \subset M \\ K \text{ compact}}} H^i(M, M - K) \to H_{n-i}(M)$$ where the partial ordering is induced

by inclusion. Let $H^i_c(M) = \varinjlim_{\substack{K \subset M \\ K \text{ compact}}} H^i(M, M - K)$, called the *cohomology of M*

with compact support. An element of $H^*_c(M)$ is represented by a singular cochain which vanishes outside of some compact set.

Theorem 6.3.1 (Poincaré Duality) *Let M be an oriented manifold (or suppose that $(\mathbb{Z}/2\mathbb{Z})$-coefficients are used). Then $D : H^i_c(M) \to H_{n-i}(M)$ is an isomorphism for all i.*

Outline of Proof

1) Check that the theorem holds for $M = \mathbb{R}^n$.
2) Use Mayer-Vietoris to show that if $M = U \cup V$ where U and V are open subsets of M such that the theorem holds for U, V, and $U \cap V$, then the theorem holds for M.
3) If M is the union of a nested family of open sets U_j such that the theorem holds for each U_j, check that the theorem holds for M.
4) Show that the theorem holds when M is an open subset of $\mathbb{R}^n$ by writing M as the nested union of open sets each of which is a finite union of open convex sets and applying steps (1), (2), and (3).
5) Do the general case by using Zorn's Lemma to find a maximal open subset U of M such that the theorem holds for U and obtain a contradiction, if $U \neq M$, by selecting a Euclidean neighbourhood V of a point in $M - U$ and using (2). Step (4) needs to be applied here to show that the theorem holds for $U \cap V$. $\qquad\square$

The key technical point appears in step (2), where one has to know for compact subsets $K \subset U$, $L \subset V$ that there is a commutative diagram

$$
\begin{array}{ccc}
H^{q-1}\big(M, M - (K \cup L)\big) & \xrightarrow{\quad D_{K \cup L} \quad} & H_{n-q+1}(M) \\
\Big\downarrow{\scriptstyle \Delta^*} & & \Big\downarrow{\scriptstyle \Delta^*} \\
H^q\big(M, M - (K \cap L)\big) \xrightarrow{\;\cong\;} H^q\big(U \cap V, (U \cap V) - (K \cap L)\big) & \xrightarrow{D_{K \cap L}} & H_{n-q}(U \cap V)
\end{array}
$$

where Δ^* is the connecting map in the Mayer-Vietoris sequence. This is a long exercise in cap products. This is a long exercise in cap products.

Remark Under the sign conventions used here the diagram strictly commutes, although under other sign conventions (Whitehead, Greenberg-Harper, Spanier) it commutes only up to sign.

Proposition 6.3.2

a) $H^*(\mathbb{R}P^n; \mathbb{Z}/2\mathbb{Z}) \cong (\mathbb{Z}/2\mathbb{Z})[t]/(t^{n+1})$ with $|t| = 1$ for $n \leq \infty$.
b) $H^*(\mathbb{C}P^n) \cong \mathbb{Z}[t]/(t^{n+1})$ with $|t| = 2$ for $n \leq \infty$. $\qquad\square$

Part 2

Homotopy Theory

CHAPTER 7

Higher Homotopy Theory

1 Fibrations and Cofibrations

One of the features of a fibre bundle $p : X \to B$ (Chapter 9) is that the "fibres",
$F_b = p^{-1}(b)$, are homeomorphic for all points b in a common path component of B.
From the homotopy point of view, the key property of a fibre bundle is that for any
pointed space W, there is an exact sequence $[W, F_b] \to [W, X] \to [W, B]$, where b is
the basepoint of B. Of course, from a homotopy viewpoint, having homeomorphic
fibres and the rest of the rigid structure of a fibre bundle is overkill. In this section
we will study a generalization (and its dual) of the concept of fibre bundle in
which the fibres over points in a common path component are not homeomorphic
but merely homotopy equivalent, and although the overall structure is much less
rigid than that of a fibre bundle, it is still sufficient to give the exact sequence
$[W, F_b] \to [W, X] \to [W, B]$.

A map $p : E \to B$ is said to have the *homotopy lifting property* with respect
to a space Y if, given a homotopy $H : Y \times I \to B$ and a lift $f' : Y \to E$ of H_0,
there exists a lift $H' : Y \times I \to E$ of H such that $H'_0 = f'$. A surjection $p : E \to B$
is called a (*Hurewicz*) *fibration* if it has the homotopy lifting property with respect
to Y for all Y. Covering projections are examples of fibrations. As with covering
projections, B is called the *base space* of the fibration, E the *total space*, and the
inverse image $p^{-1}(b)$ of a point $b \in B$ is called the *fibre* of p over b. If we are dealing
with pointed spaces and maps, the fibre over the basepoint of B is called the fibre
of p. The phrase "$F \longrightarrow E \xrightarrow{\ p\ } B$ is a fibration sequence" is used to mean that p
is a fibration between pointed spaces with fibre F. A projection map $F \times B \to B$
from a product is the canonical example of a fibration and is sometimes called a
"trivial fibration".

Let $p : X \to B$ and $q : Y \to B$ be fibrations. If there exist maps $f : X \to Y$
and $g : Y \to X$ and homotopies $K : gf \simeq 1_X$ and $L : fg \simeq 1_Y$ such that $p = qf$,
$q = pg$, $pK(x, t) = p(x)$ for all $x \in X$, $t \in I$, and $qL(y, t) = q(y)$ for all $y \in Y$,
$t \in I$ then p and q are called *fibre homotopy equivalent* and f and g are called *fibre
homotopy equivalences*.

The *homotopy category* is the category whose objects are pointed topological
spaces and whose morphisms are equivalence classes of pointed maps under the
relation of homotopy rel $\{*\}$.

We say that two diagrams of topological spaces are "equivalent up to homotopy"
if there is a natural equivalence between them in the homotopy category. A sequence
$W \to X \to Y$ will be called a *homotopy fibration* if it is equivalent up to homotopy
to a fibration sequence. Explicitly, $W \to X \to Y$ is a homotopy fibration if there
exists a homotopy commutative diagram

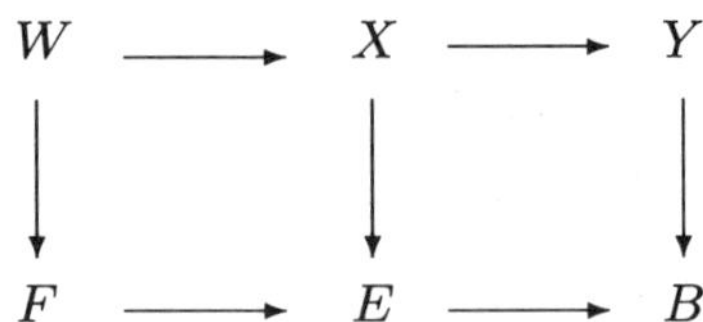

in which the vertical maps are homotopy equivalences and the bottom row is a fibration sequence.

We say that a space B has *numerable category* if it has a numerable cover $\{U_j\}_{j \in J}$ (see Definition 2.5.5) such that the inclusion map $U_j \hookrightarrow B$ is null homotopic for all j. (See Section 7.10 for motivation behind the name.) All CW-complexes have numerable category.

The basic properties of fibrations are given in the following theorems. It follows from them that a fibration over a base with numerable category has a cover by open sets over each of which the restriction of the fibration is fibre homotopy equivalent to a product fibration; this is the analogue of the local triviality condition imposed on fibre bundles.

Proposition 7.1.1

a) *The composition of fibrations is a fibration.*

b) *Let $p : E \to B$ a fibration and let $f : A \to B$ be any map. Then the induced map $p' : P \to A$ from the pullback P of p and f to A is a fibration.* $\square$

Theorem 7.1.2 *Let $p : E \to B$ be a fibration and let $f,\ \hat{f} : X \to B$ be homotopic. Then the fibrations over X induced from p by taking pullbacks with f and $\hat{f}$ are fibre homotopy equivalent.*

Proof By writing each of the maps $f,\ \hat{f}$ as a composite with the homotopy, we are reduced to the special case where the base has the form $X \times I$ and the maps $f,\ \hat{f}$ are the inclusions $f = i_0,\ \hat{f} = i_1$ of X at the two ends. Let $E_s = p^{-1}(X \times \{s\})$ denote the pullback of p with i_s. Define $h : X \times I \times I \times I \to X \times I$ by $h(x, r, s, t) = \big(x, (1 - t)r + st\big)$. Applying the homotopy lifting property to the commutative diagram

$$
\begin{array}{ccc}
E \times I & \xrightarrow{\ \ \pi'\ \ } & E \\[2mm]
\Big\downarrow{\scriptstyle E \times i_0} & & \Big\downarrow{\scriptstyle p} \\[2mm]
E \times I \times I & \xrightarrow{\ h \circ (p \times I \times I)\ } & X \times I
\end{array}
$$

(where π' denotes projection onto the first factor) gives a map $F : E \times I \times I \to E$ making both resulting triangles commute. Set $K = F_1 : E \times I \to E$ (where F_1 means $F|_{E \times I \times \{1\}}$, following our usual notation for homotopies).

$F(e, \pi''pe, t)$ for $0 \le t \le 1$ provides a homotopy from 1_E to the map $k(e) = K(e, \pi''pe)$ and satisfies $pF(e, \pi''pe, t) = pe$ for all t (where π'' denotes projection onto the second factor). Notice that $pK(e, s) = (\pi'pe, s)$ and so in particular $K(e, s)$ belongs to E_s. Define $\alpha : E_0 \to E_1$ by $\alpha(e) = K(e, 1)$ and $\beta : E_1 \to E_0$

by $\beta(e) = K(e, 0)$. Then $E_0 \times I \to E_0$ given by $(e, s) \mapsto K\big(K(e, 1-s), 0\big)$ gives a homotopy from $\beta \circ \alpha$ to $k \circ k|_{E_0}$ covering the constant homotopy $1_X \simeq 1_X$. Similarly $(e, s) \mapsto K\big(K(e, s), 1\big)$ gives a homotopy of $\alpha \circ \beta$ to $k \circ k|_{E_1}$ covering this constant homotopy. Therefore α and β are inverse fibre homotopy equivalences. $\square$

Notice that it follows from Theorem 7.1.2 that if $p : E \to B$ is a fibration with $B \approx *$ then the inclusion $p^{-1}(*) \hookrightarrow E$ is a homotopy equivalence.

Proposition 7.1.3 *Let $p : E \to B$ be a fibration and let $\gamma : I \to B$ be a path from b_0 to b_1. Let $\lambda : p^{-1}(b_0) \times I \to E$ be any lift of the composite $p^{-1}(b_0) \times I \xrightarrow{\pi''} I \xrightarrow{\gamma} B$ extending the inclusion $p^{-1}(b_0) \hookrightarrow E$. (Note: such a lift exists by the homotopy lifting property.) Then the composite $p^{-1}(b_0) \xrightarrow{i_1} p^{-1}(b_0) \times I \xrightarrow{\lambda} p^{-1}(b_1)$ is a homotopy equivalence. In particular, the fibres over any points in the same path component of B are homotopy equivalent. Furthermore, if $\gamma \simeq \gamma'$ (rel $\{0, 1\}$) and λ' is a similar lift of γ' then $\lambda \circ i_1 \simeq \lambda' \circ i_1$, so there is a well defined homotopy class $[g]_! : p^{-1}(b_0) \approx p^{-1}(b_1)$.* $\square$

Theorem 7.1.4 *Let $p : X \to B$ and $q : Y \to B$ be fibrations and let $f : X \to Y$ satisfy $qf = p$. Suppose that B has numerable category. Then the following are equivalent:*

a) *the induced map $f : p^{-1}(b) \to q^{-1}(b)$ is a homotopy equivalence for all $b \in B$;*

b) *f is a homotopy equivalence;*

c) *f is a fibre homotopy equivalence.*

See [May2, Theorems 1.5 and 2.6] for details. $\square$

Corollary 7.1.5 *Let $F \xrightarrow{j} E \xrightarrow{p} B$ be a fibration in which B has numerable category. Then there exists a (homotopy retraction) map $r : E \to F$ such that $rj \simeq 1_F$ if and only if there exists $r : E \to F$ such that $E \xrightarrow{(r, p)} F \times B$ is a homotopy equivalence.* $\square$

The next corollary follows from Corollary 7.5.6 and the 5-Lemma, although most of it could be deduced from the theorems already presented.

Corollary 7.5.10 *Let*

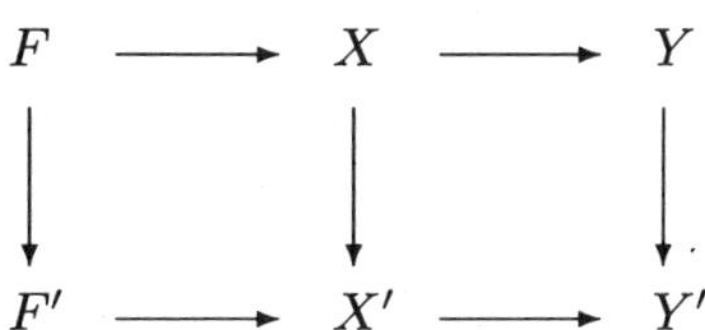

be a homotopy commutative diagram in which the rows are homotopy fibrations and X', Y, and Y' have the homotopy type of connected CW-complexes. If two of the vertical maps are homotopy equivalences then so is the third.

Contrasting Corollary 7.1.5 to Proposition 4.1.4, note that to conclude that $E \approx F \times B$ it is not sufficient to have a splitting $s : B \to E$ such that $ps = 1_B$. The situation is analogous to that in the category of groups, not that in abelian groups. Specifically, let N be a normal subgroup of a group G. If the inclusion

$N \hookrightarrow G$ is a split monomorphism then $G \cong N \times G/N$, however if $G \to G/N$ is a split epimorphism then it does not follow that G is a product group; G might be a semi-direct product of N and G/N.

The following theorem is due to Hurewicz. It strengthens the motivation for considering fibrations as generalized (locally trivial) fibre bundles.

Theorem 7.1.6 (Local Fibration is a Fibration) *Let $p : E \to B$. Suppose that there exists an open cover $\{V_j\}_{j \in J}$ of B such that the induced map $p : p^{-1}(V_j) \to V_j$ is a fibration for all j and such that $\{V_j\}_{j \in J}$ has a numerable refinement. Then p is a fibration.*

See [May2, Theorem 3.8]. $\qquad\qquad\qquad\qquad\qquad\qquad\qquad\qquad\qquad\qquad\qquad\qquad$ $\square$

Dualizing the concept of fibration gives the following. An inclusion $i : A \hookrightarrow X$ is said to have the *homotopy extension property* with respect to a space Y if given a homotopy $H : A \times I \to Y$ and an extension $f' : X \to Y$ of H_0, there exists an extension $H' : X \times I \to Y$ of H such that $H'_0 = f'$. An inclusion $i : A \hookrightarrow X$ is called a *cofibration* if it has the homotopy extension property with respect to Y for all Y. If (X, x_0) is a pointed space and the inclusion $\{x_0\} \hookrightarrow X$ is a cofibration then x_0 is called a *nondegenerate basepoint*. If A is a subspace of X we write X/A for the space $X/\sim$ where $a \sim a'$ for all $a \in A$. If $i : A \hookrightarrow X$ is a cofibration, X/A is called the *cofibre* of i and we sometimes say that $A \xrightarrow{i} X \longrightarrow C$ is a cofibration to indicate that C is the cofibre of i. The notation X/A unfortunately duplicates the notation introduced earlier for the space of orbits of the action of a group A on X and we must rely on the context to determine which is meant.

It is sometimes convenient to replace a pointed space X by one with a nondegenerate basepoint obtained as follows. Set $X' = X \vee I$ with the basepoint of X' chosen to be the end of I which is not attached to X. The pointed space X' has the same homotopy and homology groups as X, and has a nondegenerate basepoint. This replacement is sometimes called the "whisker trick".

Let A be a subspace of X and let $i : A \hookrightarrow X$ be the inclusion map. If there exists $r : X \to A$ such that $ri = 1_A$ then A is called a *retract* of X. If there exists $r : X \to A$ such that $ri \simeq 1_A$ then A is called a *homotopy retract* of X. A retraction $r : X \to A$ is called a *deformation retraction* if $ir \simeq 1_X$. A retraction $r : X \to A$ is called a *strong deformation retraction* if $ir \simeq 1_X$ (rel A).

Proposition 7.1.7

   a) *An inclusion $A \hookrightarrow X$ is a cofibration if and only if the inclusion $(A \times I) \cup (X \times 0) \hookrightarrow X \times I$ has a retraction.*
   b) *The composition of cofibrations is a cofibration.*
   c) *Let $i : A \to X$ be a cofibration and let $f : A \to B$ be any map. Then the induced map $i' : B \to Q$ from B to the pushout Q of i and f, is a cofibration.*
   $\qquad\qquad\qquad\qquad\qquad\qquad\qquad\qquad\qquad\qquad\qquad\qquad\qquad\qquad\qquad\qquad$ $\square$

The dual of Theorem 7.1.2 is

Theorem 7.1.8 *Let $j : A \to X$ be a cofibration with A closed in X and let $f, g : A \to Y$ be homotopic. Then the pushouts of j with f and g are homotopy equivalent* rel Y.

Proof Let $H : A \times I \to Y$ be a homotopy from f to g. Let

$$
\begin{array}{ccc}
A \xrightarrow{\ f\ } Y & \quad A \xrightarrow{\ g\ } Y & \quad A \times I \xrightarrow{\ H\ } Y \\
\downarrow{j} \qquad \downarrow{p_f} & \downarrow{j} \qquad \downarrow{p_g} \quad \text{and} & \downarrow{j \times I} \qquad \downarrow{p} \\
X \xrightarrow{\ i_f\ } Q_f & \quad X \xrightarrow{\ i_g\ } Q_g & \quad X \times I \xrightarrow{\ i\ } Q
\end{array}
$$

be pushouts. To show $Q_f \approx Q_g$, by symmetry it suffices to show $Q_f \approx Q$. The maps $i_0 : Z \to Z \times I$, for $Z = A$ and X, yield a map from the first square to the third thus inducing a map $\eta : Q_f \to Q$ satisfying $\eta p_f = p$ and $\eta i_f = ii_0$. By the homotopy extension property, there exists $J : X \times I \to Q_f$ such that $J(j \times I) = p_f H$ and $J_0 = i_f$. The first equality yields an induced map $\phi : Q \to Q_f$ such that $\phi p = p_f$ and $\phi i = J$. $\phi \eta p_f = \phi p = p_f$ and $\phi \eta i_f = \phi ii_0 = Ji_0 = J_0 = i_f$ and so $\phi \eta = 1_{Q_f}$. Let $\tilde{I} = (I \times 0) \cup (0 \times I) \cup (I \times 1) \subset I \times I$. Choose a self-homeomorphism α of $I \times I$ such that $\alpha(\tilde{I}) = 0 \times I$ and let $\beta = \alpha^{-1}$. Let $r : X \times I \to (X \times 0) \cup (A \times I)$ be a retraction (see Proposition 7.1.7a). Set

$$
\tilde{r} = (1_X \times \beta)(r \times 1_I)(1_X \times \alpha) : X \times I \times I \to (X \times \tilde{I}) \cup (A \times I \times I).
$$

Then $\tilde{r}$ is a retraction. Define $M : (X \times \tilde{I}) \cup (A \times I \times I) \to Q$ by $M(x, s, 0) = i(x, s)$, $M(X, 0, t) = i(x, 0)$, $M(x, s, 1) = \eta J(x, s)$ and $M(a, s, t) = i(a, s)$ for all $a \in A$, $x \in X$, $s, t \in I$. The definitions imply that M has been defined consistently on the overlapping subsets, and since A is closed in X these overlaps are all closed subsets of $(X \times \tilde{I}) \cup (A \times I \times I)$ and so M is continuous. Let $K = M\tilde{r} : X \times I \times I \to Q$ and let $L : Q \times I \to Q$ be defined by $L(p_f y, t) = p_f y$ and $L(i(x, s), t) = K(x, s, t)$, which is well defined and continuous since it is consistent on the image of $A \times I \times I$. Then $L_0 p_f = p_f$ and $L_0 i(x, s) = K(x, s, 0) = M\tilde{r}(x, s, 0) = M(x, s, 0) = i(x, s)$ for all $x \in X$, $s \in I$. Therefore $L_0 = 1_Q$. $L_1 p_f = p_f$ and $L_1 i(x, s) = K(x, s, 1) = M\tilde{r}(x, s, 1) = \eta J(x, s) = \eta \phi i(x, s)$ for all $x \in X$, $s \in I$. Consequently $L_1 = \eta \phi$. Thus $L : 1_Q \simeq \eta \phi$ and so η and ϕ are inverse homotopy equivalences between Q_f and Q. $\qquad\qquad\square$

Proposition 7.1.9

a) *Let $i : A \hookrightarrow X$ be a cofibration. Then i has a retraction if and only if i has a homotopy retraction.*

b) *Let $i : A \hookrightarrow X$ be a cofibration where A is closed in X. Then i is a homotopy equivalence if and only if i has a strong deformation retraction.*

c) *Let $A \hookrightarrow X$ be a cofibration where A is contractible and closed in X. Then the quotient map $X \to X/A$ is a homotopy equivalence.* $\qquad\square$

Theorem 7.1.10 *Let A be a closed subspace of X. Then the following are equivalent:*

1) *the inclusion $A \hookrightarrow X$ is a cofibration;*
2) *there exist continuous functions $u : X \to I$ and $h : X \times I \to X$ such that:*
    (i) $A = u^{-1}(0)$;
    (ii) $h(x, 0) = x$ for all $x \in X$;
    (iii) $h(a, t) = a$ for all $a \in A$, $t \in I$;
    (iv) $h(x, 1)$ lies in A for all $x \in u^{-1}([0, 1))$.

See Steenrod [St1].   □

Notice that the conditions of part (2) imply that h is a strong deformation retraction from $u^{-1}([0,1))$ to A. If $A \subset X$ satisfies the conditions of the preceding theorem we say that A is a *neighbourhood deformation retract* of X or that (X, A) is an NDR-pair.

Intuition suggests that $H_*(X, A)$ should be related in some way to $H_*(X/A)$.

Proposition 7.1.11 *Let* (X, A) *be an* NDR-*pair. Then* $H_*(X, A) \cong \tilde{H}_*(X/A)$.

Proof Set $W = u^{-1}([0,1))$. Then h yields a strong deformation retraction from W to A and W/A is contractible. Combining excisions with isomorphisms coming from long exact homology sequences gives

$$\tilde{H}_*(X/A) \cong H_*(X/A, W/A) \cong H_*(X - A, W - A) \cong H_*(X, W) \cong H_*(X, A).$$

□

Thus there is a long exact homology sequence for an NDR-pair.
The next theorem is demonstrated inductively over the cells.

Theorem 7.1.12 *Let* A *be a relative subcomplex of a relative* CW-*complex* X. *Then* (X, A) *is an* NDR-*pair.*   □

The following is dual to Corollary 7.1.5 and Corollary 7.5.10.

Corollary 7.5.11

a) *Let* $A \xrightarrow{i} X \xrightarrow{c} Y$ *be an* NDR-*pair sequence of pointed spaces (i.e., a cofibration sequence in which* (X, A) *is a pointed* NDR-*pair) in which each of the spaces is simply connected and has the homotopy type of a* CW-*complex. Suppose that there exists a (homotopy splitting) map* $s : Y \to X$ *such that* $cs \simeq 1_Y$. *Then* $A \vee Y \xrightarrow{i \perp s} X$ *is a homotopy equivalence.*

b) *Let*

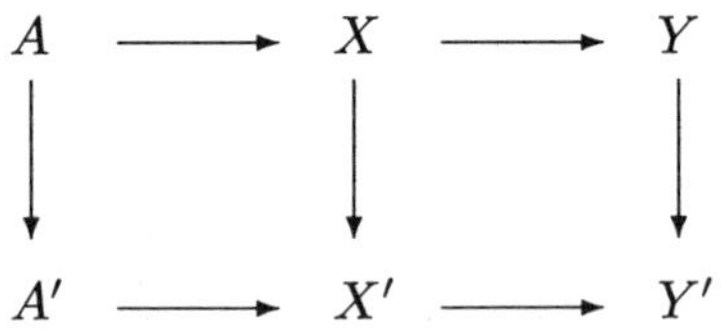

be a homotopy commutative diagram in which the rows are equivalent up to homotopy to NDR-*pair sequences and each of the spaces is simply connected and has the homotopy type of a* CW-*complex. If two of the vertical maps are homotopy equivalences then so is the third.*

This follows from Corollary 7.5.7 by application of the 5-Lemma. In contrast to Corollary 7.5.10 which holds for any connected CW-complex, the dual statements of Corollary 7.5.11 are false without the simply connectedness hypothesis. For a counterexample, see Example 7.4.7.

Proposition 7.1.13 *Let* $A \to X$ *be a cofibration. Then for any space* Y:

a) *if* A *and* X *are compactly generated then the induced map* $(Y^X)_{\mathcal{K}} \to (Y^A)_{\mathcal{K}}$ *has the homotopy lifting property with respect to* Z *for all* Z *and is therefore a fibration if it is surjective;*

b) *if A and X are locally compact then the induced map $Y^X \to Y^A$ has the homotopy lifting property with respect to Z for all Z and is therefore a fibration if it is surjective.* $\qquad\square$

Of critical importance in homotopy theory is the fact that every map can be "turned into a fibration" according to the following theorem.

Theorem 7.1.14 *Let $f : X \to Y$ be a map of pointed spaces which induces a surjective map on path components. Then there exists a factorization $f = p\phi$ where $\phi : X \to X'$ is a homotopy equivalence and $p : X' \to Y$ is a fibration.*

An explicit construction satisfying the conditions is as follows. Let $P^f \to X$ be the pullback of f with the map $ev_0 : Y^I \to Y$ given by $ev_0(\omega) = \omega(0)$, which is both a fibration and a homotopy equivalence. The composite $P^f \longrightarrow Y^I \overset{ev_1}{\longrightarrow} Y$ is a fibration and the map $\phi^f : X \to P^f$ given by $\phi^f(x) = (x, \omega)$ where $\omega(t) = f(x)$ for all t is a homotopy equivalence. P^f is called the *mapping path fibration* of f.

If Y is path connected, applying this construction to the map $* \to Y$ gives a fibration $ev : PY \to Y$ with contractible total space where $PY = \{\omega : I \to Y \mid \omega(0) = *\}$ and $ev(\omega) = \omega(1)$. PY is called the *path space* over Y. The fibre of ev is called the *loop space* of Y, denoted ΩY. Explicitly

$$\Omega Y = ev^{-1}(*) = \{\omega : I \to Y \mid \omega(0) = \omega(1) = *\} = \mathrm{Map}_*(S^1, Y).$$

Of course, this construction makes sense for any pointed space Y, but is determined by the component of Y containing the basepoint. ΩY becomes an H-space under the operation which multiplies paths in the same way that elements of $\pi_1(Y)$ are multiplied. (See Section 7.2 for basic definitions concerning H-spaces.) The same operation also induces a homotopy action $\Omega Y \times PY \to PY$ of ΩY on PY.

If the fibration $q : P \to W$ is the pullback of $ev : PX \to X$ and f for some space X and some map $f : W \to X$ then q is called a *principal fibration*. In this case, there is an induced homotopy action $\Omega X \times P \to P$. We noted earlier that, for an arbitrary fibration $F \overset{j}{\longrightarrow} E \overset{p}{\longrightarrow} B$, a right homotopy splitting of p does not imply that $E \approx F \times B$, however in the case of a principal fibration, a right homotopy splitting of p implies a left homotopy splitting of j, so the situation reduces to that of Corollary 7.1.5.

Beginning from a fibration $f : X \to Y$ as above we can form the mapping fibration sequence $F^f \overset{j_0}{\longrightarrow} P^f \overset{f'}{\longrightarrow} Y$ where $f = f'\phi$ and $\phi : X \to P^f$ is a homotopy equivalence with homotopy inverse $\psi : P^f \to X$ given by $\psi(x, \omega) = x$. Let $f_1 = \psi j_0 : F^f \to X$, let F be the fibre of f, and let $\tilde{P}Y = \{\omega : I \to Y \mid \omega(1) = *\}$ be an "inverted" PY. In the commutative diagram

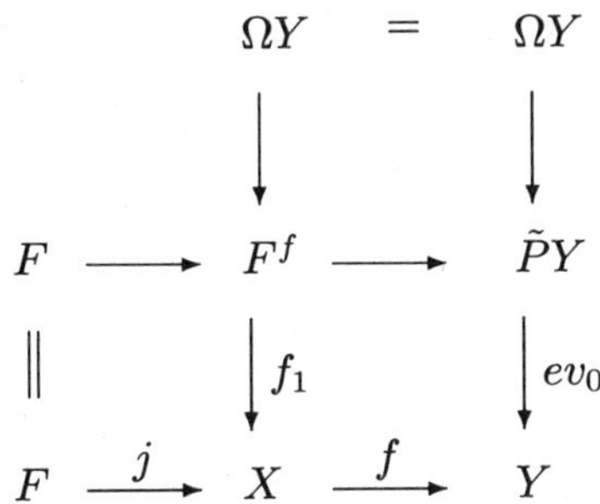

the bottom right square is a pullback square, the vertical maps and horizontal sequences are fibration sequences, and the map $F \to F^f$ is a homotopy equivalence by Theorem 7.1.2. Applying this procedure again to the fibration $f_1 : F^f \to X$ gives a mapping fibration sequence $F^{f_1} \longrightarrow P^{f_1} \xrightarrow{f_1'} X$ and a diagram

$$
\begin{array}{ccc}
\Omega X & = & \Omega X \\
\downarrow{\scriptstyle i} & & \downarrow \\
\Omega Y \xrightarrow{\ \eta\ } F^{f_1} & \longrightarrow & \tilde{P}X \\
\| \qquad\qquad \downarrow{\scriptstyle f_2} & & \downarrow{\scriptstyle ev_0} \\
\Omega Y \longrightarrow F^f & \xrightarrow{\ f_1\ } & X
\end{array}
$$

Lemma 7.1.15 $i \simeq \eta \circ (\Omega f)^{-1}$.

Note $(\Omega f)^{-1}$ is defined by $(\Omega f)^{-1}(\omega)(t) = f(\omega(1-t))$ as discussed at the beginning of Section 7.2.

Proof From the definitions, $F^{f_1} = \{(\alpha, \omega, x) \mid x \in X,\ \alpha : I \to X,\ \omega : I \to Y,\ \alpha(0) = x,\ \alpha(1) = *,\ \omega(0) = f(x),\ \omega(1) = *\}$ and the maps i and η are given by $i(\alpha) = (\alpha, c_*, *)$ and $\eta(\omega) = (c_*, \omega, *)$ where c_* denotes the constant path at $*$. Since $\tilde{P}X$ is contractible, the fibration $\Omega Y \xrightarrow{\ \eta\ } F^{f_1} \longrightarrow \tilde{P}X$ shows that η is a homotopy equivalence (Theorem 7.1.2). We will show that an explicit homotopy inverse is given by $\phi : F^{f_1} \to \Omega Y$ where $\phi(\alpha, \omega, x) = f(\alpha^{-1}) \cdot \omega$. Note that $f(\alpha^{-1})(1) = f(x) = \omega(0)$ so $f(\alpha^{-1}) \cdot \omega$ makes sense. It is clear that $\phi\eta \simeq 1_{\Omega Y}$. Intuitively the required homotopy $H : F^{f_1} \times I \to F^{f_1}$ from $1_{F^{f_1}}$ to $\eta\phi$ gradually reduces the distance travelled by the curve in the first component while taking the remainder of the curve and adjoining its image under f to the second component. Explicitly, set $H\big(\alpha, (\omega, x), t\big) = \Big(\alpha_{t,1}, \big(f(\alpha_{0,t}^{-1}) \cdot \omega, \alpha(t)\big)\Big)$ where for $0 \le a \le b \le 1$, $\alpha_{a,b}$ is the portion of the path α joining $\alpha(a)$ to $\alpha(b)$ reparametrized; i.e., $\alpha_{a,b}(s) = \alpha\big((1-s)a + sb\big)$. Since it is clear that $\phi i \simeq (\Omega f)^{-1}$, the lemma follows. $\qquad\square$

Continuing the above procedure produces a sequence of mapping fibrations $F^{f_n} \longrightarrow P^{f_n} \xrightarrow{f_n'} F^{f_{n-2}}$, and replacing P^{f_n} by the homotopy equivalent space $F^{f_{n-1}}$ gives an infinite sequence $\ldots \to F^{f_n} \to F^{f_{n-1}} \to F^{f_{n-2}} \to \ldots \to F^{f_1} \to F \to X \to Y$ in which every triple of consecutive maps forms a homotopy fibration sequence. Replacing by means of the homotopy equivalences $\eta : \Omega F^{f_{n-3}} \to F^{f_n}$ and taking into account Lemma 7.1.15 gives the equivalent sequence

$$
\ldots \longrightarrow \Omega^k F \xrightarrow{(\Omega^k j)^{(-1)^k}} \Omega^k X \xrightarrow{(\Omega^k f)^{(-1)^k}} \Omega^k Y \xrightarrow{\ \partial\ } \Omega^{k-1} F \longrightarrow \ldots
$$
$$
\longrightarrow \Omega X \xrightarrow{(\Omega f)^{-1}} \Omega Y \xrightarrow{\ \partial\ } F \xrightarrow{\ j\ } X \xrightarrow{\ f\ } Y.
$$

For an arbitrary map $f : X \to Y$, applying the above procedure to the map $f' : X' \to Y$ obtained by turning f into a fibration and then replacing X' by the homotopy equivalent X at the end gives the same sequence. If one prefers, one can use the self-homotopy-equivalence $\omega \to \omega^{-1}$ on some of the loop spaces together with replacement of some of the connecting maps ∂ to produce an equivalence up to homotopy of the above diagram with one in which no inverses appear.

We now consider the dual situation. Any map can be turned into a cofibration.

Theorem 7.1.16 *Let $f : A \to X$ be a map of pointed spaces. Then there exists a factorization $f = \phi i$ where $i : A \to X'$ is a cofibration and $\phi : X' \to X$ is a homotopy equivalence.*

Let $f : A \to X$ be a map of pointed spaces. An explicit construction satisfying the conditions is as follows. Let $X \to M_f$ be the pushout of f and the map $i_0 : A \to (A \times I)/(* \times I)$ given by $i_0(a) = (a, 0)$, which is both cofibration and a homotopy equivalence. Then the composite $A \xrightarrow{i_1} A \times I \to M_f$ is a cofibration and the map $\phi_f : M_f \to X$ given by $\phi_f(x) = x$ and $\phi_f(a, t) = f(a)$ is a homotopy equivalence whose homotopy inverse $\psi_f : X \to M_f$ is the inclusion map in the pushout defining M_f. M_f is called the *(reduced) mapping cylinder* of f and $C_f = M_f/(A \times 1)$ is called the *(reduced) mapping cone* on f. The unreduced mapping cylinder and mapping cone can be defined by replacing $(A \times I)/(* \times I)$ by $A \times I$ in each definition. If the basepoint of A is nondegenerate then these are homotopy equivalent to the reduced versions, however we use the reduced spaces because with the unreduced spaces it is not possible to choose a basepoint for the mapping cylinder in such a way that $i : A \hookrightarrow M_f$ and ψ_j are both pointed maps.

Applying this construction to the map $A \to *$ gives a cofibration $i_1 : A \to CA$ where CA is the contractible space $CA = (A \times I)/((A \times 0) \cup (* \times I))$ called the *(reduced) cone* on A. The cofibre of i_1 is called the *(reduced) suspension* of A, denoted SA. We also have $CA = A \wedge I$ and $SA = A \wedge S^1$ where $X \wedge Y = (X \times Y)/(X \vee Y)$. From the Künneth formula and the exact homology sequence of the pair $(X \times Y)/(X \vee Y)$ we get the reduced Künneth formula

$$0 \to \left(\tilde{H}_*(X) \otimes \tilde{H}_*(Y)\right)_n \to H_n(X \wedge Y) \to \mathrm{Tor}^R\left(\tilde{H}_*(X), \tilde{H}_*(Y)\right)_{n-1} \to 0.$$

Therefore we get the formula $H_n(SX) \cong H_{n-1}(X)$, which can also be obtained by applying Mayer-Vietoris to a cover of SX by upper and lower cones.

For any space Z, the multiplication $Z^{S^1 \vee S^1} = \Omega Z \times \Omega Z \to \Omega Z = Z^{S^1}$ described earlier is induced by the map $c : S^1 \to S^1 \vee S^1$ which pinches the equator S^0 to a point. This induces a co-H-space structure $1_A \wedge c$ on $A \wedge S^1 = SA$ and the same operation induces a homotopy co-action map $CA \to CA \vee SA$. If the cofibration $j : W \to Q$ is the pushout of $i_1 : B \to CB$ and k for some space W and some map $k : B \to W$, then j is called a *principal cofibration*. In this case, there is an induced homotopy co-action $Q \to Q \vee SB$. If $A \xrightarrow{j} X \xrightarrow{q} Y$, is a principal cofibration, a left homotopy splitting of j implies a right homotopy splitting of q.

Let $\mathcal{T}op_*$ denote the category of pointed topological spaces and basepoint preserving maps.

Proposition 7.1.17 $\mathcal{T}op_*(SX, Y) = \mathcal{T}op_*(X, \Omega Y)$ *for pointed spaces X and Y. In other words, $S(\ )$ and $\Omega(\ )$ are adjoint functors from the category of pointed topological spaces to itself.* $\square$

If $j : A \to X$ is a cofibration with cofibre C in which A is closed in X, the dual of the procedure described earlier for fibrations yields an infinite sequence

$$A \xrightarrow{j} X \xrightarrow{q} C \xrightarrow{\delta} SA \xrightarrow{(Sj)^{-1}} SX \longrightarrow \dots$$

$$\longrightarrow S^{k-1}C \xrightarrow{\delta} S^k A \xrightarrow{(S^k j)^{(-1)^k}} S^k X \xrightarrow{(S^k q)^{(-1)^k}} C \longrightarrow \dots.$$

in which each triple of consecutive maps is a homotopy cofibration sequence. In the more general situation of an arbitrary map $f : A \to X$ from a space A with nondegenerate basepoint, we can consider instead the cofibration map $A \to M_f$. Then A will be a closed subset of M_f and so the same sequence is obtained by applying the above procedure to $A \to M_f$.

If one draws pictures in which $Y \cup CB$ is represented by a cone over the subset B and considers the homotopy equivalence $Y \cup CB \to Y/B$ which collapses the cone, the appearance of SA and SX in this sequence is easy to visualize. As in the case of fibrations, the sequence can be replaced by an equivalent sequence up to homotopy in which the minus signs have been eliminated.

2 Group Structures on Sets of Homotopy Classes

In this section we will assume that all the spaces and maps are pointed and that all homotopies are relative to the basepoint.

Let X be a (pointed) topological space. If G is a topological group (regarded as a pointed space by choosing the identity $e \in G$ as the basepoint) then the group structure on G induces a group structure on $[X, G]$. However it is clear that all that is really needed from G is a "group up to homotopy".

An *H-space* (or *Hopf space*) consists of a pointed space (H, e) together with a continuous map $m : H \times H \to H$ such that $mi_1 \simeq 1_H$ and $mi_2 \simeq 1_H$ where $i_1(x) = (x, e)$ and $i_2(x) = (e, x)$. The multiplication m is often ignored in the notation and we refer to the H-space H. Occasionally this expression is also used to indicate a space for which an H-space structure exists. An H-space H is called *homotopy associative* if $m \circ (1_H \times m) \simeq m \circ (m \times 1_H)$. If H is an H-space, a map $c : H \to H$ is called a *homotopy inverse* for H if $m \circ (1_H, c) \simeq *$ and $m \circ (c, 1_H) \simeq *$. A homotopy associative H-space together with a homotopy inverse c for H is called an *H-group*. An H-space is called *homotopy abelian* if $mT \simeq m$, where $T : H \times H \to H \times H$ is given by $T(x, y) = (y, x)$. A map $f : X \to Y$ between H-spaces is called an *H-map* if $m_Y \circ (f \times f) \simeq f \circ m_X$.

Proposition 7.2.1 *Let H be an H-group. Then for all W, the operation $([f], [g]) \to [m \circ (f, g)]$ gives a natural group structure on $[W, H]$. An H-map $f : H \to H'$ induces a group homomorphism $f_\# : [W, H] \to [W, H']$. If H is homotopy abelian then the group $[W, H]$ is abelian.* $\square$

Two H-spaces structures m, m' on X are called equivalent if $m \simeq m' : X \times X \to X$. Equivalent H-space structures on X result in the same group structure on $[Y, X]$. If X is an H-space, a homotopy equivalence $\phi : X \to X'$ induces an H-space structure on X'. If $f : X \to Y$ is an H-map and $f = \phi f'$ with ϕ a homotopy equivalence and f' a fibration then f' becomes an H-map when using this H-space structure on X'.

It is easy to see that a retract of an H-space is an H-space. We shall later prove the following stronger statement for connected CW-complexes.

Theorem 7.9.5 *A connected CW-complex X is an H-space if and only if the canonical map $X \to \Omega SX$ has a left homotopy splitting.*

A retract of a homotopy associative H-space need not be homotopy associative. Indeed if that were true then Theorem 7.9.5 would imply that for connected CW-complexes every H-space would be homotopy associative, since ΩY is homotopy associative for any Y. However it is not hard to check that a retract of a homotopy commutative H-space is homotopy commutative.

In Section 7.5 we will see

Theorem 7.5.12

a) *Let (H, m) be an H-space with nondegenerate basepoint e. Then there exists an equivalent multiplication m' on H such that e forms a strict identity for m'. That is, $m'(e, x) = m'(x, e) = x$ for all $x \in H$.)*

b) *Let X and Y have the homotopy type of connected CW-complexes and let $f : X \to Y$ be a fibration which is also an H-map between H-spaces with strict nondegenerate identities. Then there exists an equivalent multiplication m' on X such that m' has a strict identity and f is strictly multiplicative. (f is strictly multiplicative means $m_Y \circ (f \times f) = f \circ m'$.)*

Corollary 7.5.13 *Let*

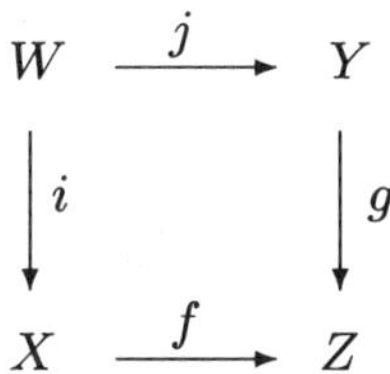

be a pullback square in which at least one of f, g is a fibration. Suppose that X, Y, and Z are H-spaces having the homotopy types of connected CW-complexes and that f and g are H-maps. Then there is an H-space structure on W such that i and j are H-maps.

A natural group structure on $[X, Y]$ can also come from extra structure on X instead of coming from an H-space structure on Y. A *co-H-space* consists of a space X together with a continuous map $\psi : X \to X \vee X$ such that $(1_X \perp *) \circ \psi \simeq 1_X$ and $(* \perp 1_X) \circ \psi \simeq 1_X$ where $f \perp g : A \vee B \to Z$ denotes the map induced by $f : A \to Z$, $B \to Z$. Dualizing the other notions in the obvious way gives the concepts of *homotopy (co)associative*, *homotopy inverse*, *co-H-group*, and *homotopy (co)abelian* for a co-H-space, and of a *co-H-map* between co-H-spaces.

Proposition 7.2.2 *Let X be a co-H-group. Then for all Y, the operation $([f], [g]) \to [(f \perp g) \circ \psi]$ gives a natural group structure on $[X, Y]$. A co-H-map $f : X \to X'$ induces a group homomorphism $f^\# : [X', Y] \to [X, Y]$. If X is homotopy co-abelian then the group $[X, Y]$ is abelian.* $\qquad\square$

The following two theorems are dual to Theorem 7.9.5 and Corollary 7.5.13.

Theorem 7.9.6 *A connected CW-complex X is a co-H-space if and only if the canonical map $S\Omega X \to X$ has a right homotopy splitting.*

Theorem 7.9.8 *Let*

$$
\begin{array}{ccc}
X & \xrightarrow{\;f\;} & Y \\
\downarrow{\scriptstyle g} & & \downarrow{\scriptstyle p} \\
Z & \xrightarrow{\;q\;} & Q
\end{array}
$$

be a pushout square in which at least one of f, g is a cofibration. Suppose that X, Y, and Z are co-H-spaces having the homotopy types of connected CW-complexes. Suppose that f and g are co-H-maps and that X is simply connected. Then there is a co-H-space structure on Q such that i and j are co-H-maps.

Theorem 7.2.3 *Let X be a co-H-group and Y an H-group. Then the group operation on $[X,Y]$ induced by the co-H-structure on X equals the group operation induced by the H-structure on Y. The resulting group structure is abelian.*

Proof Let Δ and ∇ denote the diagonal and fold maps respectively. For any maps a, b, c, $d : X \to Y$, the equality

$$(\nabla_Y \times \nabla_Y) \circ \big((a \vee b) \times (c \vee d)\big) \circ \Delta_{X \vee X}$$
$$= \nabla_{Y \times Y} \circ \big((a \times c) \vee (b \times d)\big) \circ (\Delta_X \vee \Delta_X) : X \vee X \to Y \times Y$$

can be checked by examining the four maps from $X \to Y$ obtained by restricting to each copy of X and projecting to each copy of Y. Composing with m_Y on one end and ψ_X on the other shows that $(a * b) \bullet (c * d) = (a \bullet c) * (b \bullet d)$, where $*$ and $\bullet$ denote the group operations obtained from X and Y respectively. Therefore the theorem follows from the following easily verified lemma.

Lemma 7.2.4 *Let Z be a set with two binary operations $*$ and $\bullet$ which have a common two-sided identity element and satisfy $(a * b) \bullet (c * d) = (a \bullet c) * (b \bullet d)$ for all a, b, c, $d \in Z$. Then the operations $*$ and $\bullet$ are equal and this common operation is abelian.* $\qquad\square$

Remark Under the conditions of Lemma 7.2.4, the operations are also associative, so the hypotheses of Theorem 7.2.3 can be weakened somewhat.

It is clear that the unreduced suspension of S^n is homeomorphic to S^{n+1} and it follows that $SS^n \approx S^{n+1}$ but in fact a stronger statement is true.

Proposition 7.2.5 $SS^n \cong S^{n+1}$ *for all $n \geq 0$.*

Proof $SS^n \cong S\big(I^n/\partial(I^n)\big) \cong I^{n+1}/\partial(I^{n+1}) \cong S^{n+1}$. $\qquad\square$

We previously defined $\pi_n(X) = [S^n, X]$. For $n \geq 1$ we give it the natural group structure coming from the homeomorphism $S^n \cong SS^{n-1}$. It follows from Theorem 7.2.3 and Proposition 7.1.17 that $\pi_n(X)$ is abelian for $n \geq 2$. By the lifting theorem, if X has a simply connected (thus universal) covering space $\tilde{X}$ then the covering projection induces an isomorphism $\pi_n(\tilde{X}) \to \pi_n(X)$ for $n \geq 2$. For a basepointed topological pair (X, A), we set $\pi_n(X, A) = [(D^n, S^{n-1}), (X, A)]$. For $n \geq 2$ we give it the natural group structure induced by the map $(D^n, S^{n-1}) \to (D^n, S^{n-1}) \vee (D^n, S^{n-1})$ which pinches the equator to a point. It is easy to see that $\pi_n(X, *) \cong \pi_n(X)$.

Proposition 7.2.6 *For a basepointed pair X, A there is a (long) exact sequence*

$$\ldots \longrightarrow \pi_n(A) \xrightarrow{j_\#} \pi_n(X) \xrightarrow{q_\#} \pi_n(X, A) \xrightarrow{\partial} \pi_{n-1}(A) \longrightarrow \ldots \longrightarrow \pi_0(A) \xrightarrow{j_\#} \pi_0(X)$$

where $j : A \to X$ is the inclusion, $q_\#(g)$ is the composition $D^n \twoheadrightarrow S^n \xrightarrow{g} X$, and $\partial(f) = f|_{S^{n-1}}$.

All but the last three terms in the sequence are groups and all but the last six are abelian groups.

Let $f : S^n \to W$ represent an element of $\pi_n(W, w_0)$. Since $* \to S^n$ is a cofibration, any path γ in W from w_0 to w_1, regarded as a homotopy between its endpoints, can be extended to a homotopy $H : S^n \times I \to W$ such that $H_0 = f$. The homotopy class of H_1 in $\pi_n(W, w_1)$ is uniquely determined by that of f and the homotopy class rel $\{0, 1\}$ of γ, and is written $[\gamma] \cdot [f]$. In particular, if γ is a closed path then $[\gamma] \cdot [f]$ is again in $\pi_n(W, w_0)$ and the assignment $\mu([\gamma], [f]) = [\gamma] \cdot [f]$ yields an action of $\pi_1(W, w_0)$ on $\pi_n(W, w_0)$. More generally, let (X, A, a_0) be a basepointed pair with inclusion $j : A \hookrightarrow X$. Given a pointed map $f : (D^n, S^{n-1}) \to (X, A)$, any path γ joining a_0 to a_1 in A can be extended using the cofibration $* \to S^{n-1}$ to a homotopy $H : S^{n-1} \times I \to A$ such that $H_0 = f|_{S^{n-1}}$, and then $(j \times 1_I) \circ H$ can be extended using the cofibration $S^{n-1} \to D^n$ to a homotopy $J : D^n \times I \to X$ such that $J_0 = g$. This yields an action of $\pi_1(A, a_0)$ on $\pi_n(X, A, a_0)$. Through the map $\pi_1(A, a_0) \xrightarrow{j_\#} \pi_1(X, a_0)$, $\pi_1(A, a_0)$ also acts on $\pi_n(X, a_0)$ and all the maps in the exact homotopy sequence of the pair (X, A) commute with the action of $\pi_1(A)$.

Proposition 7.2.7 *Let $F \xrightarrow{j} E \xrightarrow{p} B$ be a fibration. Then $[W, F] \xrightarrow{j_\#}$ $[W, E] \xrightarrow{p_\#} [W, B]$ is an exact sequence of pointed sets for any space W. (That is, $\operatorname{Im} j_\# = \{x \in [W, F] \mid p_\#(x) = *\}$.)* $\qquad\square$

Applying the preceding proposition to the infinite sequence of mapping fibrations (preceding Theorem 7.1.16) and using adjointness gives

Corollary 7.2.8 *Let $F \longrightarrow E \longrightarrow B$ be a fibration. Then for any space W there is a long exact sequence*

$$\ldots \to [S^n W, F] \to [S^n W, E] \to [S^n W, B] \to [S^{n-1} W, F] \to$$
$$\ldots \to [W, F] \to [W, E] \to [W, B].$$

$\qquad\square$

In the case $W = S^1$ this sequence is equivalent to the long exact sequence of a pair in the following sense. Let (X, A) be a topological pair with inclusion map $j : A \hookrightarrow X$, let $A \xrightarrow{\phi} A' \xrightarrow{p} X$ be a factorization of j as a homotopy equivalence ϕ followed by a fibration p, and let F be the fibre of p. Then there is an isomorphism $\pi_{n-1}(F) \to \pi_n(X, A)$ such that

$$
\begin{array}{ccccccccccc}
\pi_n(F) & \longrightarrow & \pi_n(A') & \xrightarrow{p_\#} & \pi_n(X) & \xrightarrow{\partial} & \pi_{n-1}(F) & \longrightarrow & \ldots & \longrightarrow & \pi_1(A') & \longrightarrow & \pi_1(X) \\
\downarrow{\scriptstyle\cong} & & \downarrow{\scriptstyle\cong} & & \downarrow{\scriptstyle\cong} & & \downarrow{\scriptstyle\cong} & & & & \downarrow{\scriptstyle\cong} & & \downarrow{\scriptstyle\cong} \\
\pi_{n+1}(X, A) & \xrightarrow{\partial} & \pi_n(A) & \xrightarrow{j_\#} & \pi_n(X) & \longrightarrow & \pi_n(X, A) & \longrightarrow & \ldots & \longrightarrow & \pi_1(A) & \longrightarrow & \pi_1(X)
\end{array}
$$

commutes.

Given a cofibration $A \xrightarrow{j} X \longrightarrow C$ and a space V, applying the preceding corollary with $W = S^0$ to the fibration $V^C \longrightarrow V^X \xrightarrow{j^*} (V^A)'$ (where $(V^A)' = \operatorname{Im} j^* \subset V^A$) gives

Corollary 7.2.9 *Let* $A \longrightarrow X \longrightarrow C$ *be a cofibration. Then for any space* V *there is a long exact sequence*

$$\ldots \to [S^n C, V] \to [S^n X, V] \to [S^n A, V] \to [S^{n-1} A, V] \to$$
$$\ldots \to [C, V] \to [X, V] \to [A, V].$$
$\square$

3 Homology of a Loop-suspension

For any pointed space X the adjoint of the identity map 1_{SX} gives a map $\alpha_X : X \to \Omega SX$. Explicitly, $\alpha_X(x)(t) = (x, t)$ for all $x \in X$, $t \in I$. Given a commutative ring R and an R-module M, let $T(M)$ denote the tensor algebra over R on M.

Theorem 7.3.1 (Bott-Samelson) *Let* R *be a PID and let* X *be a connected space such that* $H_*(X; R)$ *is torsion free. Then* $H_*(\Omega SX; R) \cong T(\tilde{H}_*(X; R))$ *and* $\alpha : X \to \Omega SX$ *induces the canonical map from* $\tilde{H}_*(X; R)$ *to* $T(\tilde{H}_*(X; R))$.

Proof Throughout this proof $H_*(\ )$ will denote homology with coefficients in R. Applying the whisker trick if necessary (see discussion preceding Proposition 7.1.7), we may assume that X has a nondegenerate basepoint. Let S^+X and S^-X be the upper and lower cones respectively in SX. Let E^+, E^-, and Y denote the pullbacks of $PSX \to SX$ sitting over S^+X, S^-X, and X respectively. Specifically, they are the spaces of paths in SX which begin at the basepoint and end in S^+X, S^-X, and X respectively.

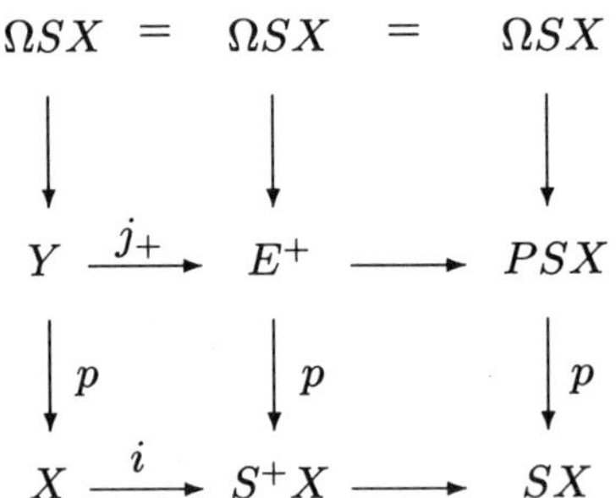

Define $u_+ : S^+X \to E^+$ and $u_- : S^-X \to E^-$ by $u_+(x,s)(t) = \big(x, 1 - (1-s)t\big)$ and $u_-(x,s)(t) = (x, st)$. Observe that u_+ and u_- restrict to maps $\tilde{u}_+$ and $\tilde{u}_-$ taking X to Y. Let $g_\pm : E^\pm \to \Omega SX$ be given by $g_\pm(\omega) = \omega \cdot u_\pm\big(\omega(1)\big)^{-1}$. Then $g_+ : E^+ \to \Omega SX$, $g_- : E^- \to \Omega SX$, $(g_+ j_+, p) : Y \to \Omega SX \times X$, and $(g_- j_-, p) : Y \to \Omega SX \times X$ are all homotopy equivalences by Theorem 7.1.2. The composite $g_+ j_+ \tilde{u}_-$ equals α, while $g_- j_- \tilde{u}_-$ is null homotopic since it factors through the contractible space S^-X.

Let $A = H_*(\Omega SX)$ and let $V = H_*(X)$ and $\tilde{V} = \tilde{H}_*(X)$. Since V is torsion free, $H_*(Y) \cong A \otimes V$. The action map $\mu : \Omega SX \times PSX \to PSX$ given by $\mu : (\omega, \gamma) = \omega \cdot \gamma$ restricts to maps $\Omega SX \times \Omega SX \to \Omega SX$, $\Omega SX \times E^\pm \to E^\pm$, and $\Omega SX \times Y \to Y$. These maps give A its algebra structure and they give $H_*(E^\pm)$ and $H_*(Y)$ the structure of left A-modules such that $(g_+)_*$, $(g_-)_*$, $(j_+)_*$, and $(j_-)_*$ are A-module homomorphisms.

Since $(S^-X, X) \to (SX, S^+X)$ is a combination of excision and homotopy equivalence (making use of our nondegenerate basepoint), taking inverse images under p gives a combination of excision and homotopy equivalence $(E^-, Y) \to (PSX, E^+)$ and thus a homology isomorphism $H_*(E^-, Y) \cong H_*(PSX, E^+)$.

$$
\begin{array}{ccccccccc}
\tilde{H}_{q+1}(E^-) & \xrightarrow{0} & H_{q+1}(E^-, Y) & \xrightarrow{\partial} & \tilde{H}_q(Y) & \xrightarrow{(j_-)_*} & \tilde{H}_q(E^-) & \xrightarrow{0} & H_q(E^-, Y) \\
\downarrow{\scriptstyle\cong} & & \downarrow{\scriptstyle\cong} & & \downarrow{\scriptstyle (j_+)_*} & & \downarrow & & \\
\tilde{H}_{q+1}(PSX) & \longrightarrow & H_{q+1}(PSX, E^+) & \xrightarrow{\cong} & \tilde{H}_q(E^+) & \longrightarrow & \tilde{H}_q(PSX) & &
\end{array}
$$

is a map of exact sequences of A-modules and so the theorem follows from the following easily checked lemma applied with $\eta = \alpha_*$,

$$
f = (g_+ j_+, p) \circ \partial \big((j_+)_* \partial \big)^{-1} \circ (g_+)_*^{-1},
$$

$$
j = (g_-)_* \circ (j_-)_* \circ (g_+ j_+, p)_*^{-1}, \quad \text{and} \quad q = (g_+)_* \circ (j_+)_* \circ (g_+ j_+, p)_*^{-1}.
$$

Lemma 7.3.2 *Let A be a connected graded algebra (see beginning of Chapter 10) over a PID R and V a connected graded module which is torsion-free over R. Let $\eta : V \to A$ be an R-module homomorphism. Suppose that there is an exact sequence of graded A-modules $R \longrightarrow A \xrightarrow{f} A \otimes V \xrightarrow{j} A \longrightarrow R$, where A acts on $A \otimes V$ by left multiplication on A and j projects V onto $V_0 = R$. Let $q : A \otimes V \to A$ be an A-module left splitting of f such that the composite $V \xrightarrow{\cong} R \otimes V \longrightarrow A \otimes V \xrightarrow{q} A$ is η. Then $A \cong T(\tilde{V})$ as an algebra with η corresponding to the canonical inclusion.* $\square$

Corollary 7.3.3 $H_*(\Omega S^n) \cong T(x)$ *where* $|x| = n - 1$. $\square$

4 Relation between Homotopy Groups and Homology Groups

In this section and the next we shall consider the Hurewicz and CW-approximation theorems. These topics are interrelated in that some proofs may be easier or harder depending on which of the other facts have already been proved. We wish to establish, in some order, all of the following facts (some of which involve definitions not yet given but which will appear in the relevant places):

1) For a CW-complex X, the pair $\big(X, X^{(n)}\big)$ is n-connected.
2) Cellular Approximation Theorem (7.5.1).
3) CW-Approximation Theorem (7.5.8).
4) A weak homotopy equivalence between CW-complexes is a homotopy equivalence
5) The inclusion map from an Eilenberg sub-chain-complex is a chain homotopy equivalence (7.4.3).
6) A weak homotopy equivalence induces an isomorphism on homology (7.4.6).
7) Hurewicz Theorem for wedges of spheres (7.3.1).
8) Hurewicz Theorem (7.4.4).
9) Relative Hurewicz Theorem (7.4.5).
10) Ganea's Theorem (7.7.3).

Two possible strategies are:

Plan A

- Give direct proofs of (1), (5), (7), and (10).
- Use (1) to prove (2), then (3), then (4).
- Drop out (6) as a corollary of (5).
- Use (4), (6), and (7) to prove (8).
- Use (8) and (10) to prove (9).

Plan B

- Give direct proofs of (5), (7), and (10).
- Use (5) and (7) to prove (9) and (8).
- Drop out (1) as a corollary of (9).
- Use (1) to prove (2), then (3), then (4).
- Drop out (6) as a corollary of either (5) or (9).

The plans have many parts in common. The main difference is whether one goes first for the CW-approximation theorems (Plan A) or the Hurewicz theorems (Plan B), with the proof of the second made easier by making use of the first. In Plan A, one works hard to establish (1), using barycentric subdivision and essentially duplicating parts of the proof that singular homology satisfies excision. This pays off in the proofs of the Hurewicz theorems, which are made easier by reducing to the case of CW-complexes. Under Plan B, one works harder to prove the Hurewicz theorems, but then (1) becomes a triviality by passing to homology and, in effect, using excision. These notes illustrate Plan B.

A pointed space X is called *k-connected* if $\pi_j(X)$ is trivial for $j \leq k$. More generally, a pair (X, A) is called *k-connected* if $\pi_j(X, A)$ is trivial for $j \leq k$.

Let $(\Delta^m)^{(k)}$ denote the k-skeleton of Δ^m, the union of the faces of dimension less than or equal to k. Observe that $(\Delta^m)^{(m-1)} \cong S^{m-1}$ and $(\Delta^m)^{(m-2)} \approx \bigvee_{k=0}^{m-1} S^{m-2}$. Let $S_m^{(k,A)}(X)$ be the free abelian group on

$$\{f : \Delta^m \to X \mid f|_{(\Delta^m)^{(0)}} = * \text{ and } f|_{(\Delta^m)^{(k)}} \subset A\}.$$

Then $S_*^{(k,A)}(X)$ forms a subcomplex of $S_*(X)$, called an Eilenberg subcomplex, whose homology will be denoted $H_*^{(k,A)}(X)$. Let $S_*^{(k,A)}(X, A)$ be

$$S_*^{(k,A)}(X) \Big/ \left(S_*^{(k,A)}(X) \cap S_*(A) \right).$$

Theorem 7.4.1 *If X is path connected and (X, A) is n-connected then the inclusion $\alpha : S_*^{(n,A)}(X) \hookrightarrow S_*(X)$ is a chain homotopy equivalence.*

The proof makes use of the following lemma.

Lemma 7.4.2 *For each $\sigma : \Delta^q \to X$ there is a map $P(\sigma) : \Delta^q \times I \to X$ such that:*

1) $P(\sigma)_0 = \sigma$;
2) $P(\sigma)_1$ *lies in* $S_*^{(n,A)}(X)$ *and if* $\sigma \in S_*^{(n,A)}(X)$ *then* $P(\sigma)_1 = \sigma$;
3) $P(\sigma) \circ (\delta^j \times I) = P(\sigma \circ \delta^j)$, *where* $\delta^j : \Delta^{q-1} \to \Delta^q$ *is the inclusion of the jth face.*

Proof For σ of degree 0, $P(\sigma)$ can be chosen with the desired properties since X is path connected. To define $P(\sigma)$ for $|\sigma| = q$, suppose by induction that $P(\sigma')$ has been defined for σ' with $|\sigma'| < q$. If $\sigma \in S_*^{(n,A)}(X)$ then set $P(\sigma)(x,t) = \sigma(x)$ for all t. For $\sigma \notin S_*^{(n,A)}(X)$, let $f_\sigma : (\Delta^q \times 0) \cup ((\Delta^q)^{(q-1)} \times I) \to X$ be the unique map satisfying conditions (1) and (3). Consider first the case $q \leq n$. Since there exists a homeomorphism $\phi : D^q \times I \to \Delta^q \times I$ such that

$$\phi(D^q \times 0) = (\Delta^q \times 0) \cup \left((\Delta^q)^{(q-1)} \times I\right),$$
$$\phi(S^{q-1} \times 0) = (\Delta^q)^{(q-1)} \times 1, \quad \text{and}$$
$$\phi\left((S^{q-1} \times I) \cup (D^q \times 1)\right) = \Delta^q \times 1,$$

the fact that (X, A) is n-connected implies that there is a homotopy $H : \Delta^q \times I \to X$ such that the restriction of H to the domain of f_σ is f_σ and $H_1(\Delta^q) \subset A$. Choose such a homotopy and set $P(\sigma) = H$. Finally, in the case $q > n$, by the homotopy extension property f_σ extends to a map $\Delta^q \times I \to X$, and any such extension is a suitable choice for $P(\sigma)$. $\qquad\square$

Proof of Theorem 7.4.1 Define $s : S_q(X) \to S_{q+1}^{(n,A)}(X)$ on generators by $s(\sigma) = P(\sigma)_* D_{\Delta^q}(1_{\Delta^q})$, where D_Y is a natural chain homotopy $D_Y : S_*(Y) \to S_{*+1}(Y \times I)$ between $(i_0)_*$ and $(i_1)_*$ as in Theorem 5.2.2. Then $r(\sigma) = P(\sigma)_1$ yields a chain retraction of α, and $s : \alpha r \simeq 1_{S_*(X)}$. $\qquad\square$

Since $S_*^{(k,A)}(X) \cap S_*(A) = S_*^{(k,A)}(A)$,

Corollary 7.4.3 *If X is path connected and (X, A) is n-connected then the inclusion $\alpha : S_*^{(n,A)}(X, A) \hookrightarrow S_*(X, A)$ is a chain homotopy equivalence.* $\qquad\square$

For each n pick a generator ι_n of $H_n(S^n)$. Since a choice of ι_n corresponds to an orientation on S^n, these can be chosen consistently by using the orientation ι_{n-1} on the equator of S^n to induce the orientation on S^n. For a pointed space X, define the *Hurewicz homomorphism*, $h_X : \pi_n(X) \to H_n(X)$, by $h_X([f]) = f_*(\iota_n)$. This is a well defined map of sets by the homotopy axiom. Let $\psi : S^n \to S^n \vee S^n$ be the standard co-H-space structure on S^n, and let $\iota'_n = i'_*(\iota_n)$ and $\iota''_n = i''_*(\iota_n)$ be the canonical generators of $H_n(S^n \vee S^n)$, where i' and i'' are the inclusions of S^n as the summands of $S^n \vee S^n$. Then

$$h_X([f][g]) = h_X\left([(f \perp g)\psi]\right) = (f \perp g)_* \psi_*(\iota_n) = (f \perp g)_*(\iota'_n + \iota''_n) =$$
$$f_*(\iota_n) + g_*(\iota_n) = h_X([f]) + h_X([g])$$

and so h_X is a homomorphism. Similarly, for a pair (X, A) we can define the relative Hurewicz homomorphism $h_{(X,A)} : \pi_n(X, A) \to H_n(X, A)$ by $h_{(X,A)}([f]) = f_*(\tilde{\iota}_n)$ where $\tilde{\iota}_n \in H_n(D^n, S^{n-1})$ corresponds to our chosen generator ι_{n-1} under the isomorphism from the long exact homology sequence. In fact, if we replace (D^n, S^{n-1}) by the homeomorphic pair $\left(\Delta^n, (\Delta^n)^{(n-1)}\right)$ then we can choose $\tilde{\iota}$ to be the class represented on the chain level by the identity map.

It is easy to see that these homomorphisms map the homotopy exact sequence of a pair to its homology exact sequence.

$$\pi_{n+1}(X, A) \xrightarrow{\partial} \pi_n(A) \to \pi_n(X) \to \pi_n(X, A) \to \ldots \to \pi_1(A) \to \pi_1(X)$$
$$\downarrow{h_{(X,A)}} \qquad \downarrow{h_A} \qquad \downarrow{h_X} \qquad \downarrow{h_{(X,A)}} \qquad \downarrow{h_A} \qquad \downarrow{h_X}$$
$$H_{n+1}(X, A) \xrightarrow{\partial} H_n(A) \to H_n(X) \to H_n(X, A) \to \ldots \to H_1(A) \to H_1(X)$$

For a group G, let the *abelianization* of G be $G_{ab} = G/[G, G]$, the quotient of G by its commutator subgroup. For $n > 1$, $\pi_n(X)$ is already abelian and for $n = 1$

the fact that $H_1(X)$ is abelian means that h_X induces a map $(h_X)_{ab} : \big(\pi_1(X)\big)_{ab} \to H_1(X)$. The Hurewicz theorem says that (with appropriate modification for $n = 1$) the least nonvanishing homotopy group is isomorphic to the least nonvanishing homology group.

Theorem 7.4.4 (Hurewicz Theorem) *Suppose X is $(n-1)$-connected. If $n > 1$ then $h_X : \pi_n(X) \to H_n(X)$ is an isomorphism and if $n = 1$ then $(h_X)_{ab} : \big(\pi_1(X)\big)_{ab} \to H_1(X)$ is an isomorphism.*

Theorem 7.4.5 (Relative Hurewicz Theorem) *Suppose (X, A) is $(n-1)$-connected where X and A are connected and $n \geq 2$. Then the induced map $\tilde{h}_{(X,A)} : \pi_n(X, A)/N \to H_n(X, A)$ is an isomorphism, where N is the normal subgroup of $\pi_n(X, A)$ generated by $\{(\gamma \cdot f)f^{-1} \mid f \in \pi_n(X, A), \gamma \in \pi_1(A)\}$.*

Proof It is easy to see that $h_{(X,A)}$ annihilates N. The cases $n = 1$ of the absolute theorem and $n = 2$ of the relative theorem can be checked directly from the definitions. For larger n it suffices to consider the relative theorem since $\pi_n(X) = \pi_n(X, *)$ for $n > 1$. Suppose by induction that the theorem has been proved for all spaces for $k < n$.

$h_{S^n}([1_{S^n}])$ gives a generator of $H_n(S^n)$ so h_{S^n} is surjective in degree n. However the Bott-Samelson Theorem (Theorem 7.3.1) and our induction hypothesis imply that $\pi_n(S^n) \cong \pi_{n-1}(\Omega S^n) \cong H_{n-1}(\Omega S^n) \cong \mathbb{Z}$, and so h_{S^n} is an isomorphism in degree n. Let $K = \Delta^{n+1}$ and observe that $K^{(n)} \cong S^n$ and that $K^{(n-1)}$ is homotopy equivalent to a wedge of $(n-1)$-spheres. Therefore the commutative diagram with exact rows

$$\pi_n\left(K^{(n-1)}\right) \xrightarrow{\ 0\ } \pi_n\left(K^{(n)}\right) \longrightarrow \pi_n\left(K^{(n)}, K^{(n-1)}\right) \longrightarrow \pi_{n-1}\left(K^{(n-1)}\right) \longrightarrow 0$$

$$\downarrow \qquad\qquad \cong\downarrow \qquad\qquad\qquad \downarrow \qquad\qquad\qquad \cong\downarrow \qquad\qquad$$

$$0 \qquad \longrightarrow H_n\left(K^{(n)}\right) \longrightarrow H_n\left(K^{(n)}, K^{(n-1)}\right) \longrightarrow H_{n-1}\left(K^{(n-1)}\right) \longrightarrow 0$$

shows that the theorem holds for $\left(K^{(n)}, K^{(n-1)}\right)$.

Define $\psi : S_n^{(n-1, A)}(X, A) \to \pi_n(X, A)/N$ to be the map taking each generator to its homotopy class. We wish to show that ψ induces a well defined map on $H_n^{(n-1, A)}(X, A)$. Let $\tau : \Delta^{n+1} \to X$ be a generator of $S_{n+1}^{(n-1, A)}(X, A)$. Since $\psi(\partial \tau) = \sum_{j=0}^{n+1}(-1)^j[\tau \circ \delta^j]$, we need to show that $\sum_{j=0}^{n+1}(-1)^j[\tau \circ \delta^j] = 0$ in $\pi_n(X, A)/N$. We plan to use $\tau_\# : \pi_n\left(K^{(n)}, K^{(n-1)}\right) \to \pi_n(X, A)$ to transfer this problem from (X, A) to the pair $\left(K^{(n)}, K^{(n-1)}\right)$.

For $k > 0$, choose ϵ_0 as the basepoint of Δ^k and and let $\epsilon_{1,0}$ denote the path from ϵ_1 to ϵ_0. For $j > 0$, $[\tau \circ \delta^j]$ is the image of $[\delta^j]$ under $\tau_\#$. The corresponding statement for $j = 0$ does not make sense since δ^0 is not a basepoint preserving map. However $\epsilon_{1,0} \cdot \delta^0$ is a basepoint preserving map and $[\tau \circ \delta^0]$ differs from $\tau_\#\big([\epsilon_{1,0} \cdot \delta^0]\big)$ by an element of N. (This depends on the fact that $\tau \circ \epsilon_{1,0}$ is a loop in A by definition of the Eilenberg subcomplex.) Therefore it suffices to establish the equality $[\epsilon_{1,0} \cdot \delta^0] + \sum_{j=1}^{n+1}(-1)^j[\delta^j] = 0$ in $\pi_n\left(K^{(n)}, K^{(n-1)}\right)$, known as the "Homotopy Addition Theorem". However since we already have the Hurewicz theorem for this special case, it suffices to check the corresponding equation on homology after applying the Hurewicz map, and this becomes simply $[\sum_{j=0}^{n+1}(-1)^j \delta^j] = [\partial(1_{\Delta^{n+1}})] = 0$.

Therefore ψ induces a map $\bar\psi : H_n^{(n-1,A)}(X,A) \to \pi_n(X,A)/N$. Since (X,A) is $(n-1)$-connected, we have inverse chain homotopy equivalences $\alpha : S_*^{(n-1,A)}(X,A) \hookrightarrow S_*(X,A)$ and $r : S_*(X,A) \to S_*^{(n-1,A)}(X,A)$. From the definitions, $h_{(X,A)}\bar\psi = \alpha_*$, and for $f : (D^n, S^{n-1}) \to (X,A)$, $\psi r_* f_*(\iota)$ is homotopic to f by means of $P\big(f_*(\iota)\big)$, using the notation of Theorem 7.4.1 and Lemma 7.4.2. Therefore $h_{(X,A)}$ is an isomorphism. $\qquad\square$

A map which induces an isomorphism on $\pi_n(\ )$ for all n is called a *weak homotopy equivalence*.

Corollary 7.4.6 (Whitehead Theorem) *Let $f : X \to Y$ be a weak homotopy equivalence between path connected spaces. Then f induces an isomorphism on $H_n(\ )$ for all n. Conversely, if X and Y are simply connected, then a map $f : X \to Y$ which induces an isomorphism on $H_n(\ )$ for all n is a weak homotopy equivalence.* $\qquad\square$

Example 7.4.7 Corollary 7.4.6 fails if X and Y are not simply connected, even with the additional hypothesis that f induces an isomorphism on $\pi_1(\ )$. Consider the following situation. Let $W = S^1 \vee S^2$. Then the universal covering space $\tilde{W}$ of W is the real line with a copy of S^2 attached at each integer. Therefore $\pi_2(W) = \bigoplus_{j \in \mathbb{Z}} \mathbb{Z}$ and the action of the fundamental group shifts the factors. Let $\alpha : S^1 \to W$ and $\beta : S^2 \to W$ be the inclusions. Let $Y = D^3 \cup_{2\beta - \alpha \cdot \beta} W$. The composition $S^1 \xrightarrow{\alpha} W \hookrightarrow Y$ induces an isomorphism on $\pi_1(\ )$. It cannot induce an isomorphism on $\pi_2(\ )$ because $\pi_2(Y) \cong \varinjlim_n \mathbb{Z} \cong \mathbb{Z}[1/2]$, since the maps in this direct system are multiplication by 2. However $Y/S^1 \cong D^3 \cup_{2\bar\beta - \bar\beta} S^2 \approx *$ since $\bar\beta$ is the composite $S^2 \xrightarrow{\beta} W \longrightarrow X/S^1$ which has degree 1. Therefore $H_n(S^1) \to H_n(Y)$ is an isomorphism for all n. $\qquad\square$

There is however the following version of the Whitehead theorem.

Theorem 7.4.8 *Let $f : X \to Y$ be a map between path connected H-spaces which induces an isomorphism on $H_n(\ )$ for all n. Then f is a weak homotopy equivalence.* $\qquad\square$

A *Serre class* of abelian groups is a class $\mathcal{C}$ containing the trivial group and having the property that for any short exact sequence $0 \to A \to B \to C \to 0$, A and C are in $\mathcal{C}$ if and only if B is in $\mathcal{C}$. Let $\mathcal{C}$ be a Serre class of abelian groups. A homomorphism $f : G \to H$ of abelian groups is called a mod $\mathcal{C}$ *isomorphism* if $\operatorname{Ker} f$ and $\operatorname{CoKer} f$ are in $\mathcal{C}$.

There is a version of the Hurewicz theorem modulo a Serre class.

Theorem 7.4.9 *Let $\mathcal{C}$ be a Serre class. Suppose also that G, H in $\mathcal{C}$ implies that $G \otimes H$ and $\operatorname{Tor}(G,H)$ are in $\mathcal{C}$. Let X and A be simply connected and suppose that $n > 2$ and that $\pi_k(X,A)$ is in $\mathcal{C}$ for $k \le (n-1)$. Then $h_X : \pi_n(X,A) \to H_n(X,A)$ is a mod $\mathcal{C}$ isomorphism. In the absolute case $(A = *)$, the theorem also holds for $n = 2$.* $\qquad\square$

Corollary 7.4.10 *If X is a finite simply connected CW-complex then $\pi_n(X)$ is finitely generated for all n.* $\qquad\square$

The space $S^1 \vee S^2$ gives a counterexample if the simply connectedness hypothesis is removed.

5 Reduction to the Case of Simply Connected CW-Complexes

Theorem 7.5.1 (Cellular Approximation Theorem) *Let $f : X \to Y$ be a map between CW-complexes. Then there exists $g : X \to Y$ such that $g \simeq f$ and g is cellular.*

For the proof we need three lemmas.

Lemma 7.5.2 *Let $f : (D^n, S^{n-1}) \to (X, A)$. Then $[f] = 0$ in $\pi_n(X, A)$ if and only if there exists $J : f \simeq h$ $(\mathrm{rel}\, S^{n-1})$ where $h(D^n) \subset A$.*

Proof Suppose $[f] = 0$. Then there exists $H : D^n \times I \to X$ such that $H_0 = f$, $H_1 = *$. Define $J : D^n \times I \to X$ by

$$
J(x,t) = \begin{cases}
H\left(\frac{x}{1-t/2}, t\right) & \text{if } 0 \le \|x\| \le 1 - t/2; \\[2mm]
H\left(\frac{x}{\|x\|}, 2 - 2\|x\|\right) & \text{if } 1 - t/2 \le \|x\| \le 1.
\end{cases}
$$

Conversely, suppose $J : f \simeq h$ $(\mathrm{rel}\, S^{n-1})$ where $h(D^n) \subset A$. Set $H(x,t) = h\big((1 - t)x + t\epsilon_1\big)$, where ϵ_1 is the basepoint of S^{n-1}. Then $[f] = [h] = 0$ in $\pi_n(X, A)$. $\qquad\square$

Lemma 7.5.3 *Let $f : (X, A) \to (Y, B)$ where (Y, B) is n-connected $(n \le \infty)$ and (X, A) is a relative CW-complex with $\dim (X, A) \le n$. (That is, X is obtained from A by attaching cells of dimension less than or equal to n.) Then $f \simeq f'$ $(\mathrm{rel}\, A)$ for some f with $f'(X) \subset B$.*

Proof We will first give the proof for $n < \infty$, use that in the proof of the Lemma 7.5.4, and then return to do the case $n = \infty$, making use of Lemma 7.5.4.

Suppose by induction that $f \simeq f_{k-1}$ $(\mathrm{rel}\, A)$ where $k \le n$ and $f_{k-1}\left(X^{(k-1)}\right) \subset B$. Write $X^{(k)} = X^{(k-1)} \cup \left(\bigcup_{\alpha \in J} e_\alpha\right)$. Let $h_\alpha : D^k \to X$ be the characteristic map for e_α. Since $f_{k-1}h_\alpha(\partial D^k) \subset f\left(X^{(k-1)}\right) \subset B$, the previous lemma implies that there exists $H_\alpha : D^k \times I \to Y$ $(\mathrm{rel}\, \partial D^k)$ such that $(H_\alpha)_0 = f_{k-1}h_\alpha$ and $\mathrm{Im}(H_\alpha)_1 \subset B$. Define $H : X^{(k)} \times I \to Y$ by $H(e,t) = H_\alpha(e,t)$ for $e \in e_\alpha$, and $H(x,t) = f_{k-1}(x)$ for $x \in X^{(k-1)}$. In particular $H(a,t) = f(a)$ for all $a \in A$. By the homotopy extension property, H extends to $\tilde{H} : X \times I \to Y$. Set $f_k = \tilde{H}_1$. Then $f_k \simeq f_{k-1} \simeq f$ $(\mathrm{rel}\, A)$ and $f_k\left(X^{(k)}\right) \subset B$. Therefore $f' = f_n$ has the desired property.

Lemma 7.5.4 *Let f map $\big((X \times 0) \cup (A \times I), A \times 1\big)$ to (Y, B) where (Y, B) is n-connected and (X, A) is a relative CW-complex with $\dim (X, A) \le n < \infty$. Then there exists an extension $F : X \times I \to Y$ of f such that $F(X \times 1) \subset B$.*

Proof The homotopy extension property implies that f extends to $\phi : X \times I \to Y$. Since $\phi_1(A) = f(A \times 1) \subset B$, the previous lemma implies that there exists $K : X \times I \to Y$ $(\mathrm{rel}\, A)$ such that $K_0 = \phi_1$ and $\mathrm{Im}\, K_1 \subset B$. Let $\tilde{I} = (0 \times I) \cup (I \times 1) \cup (1 \times I)$. Define $\tilde{G} : (X \times \tilde{I}) \cup (A \times I \times I) \to Y$ by $\tilde{G}(x,0,t) = \phi(x,t)$, $\tilde{G}(x,s,1) = K(x,s)$, $\tilde{G}(x,1,t) = K(x,1)$, and $\tilde{G}(a,s,t) = f(a, \min\{s + t, 1\})$ for $x \in X$, $s, t \in I$, and $a \in A$. Since $(X \times \tilde{I}) \cup (A \times I \times I) \hookrightarrow X \times I \times I$ is a cofibration, $\tilde{G}$ extends to $G : X \times I \times I \to Y$. Set $F = G|_{X \times I \times 0}$. $\qquad\square$

Proof of Lemma 7.5.3 (continued) Consider now the case $n = \infty$. It suffices to construct for all n a map

$$f_n : \left(\left(X^{(n)} \times 0 \right) \cup \left(X^{(n-1)} \times I \right), X^{(n-1)} \times I \right) \to (Y, B)$$

such that:

1) $f_n(x, 0) = f(x)$ for all $x \in X^{(n)}$;
2) f_n is an extension of f_{n-1};
3) $f_n(a, t) = f(a)$ for all $a \in A$, $t \in I$.

Suppose by induction that f_k has already been constructed satisfying (1)–(3). The preceding lemma implies that there exists an extension $f'_{k+1} : \left(X^{(k)} \times I \right) \to Y$ such that $f'_{k+1} \left(X^{(k)} \times 1 \right) \subset B$. Extend f'_{k+1} to $f_{k+1} : \left(X^{(k+1)} \times 0 \right) \cup \left(X^{(k)} \times I \right) \to Y$ by setting $f_{k+1}(x, 0) = f(x)$. Then f_{k+1} has the desired properties and so this completes the induction. $\qquad\square$

Proof of Theorem 7.5.1 It suffices to construct for all n a map $g_n : X^{(n)} \times I \to Y$ such that:

1) $g_n|_{X^{(n)} \times 0} = f$;
2) $g_n|_{X^{(n-1)} \times I} = g_{n-1}$;
3) $g \left(X^{(n)} \times 1 \right) \subset Y^{(n)}$.

Given g_{k-1}, the existence of g_k satisfying these conditions follows from Lemma 7.5.4 and the fact that $\left(Y, Y^{(k)} \right)$ is k-connected. $\qquad\square$

Theorem 7.5.5 *Let $f : X \to Y$ be a weak homotopy equivalence. Then for any CW-complex K, the induced map $f_\# : [K, X] \to [K, Y]$ is a bijection.*

Proof Replacing Y by the mapping cylinder of f we may assume that f is an inclusion. Given $g : K \to Y$ we can construct a lift of g to X inductively over the skeletons of K using Lemma 7.5.3. Therefore $f_\#$ is surjective. Now suppose that $f_\#(g) = f_\#(h)$. Then there exists $F : K \times I \to Y$ such that $F_0 = fg$ and $F_1 = fh$. Applying Lemma 7.5.3 to $F : \left(K \times I, (K \times 0) \cup (K \times 1) \right) \to (Y, X)$ gives a homotopy $H : K \times I \to X$ such that $H_0 = F_0 = g$ and $H_1 = F_1 = h$. Therefore $f_\#$ is injective. $\qquad\square$

Corollary 7.5.6 *A map between CW-complexes is a homotopy equivalence if and only if it is a weak homotopy equivalence.* $\qquad\square$

Corollary 7.5.7 *A map between simply connected CW-complexes is a homotopy equivalence if and only if it induces an isomorphism on homology.* $\qquad\square$

Theorem 7.5.8 *(CW-Approximation Theorem) Given a topological space Y there exists a CW-complex X and a map $f : X \to Y$ such that $f_\# : \pi_n(X) \to \pi_n(Y)$ is an isomorphism for all n.* $\qquad\square$

It is clear that the mapping cylinder and mapping cone of a cellular map are CW-complexes. On the dual side, Theorem 2.7.3 is only a special case of a theorem of Milnor which implies that the homotopy fibre of a map between CW-complexes has the homotopy type of a CW-complex. (See Section 7.6 for the definitions of "homotopy fibre" and "homotopy pullback".) From this we derive the more general statement

Theorem 7.5.9 *Let*

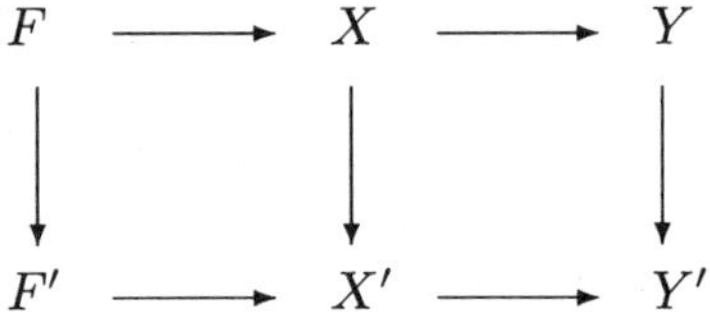

be a homotopy pullback in which X, Y, and Z have the homotopy type of CW-complexes. Then W has the homotopy type of a CW-complex.

Proof We may assume that f and g are fibrations and that the square is a pullback. Let $F_y = f^{-1}(y)$ for $y \in Y$. Since the square is a pullback, $F_y = g^{-1}(ky)$ which has the homotopy type of a CW-complex by Milnor's theorem referred to above. Applying Theorem 7.5.8 and turning the resulting map into a fibration, let $\phi : W' \to W$ be a fibration and weak homotopy equivalence where W' has the homotopy type of a CW-complex. Let $F_y' = (f\phi)^{-1}(y)$. Then F_y' also has the homotopy type of a CW-complex by Milnor's theorem. However the induced map $\phi : F_y' \to F_y$ is a weak homotopy equivalence for all $y \in Y$ and thus is a homotopy equivalence by Corollary 7.5.6. Therefore $\phi : W' \to W$ is a homotopy equivalence by Theorem 7.1.4. $\Box$

The above theorems imply

Corollary 7.5.10 *Let*

$$
\begin{array}{ccccc}
F & \longrightarrow & X & \longrightarrow & Y \\
\downarrow & & \downarrow & & \downarrow \\
F' & \longrightarrow & X' & \longrightarrow & Y'
\end{array}
$$

be a homotopy commutative diagram in which the rows are homotopy fibrations and X', Y, and Y' have the homotopy type of connected CW-complexes. If two of the vertical maps are homotopy equivalences then so is the third. $\Box$

Corollary 7.5.11

a) *Let $A \xrightarrow{i} X \xrightarrow{c} Y$ be an NDR-pair sequence of pointed spaces in which each of the spaces is simply connected and has the homotopy type of a CW-complex. Suppose that there exists a (homotopy splitting) map $s : Y \to X$ such that $cs \simeq 1_Y$. Then $A \vee Y \xrightarrow{i \perp s} X$ is a homotopy equivalence.*

b) *Let*

$$
\begin{array}{ccccc}
A & \longrightarrow & X & \longrightarrow & Y \\
\downarrow & & \downarrow & & \downarrow \\
A' & \longrightarrow & X' & \longrightarrow & Y'
\end{array}
$$

be a homotopy commutative diagram in which the rows are equivalent up to homotopy to NDR-pair sequences and each of the spaces is simply connected and has the homotopy type of a CW-complex. If two of the vertical maps are homotopy equivalences then so is the third. □

From the preceding theorems, any space X may be replaced with a *CW*-complex without affecting its homotopy or homology groups, and maps may be replaced by cellular maps. Since every *CW*-complex has a simply connected (universal) covering space (which can be chosen to have a *CW*-structure) problems concerning the fundamental group may be studied by means of this covering projection while for other purposes X may be replaced by this universal cover, a simply connected *CW*-complex.

When dealing with simply connected *CW*-complexes, we have the pleasant situation that every map inducing an isomorphism on homology groups is a homotopy equivalence. In addition, the constructions which convert maps to fibrations and cofibrations and take homotopy pullbacks and pushouts can be done using only spaces having the homotopy type of *CW*-complexes. Furthermore, any map $f : X \to Y$ between simply connected *CW*-complexes can be converted to a fibration or cofibration when desired to obtain long exact homology or homotopy sequences, and for a simply connected *CW*-pair (X, A), the homology of the pair (X, A) can be replaced by that of its (homotopy) cofibre X/A, while the homotopy of the pair can be replaced (with a dimension shift) by that of its homotopy fibre. Thus one may avoid dealing with relative homotopy and homology groups entirely if desired.

For these reasons, for the rest of these notes we shall work almost exclusively in the category of *CW*-complexes.

The remainder of this section contains a few properties of *CW*-*H*-spaces whose proofs take advantage of Corollary 7.5.6.

Theorem 7.5.12

a) *Let (H, m) be an H-space with nondegenerate basepoint e. Then there exists an equivalent multiplication m' on H such that e forms a strict identity for m'. (That is, $m'(e, x) = m'(x, e) = x$ for all $x \in H$.)*

b) *Let X and Y have the homotopy type of connected CW-complexes and let $f : X \to Y$ be a fibration which is also an H-map between H-spaces with strict nondegenerate identities. Then there exists an equivalent multiplication m' on X such that m' has a strict identity and f is strictly multiplicative. (f is strictly multiplicative means $m_Y \circ (f \times f) = f \circ m'$.)*

See [Z, following Example 1.1.6]. □

Corollary 7.5.13 *Let*

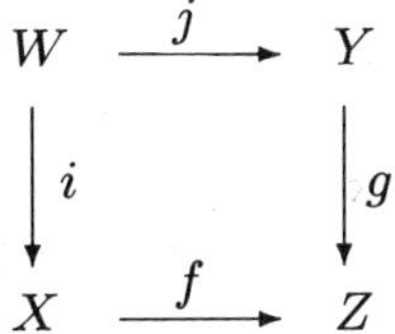

be a pullback square in which at least one of f, g is a fibration. Suppose that X, Y, and Z are H-spaces having the homotopy types of connected CW-complexes and *that f and g are H-maps. Then there is an H-space structure on W such that i and j are H-maps.*																$\square$

Remark See the proof of Theorem 7.9.8 for ideas that can be dualized to give a direct proof of Corollary 7.5.13, bypassing Theorem 7.5.12.

Proposition 7.5.14

a) *Let X be an H-space having the homotopy type of a connected CW-complex. Then X has a homotopy inverse.*

b) *Let X be a co-H-space having the homotopy type of a simply connected CW-complex. Then X has a homotopy inverse.*

Proof For part (a), suppose that X is an H-space having the homotopy type of a connected CW-complex. Define the "shearing map" sh $: X \times X \to X \times X$ by $\mathrm{sh}(x,y) = (xy, y)$. Then sh( ) induces an isomorphism on homotopy groups and so is a homotopy equivalence. Let $\gamma : X \times X \to X \times X$ be a homotopy inverse to sh( ). Define $c : X \to X$ by $c(x) = \pi'\gamma(*, x)$ (where π' denotes projection onto the first factor) and check that c is a homotopy inverse for X.

The proof of part (b) is similar.														$\square$

6 Diagrams of Exact Homotopy Sequences

In this section we will consider homotopy commutative diagrams, their conversion to strictly commuting diagrams, and the extent to which the homotopy classes of resulting induced maps depend upon choices made in the conversion.

Let

$$
\begin{array}{ccccc}
F & \longrightarrow & E & \xrightarrow{\ p\ } & B \\
\downarrow{\scriptstyle f} & & \downarrow{\scriptstyle g} & & \downarrow{\scriptstyle h} \\
F' & \longrightarrow & E' & \xrightarrow{\ p'\ } & B'
\end{array}
$$

be a commutative diagram in which the rows are fibrations. The map f between the fibres is determined by the maps g and h. A common mistake is to assume that the homotopy class of f is determined by the homotopy classes of g and h. The dual mistake would be to assume that in a commuting diagram of cofibration sequences

$$
\begin{array}{ccccc}
A & \longrightarrow & X & \longrightarrow & Y \\
\downarrow{\scriptstyle f} & & \downarrow{\scriptstyle g} & & \downarrow{\scriptstyle h} \\
A' & \longrightarrow & X' & \longrightarrow & Y'
\end{array}
$$

the homotopy class of h is determined by the homotopy classes of f and g. For a counterexample, consider the diagram

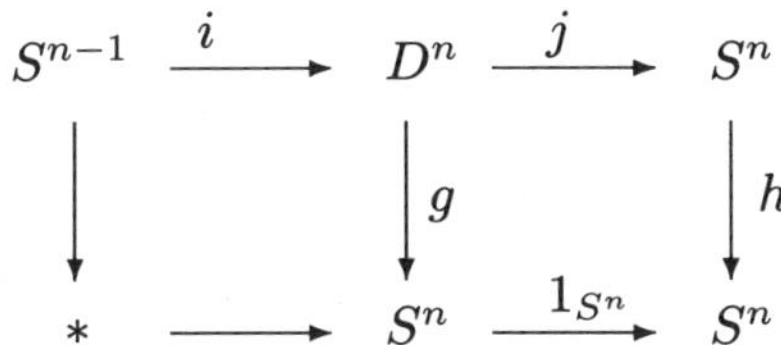

where i is the inclusion map and j is the quotient map. If $g = j$ then the diagram (strictly) commutes and the induced map h is the identity map on S^n. However if $g = *$, which is homotopic to j, then the diagram still (strictly) commutes but the induced map on S^n becomes the constant map which is not homotopic to 1_{S^n}.

Let $f : X \to Y$ be a map which is surjective on path components, where Y has numerable category. By Theorem 7.1.14, there is a homotopy fibration $F \xrightarrow{j} X \xrightarrow{f} Y$ for some space F and map $j : F \to X$. By Theorem 7.1.2 and Theorem 7.1.4, the homotopy type of F is uniquely determined by the homotopy class of f and is called the *homotopy(-theoretic) fibre* of f. A homotopy commutative square

$$
\begin{array}{ccc}
W & \xrightarrow{g'} & X \\
\downarrow{\scriptstyle f'} & & \downarrow{\scriptstyle f} \\
Y & \xrightarrow{g} & Z
\end{array}
$$

between path connected spaces is equivalent up to homotopy to a strictly commuting diagram in which the horizontal maps are fibrations and this can be used to produce a map between the homotopy fibres of g' and g which we refer to as **an** induced map of homotopy fibres. According to the preceding discussion, the homotopy class of this induced map is not uniquely determined by the original diagram, but may in general depend upon the choice of natural equivalence (in the homotopy category) from the original diagram to a commuting diagram of fibrations. However by the naturality of the mapping fibration sequence constructed in the discussion preceding Theorem 7.1.16, any such induced map γ of homotopy fibres makes the following diagram homotopy commutative.

$$
\begin{array}{ccccc}
\Omega X & \xrightarrow{\partial} & F^{g'} & \xrightarrow{j'} & W \\
\downarrow{\scriptstyle \Omega(f)} & & \downarrow{\scriptstyle \gamma} & & \downarrow{\scriptstyle f'} \\
\Omega Z & \xrightarrow{\partial} & F^{g} & \xrightarrow{j} & Y
\end{array}
$$

If g is a fibration then the homotopy commutativity of the original diagram along with the homotopy lifting property allow the construction of a map from W to the pullback of f and g creating a homotopy commutative diagram in which the maps f' and g' have been factored through this pullback. The homotopy class of the map from W to the pullback depends on the choice made when using the homotopy lifting property and this is what allows the possible variation in the homotopy class of γ.

There is a vertical/horizontal symmetry to the above which is described by the next proposition.

Proposition 7.6.1 *Let*

$$
\begin{array}{ccc}
W & \xrightarrow{\ g'\ } & X \\
\downarrow{\scriptstyle f'} & & \downarrow{\scriptstyle f} \\
Y & \xrightarrow{\ g\ } & Z
\end{array}
$$

be a homotopy commutative square in which X, Y and Z have the homotopy types of connected CW-complexes. Then the following are equivalent:

1) *there exists an induced map of fibres $F^{g'} \to F^g$ which is a homotopy equivalence;*
2) *the diagram is equivalent in the homotopy category to a pullback in which the horizontal maps are fibrations;*
3) *the diagram is equivalent in the homotopy category to a pullback in which the vertical maps are fibrations;*
4) *there exists an induced map of fibres $F^{f'} \to F^f$ which is a homotopy equivalence.* $\qquad\square$

This can be proved by turning both f and g into fibrations and making use of Corollary 7.5.6. The conditions of the preceding proposition are satisfied when, after turning either f or g into a fibration, there is an induced homotopy equivalence from W to the pullback. This motivates the following definition of "homotopy pullback".

Given any pair of maps $g : X \to Z$ and $f : Y \to Z$, there is a unique homotopy commutative square (up to equivalence of diagrams in the homotopy category) in which the maps f and g form the bottom right corner and the equivalent conditions above are satisfied. This square is called the *homotopy pullback* of f and g. It follows from Corollary 7.2.8 and Lemma 4.1.8 that a homotopy pullback yields a long exact Mayer-Vietoris sequence of homotopy groups.

Proposition 7.6.1 can be generalized as follows

Theorem 7.6.2 *Let*

$$
\begin{array}{ccc}
A_{1,1} & \xrightarrow{\ g_{1,1}\ } & A_{0,1} \\
\downarrow{\scriptstyle f_{1,1}} & & \downarrow{\scriptstyle f_{0,1}} \\
A_{1,0} & \xrightarrow{\ g_{1,0}\ } & A_{0,0}
\end{array}
$$

be a homotopy commutative diagram. Then there exists a homotopy commutative diagram (not necessarily unique up to homotopy)

$$
\begin{array}{ccccccccc}
A_{n,n} & \xrightarrow{g_{n,n}} & \cdots & \xrightarrow{g_{3,n}} & A_{2,n} & \xrightarrow{g_{2,n}} & A_{1,n} & \xrightarrow{g_{1,n}} & A_{0,n} \\
\downarrow f_{n,n} & & & & \downarrow f_{2,n} & & \downarrow f_{1,n} & & \downarrow f_{0,n} \\
\vdots & & \ddots & & \vdots & & \vdots & & \vdots \\
\downarrow f_{n,3} & & & & \downarrow f_{2,3} & & \downarrow f_{1,3} & & \downarrow f_{0,3} \\
A_{n,2} & \xrightarrow{g_{n,2}} & \cdots & \xrightarrow{g_{3,2}} & A_{2,2} & \xrightarrow{g_{2,2}} & A_{1,2} & \xrightarrow{g_{1,2}} & A_{0,2} \\
\downarrow f_{n,2} & & & & \downarrow f_{2,2} & & \downarrow f_{1,2} & & \downarrow f_{0,2} \\
A_{n,1} & \xrightarrow{g_{n,1}} & \cdots & \xrightarrow{g_{3,1}} & A_{2,1} & \xrightarrow{g_{2,1}} & A_{1,1} & \xrightarrow{g_{1,1}} & A_{0,1} \\
\downarrow f_{n,1} & & & & \downarrow f_{2,1} & & \downarrow f_{1,1} & & \downarrow f_{0,1} \\
A_{n,0} & \xrightarrow{g_{n,0}} & \cdots & \xrightarrow{g_{3,0}} & A_{2,0} & \xrightarrow{g_{2,0}} & A_{1,0} & \xrightarrow{g_{1,0}} & A_{0,0}
\end{array}
$$

in which each pair of consecutive horizontal maps and each pair of consecutive vertical maps forms a homotopy fibration.

Proof By replacing $A_{0,1}$ and $A_{1,0}$ by homotopy equivalent spaces we may assume that $f_{0,1}$ and $g_{1,0}$ are fibrations. Either $f_{1,1}$ or $g_{1,1}$ may then be replaced by a homotopic map so that the diagram strictly commutes. Let

$$
\begin{array}{ccc}
P & \xrightarrow{p} & A_{0,1} \\
\downarrow q & & \downarrow f_{0,1} \\
A_{1,0} & \xrightarrow{g_{1,0}} & A_{0,0}
\end{array}
$$

be the pullback square of $f_{0,1}$ and $g_{1,0}$. Then there is an induced map $\phi : A_{1,1} \to P$, and by replacing $A_{1,1}$ by a homotopy equivalent space we may assume that ϕ is a fibration. Define $A_{1,2}$, $A_{0,2}$, $A_{2,1}$, $A_{2,0}$, and $A_{2,2}$ to be the fibres of $f_{1,1}$, $f_{0,1}$, $g_{1,1}$, $g_{1,0}$, and ϕ respectively, with induced maps $g_{1,2}$ and $f_{2,1}$. Then

$$
\begin{array}{ccc}
A_{1,2} & \xrightarrow{g_{1,2}} & A_{0,2} \\
\downarrow f_{1,2} & & \downarrow \\
A_{1,1} & \xrightarrow{\phi} & P
\end{array}
$$

is a pullback square and so the fibre of the fibration $g_{1,2}$ equals that of ϕ. Similarly the fibre of $f_{2,1}$ is also equal to $A_{2,2}$. The rest follows by induction. $\qquad\square$

A homotopy commutative diagram of squares in which each pair of consecutive horizontal maps and each pair of consecutive vertical maps forms a homotopy fibration will be called a "diagram of homotopy fibrations".

Again we note that while the homotopy types of $A_{2,0}$, $A_{2,1}$, $A_{1,2}$ and $A_{0,2}$ are uniquely determined by the original square, the homotopy classes of the maps $f_{2,1}$, $g_{1,2}$ and the homotopy type of the space $A_{2,2}$ are not. In more detail, as shown above the original square is equivalent up to homotopy to a strictly commuting diagram of topological spaces in which the maps are fibrations. Any such strictly commuting diagram yields a map ϕ from $A_{1,1}$ to the homotopy pullback P of $f_{0,1}$ and $g_{1,0}$ and also produces induced maps of homotopy fibres $f_{2,1}$ and $g_{1,2}$. The point is that the homotopy fibres of $f_{2,1}$, ϕ, and $g_{1,2}$ have the same homotopy type, represented by the space labelled $A_{2,2}$ in the diagram. However other such strictly commuting diagrams may yield other (non-homotopic) maps and a different homotopy type $A_{2,2}$.

Combining the above with the identification of $A_{3,0}$ and $A_{0,3}$ obtained in the discussion of the mapping fibration sequence (preceding Theorem 7.1.16) gives

Theorem 7.6.3 *Let*

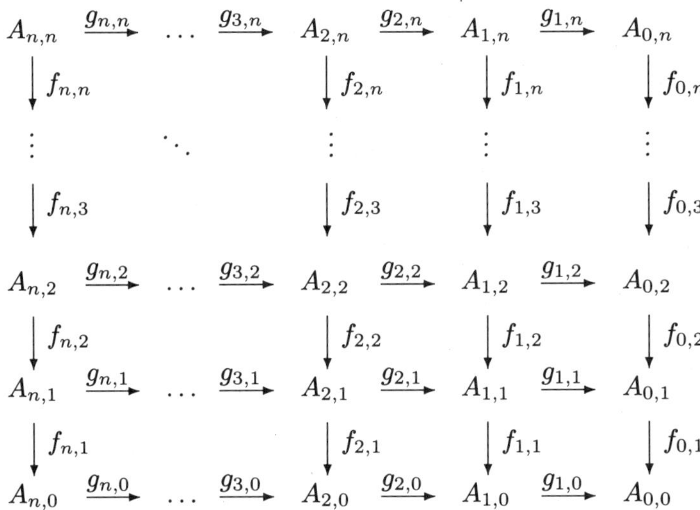

be a diagram of homotopy fibrations in which $A_{1,1}$, $A_{1,0}$, $A_{0,1}$, and $A_{0,0}$ have the homotopy type of CW-complexes. Then

    a) *$A_{i+3,j} \approx \Omega A_{i,j} \approx A_{i,j+3}$, for all i and j.*
    b) *The diagram is equivalent up to homotopy to one in which $f_{i+3,j} = \Omega f_{i,j}$ and $g_{i,j+3} = \Omega g_{i,j}$ for all i and j.*

Provided that each of the spaces has the homotopy type of a simply connected CW-complex, the preceding results for fibrations all dualize to cofibrations in the obvious way. Specifically, under these assumptions, every map determines a *homotopy(-theoretic) cofibre* uniquely up to homotopy, every homotopy commutative square produces a (non-unique) induced map of homotopy cofibres, and can be embedded (non-uniquely) in a diagram of homotopy cofibrations. This yields a definition of *homotopy pushout* and the homotopy pushout of simply connected spaces having the type of CW-complexes is again a simply connected space having the homotopy type of a CW-complex. Finally, in a diagram of cofibrations in which $A_{0,0}$, $A_{1,0}$, $A_{0,1}$, and $A_{1,1}$ have the homotopy type of simply connected CW-complexes, $A_{i+3,j} \approx SA_{i,j} \approx A_{i,j+3}$ for all i and j, and the diagram is equivalent to one in which $f_{i+3,j} = Sf_{i,j}$ and $g_{i,j+3} = Sg_{i,j}$ for all i and j.

7 Ganea's Theorem

For pointed topological spaces A and B, the (*reduced*) *join* of A and B, denoted $A * B$, is defined by $A * B = (A \times I \times B)/\sim$ where $(a, 0, b) \sim (a', 0, b)$, $(a, 1, b) \sim (a, 1, b')$, and $(*, t, *) \sim (*, t'*) \sim *$ for all $a, a' \in A$, $b, b' \in B$, and $t, t' \in I$. It is sometimes convenient to use the equivalent formulation

$$A * B = \{(a, s, b, t) \in A \times I \times B \times I \mid s + t = 1\}/\sim$$

where $(a, 0, b, t) \sim (a', 0, b, t)$, $(a, s, b, 0) \sim (a, s, b', 0)$, and $(*, s, *, t) \sim *$.

Proposition 7.7.1 *If the basepoints of A and B are closed subsets then the quotient map $A * B \to S(A \wedge B)$ is a homotopy equivalence.*

Proof This follows from Proposition 7.1.9c since $S(A \wedge B) = (A * B)/(CA \vee CB)$ where CA includes into $A * B$ via $(a, t) \mapsto (a, t, *)$ and CB includes via $(b, t) \mapsto (*, 1 - t, b)$. $\qquad\square$

Clearly the inclusion maps $A \hookrightarrow A * B$ and $B \hookrightarrow A * B$ are null homotopic.

Throughout the rest of this section, we will assume that we are working in the category of CW-complexes which in particular guarantees that our basepoints are closed.

Proposition 7.7.2 *The homotopy pushout of the projection maps $\pi' : A \times B \to A$ and $\pi'' : A \times B \to B$ is homotopy equivalent to $A * B$.*

Proof The homotopy type of the homotopy pushout can be found by taking the topological pushout after replacing $\pi' : A \times B \to A$ by the inclusion $A \times B \hookrightarrow A \times CB$. $\qquad\square$

Theorem 7.7.3 (Ganea) *Let $F \xrightarrow{\ j\ } E \xrightarrow{\ p\ } B$ be a fibration and let $\tilde{p} : E/F \to B$ be the induced map. Then there is a homotopy commutative diagram of homotopy fibrations*

$$
\begin{array}{ccccc}
F & \xrightarrow{\ j\ } & E & \xrightarrow{\ p\ } & B \\[1mm]
\downarrow{\scriptstyle *} & & \downarrow & & \| \\[1mm]
\Omega B * F & \longrightarrow & E/F & \xrightarrow{\ \tilde{p}\ } & B.
\end{array}
$$

Furthermore, up to homotopy, the composition $\Omega B * F \to E/F \to SF$ is the map $(\omega, t, f) \mapsto \big(\mu(\omega^{-1}, f), t\big)$ where $\mu : \Omega B \times F \to F$ is the action map from the induced principal homotopy fibration $\Omega B \longrightarrow F \xrightarrow{\ j\ } E$.

Proof If C is a subspace of A and B, let $A \cup_C B$ denote the quotient space of $A \amalg B$ obtained by identifying points in C. Equivalently, $A \cup_C B$ is the pushout of the inclusion maps of C into A and B, which is a homotopy pushout in the case of NDR pairs. Up to homotopy equivalence, we may replace E/F by $E \cup_F CF$, where $i : E \cup_F CF \approx E/F$ collapses the cone. The homotopy fibre of $\tilde{p}i$ is homotopy equivalent to the (topological) pullback Q of the fibration $PB \to B$ and $E \cup_F CF \to B$. The space Q is given by $Q = Q_E \cup_{Q_F} Q_{CF}$ where Q_E, Q_F, and Q_{CF} are the pullbacks of $PB \to B$ with the compositions with $E \cup_F CF \to B$ of the inclusions from E, F, and CF into $E \cup_F CF$.

The homotopy equivalence $* \hookrightarrow PB$ induces a homotopy equivalence $\eta : F \to Q_E$. Explicitly, $Q_E = \{(\omega, e) \mid \omega(1) = p(e)\}$ and $\eta(f) = (*, f)$. Since the inclusions are cofibrations, it is possible to choose an inverse equivalence $r : Q_E \to F$ such that $r(\alpha \cdot \alpha^{-1}, f) = f$ for all $\alpha \in \Omega B$, $f \in F$. The action map $\tilde{\mu} : \Omega B \times Q_E \to Q_E$ is given by $\tilde{\mu}\big(\gamma, (\omega, e)\big) = (\gamma \cdot \omega, e)$, or equivalently, $\mu : \Omega B \times F \to F$ is given by

$$\mu(\gamma, f) = r\Big(\tilde{\mu}\big(\gamma, \eta(f)\big)\Big) = r(\gamma \cdot *, f).$$

Since the composite $F \to E \to B$ is trivial, $Q_F = \Omega B \times F$ and $Q_{CF} = \Omega B \times CF$. Notice that the inclusion $Q_F \hookrightarrow Q_E$ is homotopic to the composite

$$Q_F = \Omega B \times F \xrightarrow{\Omega B \times \eta} \Omega B \times Q_E \xrightarrow{\tilde{\mu}} Q_E.$$

Let $\phi' : Q_{CF} \to \Omega B \times CQ_E$ be the homotopy equivalence $\phi'(\omega, f, t) = \big(\omega, (\omega, f), t\big)$ and let $\phi : Q_F \to \Omega B \times Q_E$ be its restriction to $t = 1$. Then $\phi \simeq \psi \circ (\Omega B \times \eta)$ where $\psi : \Omega B \times Q_E \to \Omega B \times Q_E$ is the homotopy equivalence given by $\psi\big(\omega, (\alpha, f)\big) = \big(\omega, (\omega \cdot \alpha, f)\big)$. Therefore ϕ is also a homotopy equivalence. In the commutative cube

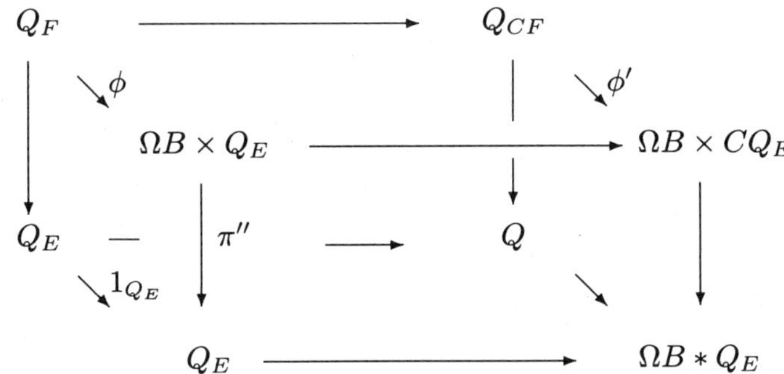

from the fact that 1_{Q_E}, ϕ, and ϕ' are homotopy equivalences, it follows that the induced map $Q \to \Omega B * Q_E$ is a homotopy equivalence, as claimed in the theorem.

The map $Q \to E \cup_F CF \to * \cup_F CF = SF$ is induced by $Q_E \to E \to *$ and $\pi'' : Q_{CF} = \Omega B \times CF \to CF$. Define $\tau' : \Omega B \times CQ_E \to CF$ by $\tau'\big(\gamma, (\omega, e), t\big) = \big(r(\gamma^{-1} \cdot \omega, e), t\big)$. Then $\tau' \circ \phi' = \pi''$ by construction of r. Therefore $Q \to SF$ factors as $Q \xrightarrow{\approx} \Omega B * Q_E \xrightarrow{\hat{\tau}} * \cup_F CF = SF$ where $\hat{\tau}$ is induced by τ' together with the trivial map $Q_E \to *$. Thus the composite $\Omega B * F \to E \cup_F CF \to SF$ is given by $(\omega, t, f) \mapsto \tau'\big(\omega, \eta(f), t\big) = \big(r(\omega^{-1} \cdot *, f), t\big) = \big(\mu(\omega^{-1}, f), t\big)$. $\qquad\square$

Remark To establish the homotopy type of Q, it would have sufficed to observe that $Q = Q_E \cup_{Q_F} Q_{CF} \approx F \cup_{\Omega B \times F} (\Omega B \times CF)$ with the map $\Omega B \times F \to F$ given by π''. The extra detail in the proof is given to demonstrate the last sentence of the theorem.

The method used in the preceding proof can be used to show

Theorem 7.7.4

a) *The homotopy fibre of the canonical inclusion $X \vee Y \to X \times Y$ is homotopy equivalent to $\Omega X * \Omega Y$.*

b) *The homotopy fibre of the fold map $X \vee X \to X$ is homotopy equivalent to $S\Omega X$.*

Proof

a) The homotopy fibre is homotopy equivalent to the pullback Q of $X \vee Y \to X \times Y$ and $P(X \times Y) \to X \times Y$. However

$$Q \approx (PX \times \Omega Y) \cup_{\Omega X \times \Omega Y} (\Omega X \times PY) \approx \Omega X * \Omega Y.$$

b) The homotopy fibre is homotopy equivalent to the pullback Q of $X \vee X \to X$ and $PX \to X$. However $Q \approx PX \cup_{\Omega X} PX \approx S\Omega X$. $\qquad\square$

Corollary 7.7.5 $\Omega(X \vee Y) \approx \Omega X \times \Omega Y \times \Omega(\Omega X * \Omega Y).$ $\qquad\square$

Assuming we are dealing with connected CW-complexes we have the following dual statement

Proposition 7.7.6 *If X and Y are connected CW-complexes then $S(X \times Y) \approx SX \vee SY \vee S(X \wedge Y).$*

Proof The principal cofibration $S(X \vee Y) \to S(X \times Y) \to S(X \wedge Y)$ has a homotopy splitting. $\qquad\square$

The same methods used in Theorem 7.7.4 yield

Theorem 7.7.7 *The homotopy fibre of the pinch map $X \vee Y \to X$ is homotopy equivalent to $(\Omega X \times Y)/(\Omega X \times *).$* $\qquad\square$

The composite of the quotient maps $A * B \to (A \times (SB))/(A \times *) \to A \wedge SB \cong S(A \wedge B)$ is a homotopy equivalence and so yields a homotopy splitting of the cofibration $SB \to (A \times (SB))/(A \times *) \to A \wedge SB$. Therefore,

Proposition 7.7.8 *If A and B are connected CW-complexes then*

$$(A \times (SB))/(A \times *) \approx SB \vee S(A \wedge B).$$ $\qquad\square$

Corollary 7.7.9

$$\Omega(X \vee SY) \approx \Omega X \times \Omega \left(\frac{\Omega X \times SY}{\Omega X \times *} \right) \approx \Omega X \times \Omega S(Y \vee (\Omega X \wedge Y)).$$ $\qquad\square$

8 Whitehead and Samelson Products

Let $f : X \to Z$ and $g : Y \to Z$ where Z is an H-group. The restriction to $X \vee Y$ of the map $X \times Y \to Z$ given by the commutator $(x, y) \mapsto f(x)g(y)f(x)^{-1}g(y)^{-1}$ is null homotopic, so there is an induced map $\langle f, g \rangle : X \wedge Y \to Z$. The sequence $[X \vee Y, Z] \to [X \times Y, Z] \to [X \wedge Y, Z]$ splits, so the homotopy class of $\langle f, g \rangle$ is uniquely determined by those of f and g and is called the *Samelson product* of f and g.

Suppose $Z = \Omega C$, and $f' : SX \to C$, $g' : SY \to C$. Let $f : X \to Z$ and $g : Y \to Z$ be the adjoints of f' and g'. Then the adjoint of $\langle f, g \rangle$ is called the *Whitehead product* and written $[f, g] : S(X \wedge Y) \to C$.

Example 7.8.1 Let ι_p and ι_q denote the canonical inclusions of S^p and S^q into $S^p \vee S^q$. The space formed by attaching a cell to $S^p \vee S^q$ by means of the Whitehead product $[\iota_p, \iota_q] : S^{p+q-1} \to S^p \vee S^q$ is $S^p \times S^q$. To see this observe that the composite $S^{p+q-1} \to S^p \vee S^q \to S^p \times S^q$ is null homotopic since its projections are $[1_{S^p}, *]$ and $[*, 1_{S^q}]$. Therefore there is an induced map $e^{p+q} \cup_{[\iota_p, \iota_q]} (S^p \vee S^q) \to S^p \times S^q$ and this map is a bijection from a compact space to a Hausdorff space. $\qquad\square$

We can use these notions to give a more precise version of Theorem 7.7.4a.

Theorem 7.7.4a' *There is a homotopy fibration*

$$S(\Omega X \wedge \Omega Y) \xrightarrow{[\beta_X, \beta_Y^{-1}]} X \vee Y \xrightarrow{i} X \times Y,$$

where β denotes the adjoint of the identity map.

Proof An explicit homotopy inverse $\bar\psi : S(A \wedge B) \to A * B$ to the homotopy equivalence of Proposition 7.7.1 is given by the following. Define $\psi : S(A \times B) \to A * B$ by

$$\psi(a, b, t) = \begin{cases} (a, 3t, *) & \text{if } 0 \le t \le 1/3; \\ (a, 2 - 3t, b) & \text{if } 1/3 \le t \le 2/3; \\ (*, 3t - 2, b) & \text{if } 2/3 \le t \le 1. \end{cases}$$

The restriction of ψ to $S(A \vee B)$ is null homotopic, and the induced map $\bar\psi : S(A \wedge B) \to A * B$ is the desired homotopy inverse.

The method of Ganea's theorem shows that the inclusion of the homotopy fibre of i into $X \vee Y$ is given by the composite

$$\Omega X * \Omega Y \approx (C\Omega X \times \Omega Y) \cup_{(\Omega X \times \Omega Y)} (\Omega X \times C\Omega Y)$$

$$\approx (PX \times \Omega Y) \cup_{(\Omega X \times \Omega Y)} (\Omega X \times PY) \xrightarrow{ev} X \vee Y$$

with $C\Omega W \to PW$ given by $(\sigma, t) \mapsto \sigma_t$, where $\sigma_t(s) = \sigma(st)$. Therefore the composite $S(\Omega X \times \Omega Y) \xrightarrow{\psi} \Omega X * \Omega Y \longrightarrow X \vee Y$ is given by

$$\psi(\omega, \gamma, t) = \begin{cases} \omega(6t) & \text{if } 0 \le t \le 1/6; \\ * & \text{if } 1/6 \le t \le 1/3; \\ \gamma(3 - 6t) & \text{if } 1/3 \le t \le 1/2; \\ \omega(4 - 6t) & \text{if } 1/2 \le t \le 2/3; \\ * & \text{if } 2/3 \le t \le 5/6; \\ \gamma(6t - 5) & \text{if } 5/6 \le t \le 1, \end{cases}$$

and so the result follows. $\qquad\qquad\qquad\qquad\qquad\qquad\qquad\qquad\qquad\square$

9 James Construction

For a pointed topological space X, let $J_k(X) = X^k/\sim$, where

$$(x_1, \dots, x_{j-1}, *, x_{j+1}, \dots, x_k) \sim (x_1, \dots, x_{j-1}, x_{j+1}, *, \dots, x_k).$$

The *James construction* on X, denoted $J(X)$ is defined by $J(X) = \varinjlim_k J_k(X)$.

There is a canonical inclusion $X = J_1(X) \hookrightarrow J(X)$. $J(X)$ is a topological monoid and any map from X to a topological monoid M extends uniquely to a morphism $J(X) \to M$ of topological monoids. That is, $X \hookrightarrow J(X)$ is universal with respect to maps from X to topological monoids.

Let $X^{(n)}$ denote the n-fold smash product of X with itself. (Unfortunately, this same notation is also customarily used to denote the n-skeleton of a CW-complex and we will have to rely on the context to determine which is meant.) From the definition, $J_k(X)/J_{k-1}(X) = X^{(k)}$. Set

$$\mathrm{FW}_k(X) = \{(x_1, x_2, \dots, x_k) \in X^k \mid x_j = * \text{ for some } j\},$$

called the kth *fat wedge* of X. There is a commutative diagram

$$
\begin{array}{ccccc}
FW_k(X) & \hookrightarrow & X^k & \longrightarrow\!\!\!\rightarrow & X^{(k)} \\
\downarrow & & \downarrow & & \| \\
J_{K-1}(X) & \hookrightarrow & J_{k(X)} & \longrightarrow\!\!\!\rightarrow & X^{(k)}.
\end{array}
$$

Proposition 7.9.1 *If X has the homotopy type of a connected CW-complex then $S\big(J(X)\big) \approx \bigvee_{k=1}^{\infty} S\big(X^{(k)}\big)$.*

Proof By Proposition 7.7.6 and induction on k the top right epimorphism in the preceding diagram has a homotopy splitting after suspending, and so the bottom does as well. $\qquad\square$

Since multiplication of paths in $\Omega(X)$ and $P(X)$ is only homotopy associative rather than strictly associative, it is convenient to introduce related spaces with strictly associative multiplications. The *Moore loop space* on X, denoted $\Omega'(X)$, is defined by

$$
\Omega'(X) = \{(s,\omega) \in \mathbb{R}^+ \times X^{\mathbb{R}^+} \mid \omega(0) = * \text{ and } \omega(t) = * \text{ for } t \ge s\}
$$

and the *Moore path space* on X, denoted $P'(X)$ is defined by

$$
P'(X) = \{(s,\omega) \in \mathbb{R}^+ \times X^{\mathbb{R}^+} \mid \omega(0) = * \text{ and } \omega(t) = \omega(s) \text{ for } t \ge s\}.
$$

We think of elements of $\Omega'(X)$ and $P'(X)$ as paths parameterized by $[0,s]$. Such paths have a strictly associative multiplication under which the product of paths of length s and s' has length $s + s'$. The inclusion $\Omega(X) \subset \Omega'(X)$ and the map $\Omega'(X) \to \Omega(X)$ which reparametrizes the paths to run from 0 to 1 are H-maps and inverse homotopy equivalences. The spaces $P'(X)$ and $P(X)$ are also homotopy equivalent, both being contractible.

By the universal property of $J(X)$, the composite $X \longrightarrow \Omega SX \xrightarrow{\approx} \Omega' SX$ extends to a map $J(X) \to \Omega' SX$.

Theorem 7.9.2 *If X has the homotopy type of a connected CW-complex then the composite $J(X) \longrightarrow \Omega' SX \xrightarrow{\approx} \Omega SX$ is a homotopy equivalence.*

Proof Consider first the special case where X is a finite CW-complex. By Proposition 7.9.1 and the Bott-Samelson Theorem, the two sides have the same homology, $T\big(\tilde{H}_*(X)\big)$, with any field as coefficients. Furthermore, the map induces an algebra map on homology whose restriction to $H_*(X)$ is the identity. Therefore the map induces an isomorphism with any field as coefficients, and so, since the two sides have finite type, it induces an isomorphism with integer coefficients. Since the map is an H-map, the Whitehead theorem implies that it is a homotopy equivalence.

For the general case, take the direct limit over all finite subcomplexes of X to see that the map induces an isomorphism on homology. $\qquad\square$

The homotopy equivalence of Proposition 7.9.1 depends on the choice of splittings. Using Theorem 7.9.2, an explicit homotopy equivalence can be described as follows. Given a map $f : X^{(m)} \to Y$, define a map $\hat{f} : J(X) \to J(Y)$ by setting

$$
\hat{f}_k|_{J_k(X)}(x_1, x_2, \dots, x_k) = \prod f(x_{i_1} \wedge x_{i_2} \wedge \cdots \wedge x_{i_m}) \in J_{\binom{k}{m}}(Y)
$$

where the product is taken of over all subsets $\{x_{i_1}, x_{i_2}, \dots, x_{i_m}\}$ of $\{x_1, x_2, \dots, x_k\}$, ordered right lexicographically. It is easy to check that the collection of maps defined this way on each $J_k(X)$ are compatible and so induce a map $\hat{f} : J(X) \to J(Y)$, called the *James combinatorial extension* of the composite $J_m(X) \twoheadrightarrow X^{(m)} \xrightarrow{f} Y$. Of special interest is the case where f is the identity map in which case the resulting map $H_m : J(X) \to J\left(X^{(m)}\right)$ is called the *mth James-Hopf invariant map*. The maps $J_k(X) \to J\left(\bigvee_{m=1}^{\infty} X^{(m)}\right)$ given by the product over $j = 1, 2, \dots, k$ (in order) of the composites $J_k(X) \hookrightarrow J(X) \xrightarrow{H_j} J\left(X^{(j)}\right) \hookrightarrow J\left(\bigvee_{m=1}^{\infty} X^{(m)}\right)$ piece together to give a map $J(X) \to J\left(\bigvee_{m=1}^{\infty} X^{(m)}\right)$, whose adjoint (after composing with the homotopy equivalence $J\left(\bigvee_{m=1}^{\infty} X^{(m)}\right) \approx \Omega S\left(\bigvee_{m=1}^{\infty} X^{(m)}\right)$ of Theorem 7.9.2) gives an explicit homotopy equivalence for Proposition 7.9.1.

From Proposition 7.9.1, Theorem 7.9.2, and Corollary 7.7.9 we get

Corollary 7.9.3 $\Omega(SX \vee SY) \approx \Omega SX \times \Omega S\left(\bigvee_{j=0}^{\infty}(X^{(j)} \wedge Y)\right).$ $\qquad\square$

It is clear that repeated application of Corollary 7.9.3 yields an infinite product decomposition of $\Omega(SX \vee SY)$, although some effort is required to keep track of the factors. The result is

Theorem 7.9.4 (Hilton-Milnor Theorem)

$$\Omega(SX \vee SY) \approx \prod_{w \in W}{}' \Omega S\left(X^{(w_x)} \wedge Y^{(w_y)}\right)$$

where W is a (vector space) basis for the free Lie algebra on x and y. $\qquad\square$

It is possible to choose a basis W such that each term is an iterated bracket in x and y, and in this formula w_x and w_y denote the number of x's and y's respectively in the iterated bracket w. The product, $\prod'$, is the direct limit of the finite subproducts, or equivalently, the result obtained by applying the compactly generated functor (see Section 2.6) to the topological product. The maps from each factor are given by loops of iterated Whitehead products.

Theorem 7.9.5 *A connected CW-complex X is an H-space if and only if the canonical map $X \to \Omega SX$ has a left homotopy splitting.*

Proof It is clear that the canonical map $X \to JX$ has a left homotopy splitting if and only if X is an H-space. $\qquad\square$

Theorem 7.9.6 *A connected CW-complex X is a co-H-space if and only if the canonical map $S\Omega X \to X$ has a right homotopy splitting.*

Lemma 7.9.7 *For any connected CW-complex X,*

$$\begin{array}{ccc} S\Omega X & \xrightarrow{\;\;f\;\;} & X \vee X \\ {\scriptstyle \beta_X}\downarrow & & \downarrow{\scriptstyle j} \\ X & \xrightarrow{\;\;\Delta\;\;} & X \times X \end{array}$$

is a homotopy pullback, where in the homotopy group $[f]$ is the adjoint of the product $[\Omega(i')][\Omega(i'')]$ of the loops on the inclusion maps.

Proof Since the diagram is homotopy commutative, there is an induced map ϕ from $S\Omega X$ to the homotopy pullback. In this case, the homotopy class of ϕ is uniquely determined because $\Omega j : \Omega(X \vee X) \to \Omega(X \times X)$ has a right homotopy splitting. Since

$$
\begin{array}{ccc}
\Omega(X \vee X) & \xrightarrow{\Omega(1 \perp -1)} & \Omega X \\[4pt]
\Big\downarrow {\scriptstyle \Omega(j)} & & \Big\| \\[4pt]
\Omega(X \times X) & \xrightarrow{\quad \tau \quad} & \Omega X
\end{array}
$$

homotopy commutes, where $\tau(\omega, \gamma) = \omega * \gamma^{-1}$, there is an induced map σ between the horizontal homotopy fibres. Since the right vertical map is the identity, after identifying the homotopy fibres we see that we have a homotopy pullback

$$
\begin{array}{ccc}
\Omega S\Omega X & \xrightarrow{\Omega(f)} & \Omega(X \vee X) \\[4pt]
\Big\downarrow {\scriptstyle \sigma} & & \Big\downarrow {\scriptstyle \Omega(j)} \\[4pt]
\Omega X & \xrightarrow{\Omega(\Delta)} & \Omega(X \times X).
\end{array}
$$

Since $\Omega(\Delta)$ has a left homotopy splitting, the fact that replacing σ by $\Omega(\beta_X)$ yields a homotopy commutative diagram implies that $\sigma \simeq \Omega(\beta_X)$. Therefore $\Omega(\phi)$ is a homotopy equivalence and thus ϕ induces an isomorphism on homotopy groups. $\qquad\square$

Proof of Theorem 7.9.6 It follows from the lemma that a homotopy lifting of the diagonal map Δ to $X \vee X$ is equivalent to a right homotopy splitting of $\beta_X : S\Omega X \to X$. $\qquad\square$

Theorem 7.9.8 *Let*

$$
\begin{array}{ccc}
X & \xrightarrow{f} & Y \\[4pt]
\Big\downarrow {\scriptstyle g} & & \Big\downarrow {\scriptstyle p} \\[4pt]
Z & \xrightarrow{q} & Q
\end{array}
$$

be a pushout square in which at least one of f, g is a cofibration. Suppose that X, Y, and Z are co-H-spaces having the homotopy types of connected CW-complexes. Suppose that f and g are co-H-maps and that X is simply connected. Then there is a co-H-space structure on Q such that p and q are co-H-maps.

Proof Let $\psi_X : X \to S\Omega X$, $\psi_Y : Y \to S\Omega Y$, and $\psi_Z : Z \to S\Omega Z$ be right homotopy splittings of β_X, β_Y, and β_Z respectively. The statements that f and g are co-H-maps are equivalents to the statements that $\psi_Y f \simeq (S\Omega f)\psi_X$ and $\psi_Z g \simeq (S\Omega g)\psi_X$. It follows from Proposition 7.5.14b that X has a homotopy inverse, so there is a homotopy cofibration sequence $X \xrightarrow{fg^{-1}} Y \vee Z \xrightarrow{p \perp q} Q$. Since the composite

$$X \xrightarrow{fg^{-1}} Y \vee Z \xrightarrow{\psi_Y \vee \psi_Z} S\Omega Y \vee S\Omega Z \xrightarrow{p \perp q} S\Omega Q$$

is null homotopic, there is an induced map $\tilde{\psi}_Q : Q \to S\Omega Q$ such that $\tilde{\psi}_Q p \simeq (S\Omega p)\psi_Y$ and $\tilde{\psi}_Q q \simeq (S\Omega q)\psi_Z$, although it need not be true that $\beta_Q \tilde{\psi}_Q \simeq 1_Q$. However since $\beta_Q \tilde{\psi}_Q \circ (p \perp q) \simeq 1_Q \circ (p \perp q)$, the principal cofibration $Y \vee Z \xrightarrow{p \perp q} Q \longrightarrow SX$ implies that there is a map $\phi : SX \to Q$ such that $1_Q \simeq \big((\beta_Q \tilde{\psi}_Q) \perp \phi\big) \circ \theta$ where $\theta : Q \to Q \vee SX$ is the co-action map. Letting $\phi' : SX \to S\Omega Q$ be the suspension of the adjoint of ϕ gives a factorization $\phi = \beta_Q \phi'$. Set $\psi_Q = (\tilde{\psi}_Q \perp \phi') \circ \theta$. Then $\psi_Q p \simeq \tilde{\psi}_Q p \simeq (S\Omega p)\psi_Y$, $\psi_Q q \simeq \tilde{\psi}_Q q \simeq (S\Omega q)\psi_Z$, and $\beta_Q \psi_Q \simeq 1_Q$, so ψ_Q yields the required co-H-structure. $\qquad\square$

Remark The dual of this method could be used to give an alternate proof of Corollary 7.5.13.

10 Lusternik-Schnirrelmann Category

The original idea behind Lusternik-Schnirrelmann category was to describe the minimum number of contractible open sets which cover X. Over the years the definition has undergone many changes including a re-indexing so that a contractible space would have category 0 rather than 1. For example, should one consider open sets or closed sets? Should the sets be required to be contractible themselves or merely contractible within X (i.e., null homotopic inclusion maps)? For "nice" spaces X these variations are not significant. The following definition is the reformulation of George Whitehead which makes it clear the concept being defined is a homotopy invariant. (That is, has the same value on homotopy equivalent spaces.)

Definition 7.10.1 The (*Lusternik-Schnirrelmann*) *category* of a pointed space X is the least n (if any) such that the iterated diagonal map $\Delta^n : X \to X^{n+1}$ factors up to homotopy through the $(n+1)$st fat wedge of X. (That is, there exists $\bar{\Delta} : X \to \mathrm{FW}_{n+1}(X)$ such that $X \xrightarrow{\bar{\Delta}} \mathrm{FW}_{n+1}(X) \hookrightarrow X^{n+1}$ is homotopic to Δ^n.)

The relation between this concept and something similar to the original concept is given by the following theorem.

Theorem 7.10.2 *Let X be a pointed topological space.*

1) *If X has category n then X has a covering by $n+1$ contractible closed subsets.*
2) *If X has a covering by $n + 1$ closed subsets whose inclusion maps are null homotopic cofibrations then X has category at most n.* $\qquad\square$

Proposition 7.10.3 *Suppose X has category n. Then the product of any $n + 1$ elements in $\tilde{H}^*(X)$ is zero.*

Proof The $(n+1)$-fold product is induced by Δ^n, but elements of $\tilde{H}^*(X)^{\otimes(n+1)}$ go to zero under $H^*(X)^{\otimes(n+1)} \to H^*\big(\mathrm{FW}_{n+1}(X)\big)$. $\qquad\square$

The Lusternik-Schnirrelmann category is one measure of the complexity of a space. Very little is known about it in general, but rational homotopy theorists have been successful in developing some properties of the LS-category of the localizations of spaces at the rationals.

Exercise 7.10.4 Find the Lusternik-Schnirrelmann category of the torus $S^1 \times S^1$.

CHAPTER 8

Simplicial Sets

1 Definitions and First Principles

In this chapter we will consider combinatorially defined objects, simplicial sets, to which we can associate a homotopy theory which is equivalent to that associated to CW-complexes. This provides an alternative to the use of CW-complexes, each theory having some technical advantages and some disadvantages. Simplicial sets show to best advantage in assigning topological spaces to objects which have a lot of algebraic structure. In these notes they will be used in Chapter 12 to construct localizations and completions of spaces.

Definition 8.1.1 A *simplicial set K* consists of a collection of sets $\{K_n\}_{n \geq 0}$ together with set maps $d_j : K_n \to K_{n-1}$ for $j = 0, \ldots, n$ and $s_j : K_n \to K_{n+1}$ for $j = 0, \ldots, n$ satisfying the simplicial identities:

$$\begin{aligned}
d_i d_j &= d_{j-1} d_i & \text{for } i < j; \\
s_i s_j &= s_{j+1} s_i & \text{for } i \leq j; \\
d_i s_j &= s_{j-1} d_i & \text{for } i < j; \\
d_i s_j &= 1_{K_n} & \text{for } i = j, j+1; \\
d_i s_j &= s_j d_{i-1} & \text{for } i > j+1.
\end{aligned}$$

A *simplicial map* $f : K \to L$ is a set map $f_n : K_n \to L_n$ which commutes with all the d_j's and s_j's. The maps d_j and s_j are called *boundary maps* and *degeneracy maps* respectively. An element $x \in K_n$ is called *degenerate* if $x = s_j y$ for some j and y. We shall sometimes use the notations $dx = (d_0 x, d_1 x, \ldots, d_n x)$ and $sx = (s_0 x, s_1 x, \ldots, s_n x)$ for $x \in K_n$ to display the boundaries and degeneracies.

Note Older books use the term "semi-simplicial set" rather than "simplicial set" for this concept, but that terminology is now obsolete.

More generally if K_n are objects in a category $\underline{C}$ and the maps d_j and s_j are morphisms in $\underline{C}$ then K is called a simplicial object in $\underline{C}$.

Example 8.1.2 Let X be a set and let $K_n = \prod_{i=0}^{n} X = X^{n+1}$. Define $d_j : K_n \to K_{n-1}$ by

$$d_j\big((x_0, x_1, \ldots, x_n)\big) = (x_0, \ldots, x_{j-1}, x_{j+1}, \ldots, x_n)$$

and $s_j : K_n \to K_{n+1}$ by

$$s_j\big((x_0, x_1, \ldots, x_n)\big) = (x_0, \ldots, x_{j-1}, x_j, x_j, x_{j+1}, \ldots, x_n).$$

$\square$

For $0 \leq j \leq n$, let $\delta^j : \Delta^{n-1} \to \Delta^n$ be the inclusion of the jth face and let $\sigma_j : \Delta^{n+1} \to \Delta^n$ be the affine map determined by

$$\sigma_j(\epsilon_i) = \begin{cases} \epsilon_i & \text{if } i \leq j; \\ \epsilon_{i-1} & \text{if } i > j. \end{cases}$$

(Thus σ_j collapses the jth face.) Given a simplicial set K, we can form a topological space $|K|$, called the *geometric realization* of K by $|K| = \coprod (K_n \times \Delta^n)/\sim$ where $(d_j x, t) \sim (x, \delta^j t)$ and $(s_j x, t) \sim (x, \sigma_j t)$. From the construction, $|K|$ has a natural CW-structure.

Example 8.1.3 Let $\Delta[k]$ be the simplicial set with

$$(\Delta[k])_n = \{(i_0, i_1, \dots, i_n) \mid i_j \in \mathbb{Z} \text{ and } 0 \le i_0 \le i_1 \le \dots \le i_n \le k\}$$

where d_j and s_j are defined by dropping and repeating the jth component. Then $|\Delta[k]| = \Delta^k$. The nondegenerate simplices of $\Delta[k]$ are the sequences without repetitions, and these correspond to the faces of Δ^k. All elements of $\Delta[k]$ can be obtained by applying sequences of boundaries and degeneracies to $(0, 1, \dots, k)$ so $\Delta[k]$ is the "free" simplicial set generated by $(0, 1, \dots, k)$ in degree k. Given any simplicial set K and any element $x \in K_k$, the assignment $f\big((0, 1, \dots, k)\big) = x$ extends uniquely to a simplicial map $\Delta[k] \to K$. We sometimes write I for the simplicial set $\Delta[1]$ and $*$ for $\Delta[0]$ and ι_k for the element $\big((0, 1, \dots, k)\big) \in (\Delta[k])_k$. $\square$

Given a simplicial set K, the *m-skeleton* $K^{(m)}$ of K is defined by

$$\left(K^{(m)}\right)_n = \{k \in K_n \mid k = s_{j_1} s_{j_2} \cdots s_{j_r} y \text{ for some } y \text{ with } |y| \le m \text{ and}$$
$$\text{some sequence } s_{j_1} s_{j_2} \cdots s_{j_r} y\}.$$

That is, $K^{(m)}$ is the subcomplex of K generated by $K_0, \dots, K_m$. The simplicial identities show that $K^{(m)}$ is closed under the maps d_i.

A *pointed simplicial set* consists of a simplicial set K together with a point $k_0 \in K_0$, or equivalently a map from $\Delta[0]$ to K. For each n, $(\Delta[0])_n$ is a singleton and so the image of this map is a point in K_n, written $*$, which can be obtained by applying any sequence of degeneracies to k_0. Since the same notation, $*$, is used for all n, the set K_n to which $*$ belongs will have to be determined from the context.

If $A \hookrightarrow B$ is an inclusion of simplicial sets, then define the quotient simplicial set by $(B/A)_n = B_n/A_n$. (That is, make all elements of A_n equivalent.) Then $|B/A| = |B|/|A|$. Define the simplicial m-sphere by $S^m = \Delta[m]/(\Delta[m])^{(m-1)}$. S^m has two nondegenerate simplices: $(0) \in (S^m)_0$ and $(0, 1, \dots, m) \in (S^m)_m$.

Let K and L be simplicial sets. The product of K and L is given by $(K \times L)_n = K_n \times L_n$ with d_j and s_j acting componentwise.

Proposition 8.1.4 *If at least one of K, L has finite type (i.e., finitely many nondegenerate elements in each degree) then $|K \times L| \cong |K| \times |L|$.* $\square$

The cell structure of the product as a CW-complex can be seen clearly in the following example.

Example 8.1.5 Let $c = (0, 1)$, $b = d_0 c = (1)$, and $a = d_1 c = (0)$ be the nondegenerate simplices of I. Then the nondegenerate simplices of $I \times I$ are: (a, a), (a, b), (b, a), (b, b) in degree 0; $(s_0 a, c)$, $(c, s_0 b)$, $(s_0 b, c)$, $(c, s_0 a)$, (c, c) in degree 1; and $(s_0 c, s_1 c)$, $(s_1 c, s_0 c)$ in degree 2.

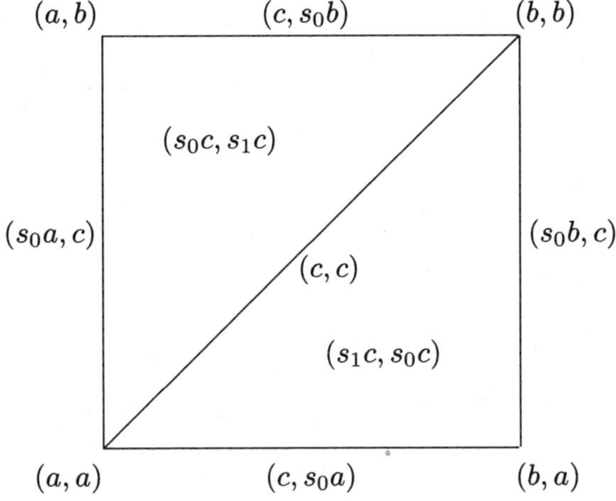

$\square$

Let $i_0 : K \to K \times I$ and $i_1 : K \to K \times I$ be the simplicial maps $i_0(k) = \big(k, (0, 0, \ldots, 0)\big)$ and $i_1(k) = \big(k, (1, 1, \ldots, 1)\big)$.

Two simplicial maps $f, g : K \to L$ are called *homotopic*, written $f \simeq g$ if there exists $H : K \times I \to L$ such that $H \circ i_0 = f$ and $H \circ i_1 = g$.

Note In general, homotopy is **not** an equivalence relation on simplicial maps.

Proposition 8.1.6 $f \simeq g : K \to L \Rightarrow |f| \simeq |g| : |K| \to |L|$. $\square$

Simplicial objects can also be defined as contravariant functors from a category Δ^* defined as follows. The objects of Δ^* are $\{(0, 1, \ldots, n)\}_{n \geq 0}$. The object $(0, 1, \ldots, n)$ of Δ^* is denoted Δ_n. The morphisms are given by

$$\Delta^*(\Delta_n, \Delta_k) = \{\text{order preserving maps from } \Delta_n \text{ to } \Delta_k\}.$$

A simplicial $\underline{C}$-object K is equivalent to a contravariant functor from Δ^* to $\underline{C}$ (with $K_n = K(\Delta_n)$) and a simplicial map is equivalent to a natural transformation of such functors.

Every morphism in Δ^* can be written uniquely as a composite of the morphisms $\delta_i : \Delta_{n-1} \to \Delta_n$ and $\sigma_i : \Delta_{n+1} \to \Delta_n$ which omit i and repeat i respectively.

Every object X in a category $\underline{C}$ determines a contravariant functor $\underline{C}(\_, X)$ and every morphism from $X \to Y$ yields a natural transformation of the corresponding functors. By examining the definitions we see that $\Delta[n] = \Delta^*(\_, \Delta_n)$ with $(\Delta[n])_k = \Delta^*(\Delta_k, \Delta_n)$. The maps δ_i and σ_i yield maps of simplicial sets $(\delta_i)_* : \Delta[n-1] \to \Delta[n]$ and $(\sigma_i)_* : \Delta[n+1] \to \Delta[n]$. As with every simplicial map from $\Delta[n-1]$, δ_i is completely determined by its value on $(0, 1, \ldots, n-1) \in (\Delta[n-1])_{n-1}$ which in this case is $(0, \ldots, i-1, i+1, \ldots, n) \in (\Delta[n])_{n-1}$. Similarly $(\sigma_i)_*$ is determined by the equation

$$(\sigma_i)_*\big((0, 1, \ldots, n+1)\big) = (0, \ldots, i-1, i, i, i+1, \ldots, n)$$

in $(\Delta[n])_{n+1}$.

Let SS denote the category of simplicial sets. For simplicial sets K and L, we define the function simplicial set L^K by $(L^K)_n = \mathrm{Hom}_{SS}(K \times \Delta[n], L)$ with $(d_i f)(k, t) = f(k, (\sigma_i)_* t)$ and $(s_i f)(k, t) = f(k, (\delta_i)_* t)$ for all $k \in K_q$, $t \in (\Delta[n])_q$, and $f \in (L^K)_n$.

Theorem 8.1.7 (Exponential Law) $M^{K \times L} \cong (M^K)^L$.

The proof is a long exercise in p, q-shuffles. See [May, Lemma 6.13]. $\square$

2 Simplicial Fibrations and Kan Complexes

Set $\Lambda^k[n]$ equal to the subcomplex of $\Delta[n]$ generated by $\{d_i((0,1,\ldots,n)) \mid i \neq k\}$. A simplicial map $f : X \to Y$ is called a *fibration* if given any diagram

$$
\begin{array}{ccc}
\Lambda^k[n] & \longrightarrow & X \\
\downarrow & & \downarrow f \\
\Delta[n] & \longrightarrow & Y
\end{array}
$$

for any k and n, there exists a map $\gamma : \Delta[n] \to X$ making the two triangles created commute. A simplicial set K is called a *Kan complex* if $K \to *$ is a fibration. Explicitly, K is a Kan complex if and only if for every k and n, given a set of n elements in K_{n-1} labelled $y_0, y_1, \ldots, y_{k-1}, y_{k+1}, \ldots, y_n$ which have the property that $d_i y_j = d_{j-1} y_i$ for $i < j$ $(i, j \neq k)$, there exists $y \in K_n$ such that $d_i y = y_i$ for $i = 0, \ldots, k-1, k+1, \ldots, n$.

Theorem 8.2.1 L *a Kan complex* $\Rightarrow L^K$ *is a Kan complex.* $\square$

Lemma 8.2.2 *Let L be a Kan complex.*

a) *If there exists a path $\omega : I \to L$ such that $\omega(0) = x$ and $\omega(1) = y$ then there exists a path $\sigma : I \to L$ such that $\sigma(0) = y$ and $\sigma(1) = x$.*

b) *If there exist paths $\omega : I \to L$ and $\tau : I \to L$ such that $\omega(0) = x$, $\omega(1) = y = \tau(0)$, $\tau(1) = z$ then there exists a path $\sigma : I \to L$ such that $\sigma(0) = x$ and $\sigma(1) = z$.*

Proof

a) Let $a = \omega((0,1)) \in L_1$. By definition of Kan complex, there exists $z \in L_2$ such that $dz = (\_, s_0 x, a)$, since the faces match. Set $b = d_0 z$. Then $db = (x, y)$ so let σ be the path determined by $\sigma((0,1)) = b$.

b) Let $a = \omega((0,1))$ and let $b = \tau((0,1))$. There exists $z \in L_2$ such that $dz = (b, \_, a)$. Let σ be the path determined by $\sigma((0,1)) = d_1 z$. $\square$

By the exponential law, a homotopy from K to L is a path in L^K and so we get the following corollary.

Corollary 8.2.3 *If L is a Kan complex then homotopy is an equivalence relation on $\mathrm{Hom}_{SS}(K, L)$ for any simplicial complex K.* $\square$

If K is a simplicial set and L is a Kan complex then let $[K, L]$ denote the set of equivalence classes of pointed simplicial maps from K to L under the homotopy relation. Set $\pi_n(L) = [S^n, L]$.

Let K be a pointed Kan complex. For $x, y \in \{k \in K_n \mid d_i k = * \text{ for all } i\}$, write $x \sim_h y$ if there exists $z \in K_{n+1}$ such that $dz = (*, *, \ldots, *, x, y, *, \ldots, *)$.

Lemma 8.2.4 *Let K be a pointed Kan complex and suppose w, x, y belong to K_n such that $d_i w = d_i x = d_i y = *$ for all $i = 0, \ldots, n$. If there exists $a \in K_{n+1}$ such that $da = (w, x, *, \ldots, *)$ and there exists $b \in K_{n+1}$ such that $db = (x, y, *, \ldots, *)$ then there exists $c \in K_{n+1}$ such that $dc = (w, y, *, \ldots, *)$. Also, for x, y as above the following are equivalent:*

a) *there exists $z \in K_{n+1}$ such that $dz = (*, \ldots, *, x, y, *, \ldots, *)$, where the x appears in position k;*

b) *there exists $z \in K_{n+1}$ such that $dz = (*, \ldots, *, y, x, *, \ldots, *)$, where the y appears in position k;*

c) *there exists $z \in K_{n+1}$ such that $dz = (*, \ldots, *, x, y, *, \ldots, *)$, where the x appears in position $k + 1$.*

Proof Given $a,\ b \in K_{n+1}$ such that $da = (w, x, *, \ldots, *)$ and $db = (x, y, *, \ldots, *)$, there exists $v \in K_{n+2}$ such that $dv = (a, \_, b, *, \ldots, *)$. Then $d(d_1 v) = (w, y, *, \ldots, *)$, so set $c = d_1 v$.

a$\Rightarrow$c) Given $z \in K_{n+1}$ such that $dz = (*, \ldots, *, x, y, *, \ldots, *)$ with the x in position k, there exists $v \in K_{n+2}$ such that

$$dv = (*, \ldots, *, \_, s_{k+1} y, z, s_k y, *, \ldots, *)$$

where the blank appears in position k. Then

$$d(d_k v) = (*, \ldots, *, x, y, *, \ldots, *),$$

where the x appears in position $k + 1$.

c$\Rightarrow$a) Given $z \in K_{n+1}$ such that $dz = (*, \ldots, *, x, y, *, \ldots, *)$ with the x in position $k + 1$, there exists $v \in K_{n+2}$ such that

$$dv = (*, \ldots, *, z, s_{k+1} y, \_, s_k y, *, \ldots, *)$$

where the blank appears in position $k + 2$. Then

$$d(d_{k+2} v) = (*, \ldots, *, x, y, *, \ldots, *),$$

where the x appears in position k.

a$\Leftrightarrow$b) Given $z \in K_{n+1}$ such that $dz = (*, \ldots, *, x, y, *, \ldots, *)$ with the x in position k, there exists $v \in K_{n+2}$ such that

$$dv = (*, \ldots, *, \_, s_{k+1} y, s_k y, z, *, \ldots, *)$$

where the blank appears in position k. Then

$$d(d_k v) = (*, \ldots, *, y, x, *, \ldots, *),$$

where the y appears in position k. $\qquad\square$

Corollary 8.2.5 $\sim_h$ *is an equivalence relation on $\{k \in K_n \mid d_i k = * \text{ for all } i\}$.* $\qquad\square$

Theorem 8.2.6 *Let $f,\ g : S^n \to K$ be pointed maps where K is a Kan complex. Then $f \simeq g$ (rel $*$) $\iff f(\iota_n) \sim_h g(\iota_n)$, where $\iota_n = (0, 1, \ldots, n) \in (S^n)_n$.* $\qquad\square$

Corollary 8.2.7 $\pi_n(K) \cong \{k \in K_n \mid d_i k = * \text{ for all } i\} / \sim_h.$ $\qquad\square$

For $[x], [y] \in \pi_n(K)$ define an element $[xy] \in \pi_n(K)$ as follows. Choose v such that $dv = (x, \_, y, *, \ldots, *)$ and set $[xy] = [d_1 v]$.

Proposition 8.2.8 *The multiplication $([x], [y]) \mapsto [xy]$ is well defined and is a group structure on $\pi_n(K)$ which is abelian if $n \geq 2$.* $\qquad\square$

Proposition 8.2.9 *Let L be a connected Kan complex. Then $\pi_1(L) \cong \pi_1(|L|)$.* $\qquad\square$

3 Simplicial Groups

Recall the definition of a simplicial group given after Definition 8.1.1.

Theorem 8.3.1 *A simplicial group is a Kan complex.*

Proof Let G be a simplicial group and let $y_0, y_1, \ldots, \widehat{y_k}, \ldots, y_n \in G_{n-1}$ be compatible. Inductively define w_j for $j = 0, \ldots, k-1$ by

$$w_0 = s_0 y_0 \quad \text{and} \quad w_j = w_{j-1}(s_j d_j w_{j-1})^{-1}(s_j y_j).$$

Next using downward induction, define w_j for $j = n, \ldots, k+1$ by

$$w_n = w_{k-1}(s_{n-1} d_n w_{k-1})^{-1}(s_{n-1} y_n) \quad \text{and} \quad w_j = w_{j+1}(s_{j-1} d_j w_{j+1})^{-1}(s_{j-1} y_j).$$

By induction on j, check that $d_i w_j = y_i$ for $i \leq j < k$. Next, by downward induction on j, check that $d_i w_j = y_i$ for $k < j \leq n$ whenever $i = 1, \ldots, k-1$ or $i \geq j$. $\qquad\square$

Let G be a simplicial group. Set $NG_n = \bigcap_{j>0} \operatorname{Ker} d_j$. Although NG does not form a simplicial subgroup of G (not closed under degeneracies), for each n, NG_n is a normal subgroup of G_n and the simplicial identities imply that $d_0(NG_n) \subset NG_{n-1}$. Even though the groups need not be abelian, $\operatorname{Im} d_0 : NG_{n+1} \to NG_n$ is a normal subgroup of $\operatorname{Ker} d_0 : NG_n \to NG_{n-1}$ and so we can form homology in the usual way by $H_n(NG, d_0) = \operatorname{Ker} d_0 / \operatorname{Im} d_0$. Choose the identity of G_0 as the basepoint $*$.

Theorem 8.3.2 $\pi_n(G) \cong H_n(NG, d_0)$.

Proof Let $f : S^n \to G$ represent an element of $\pi_n(G)$ and let x be $f(\iota_n)$. Then $d_j x = *$ for all j so x belongs to $\operatorname{Ker} d_0 : NG_n \to NG_{n-1}$. Define $\phi : \pi_n(G) \to H_n(NG, d_0)$ by $\phi(f) = x$. If $g \sim f$ then there exists z such that $dz = (x, y, *, \ldots, *)$ where $y = g(\iota_n)$. $d_0\big(z(s_0 y^{-1})\big) = (d_0 z)y^{-1} = xy^{-1}$, $d_1\big(z(s_0 y^{-1})\big) = yy^{-1} = *$ and $d_j\big(z(s_0 y^{-1})\big) = *$ for $j \geq 2$, so $xy^{-1} \in \operatorname{Im}(NG_{n+1} \to NG_n)$, and therefore ϕ is well defined. For $f, g : S^n \to G$ with $x = f(\iota_n)$, $y = g(\iota_n)$, there exists z such that $dz = (x, a, y, *, \ldots, *)$ for some a. Then $[a] = [x][y]$ and

$$d\left(z\big(s_1(y^{-1})\big)\big(s_0(ya^{-1})\big) \right) = (xya^{-1}, *, *, \ldots, *)$$

so that $xya^{-1} \in \operatorname{Im} d_0$ and therefore $\phi(x)\phi(y) = \phi(a) = \phi(xy)$. It is easy to see that ϕ is a bijection. $\qquad\square$

4 The Singular Complex

Let X be a topological space. Define a simplicial set $\operatorname{Sing}(X)$ by

$$\operatorname{Sing}(X)_n = \{\text{continuous functions from } \Delta^n \text{ to } X\}$$

with $d_j f = f \circ \delta^j$ and $s_j f = f \circ \sigma_j$.

Proposition 8.4.1 *For all spaces X, $\operatorname{Sing}(X)$ is a Kan complex.*

Proof Let $f_0, \ldots, f_{j-1}, f_{j+1}, \ldots, f_n \in (\operatorname{Sing} X)_n$ be compatible. The maps f_i piece together to give a well defined map $f : |\Lambda^k[n]| \to X$. Since the inclusion $|\Lambda^k[n]| \hookrightarrow \Delta^n$ has a retraction, f has an extension $\tilde{f} : \Delta^n \to X$. Then $d_i(\tilde{f}) = f_i$ for $i \neq j$. $\qquad\square$

Theorem 8.4.2

a) $SS(K, \mathrm{Sing}\, X) = \mathcal{T}\!op(|K|, X)$. *That is,* $|\ | \dashv \mathrm{Sing}(\ )$.

b) *The bijections*

$$\phi : SS(K, \mathrm{Sing}\, X) \to \mathcal{T}\!op(|K|, X) \ \ and \ \ \psi : \mathcal{T}\!op(|K|, X) \to SS(K, \mathrm{Sing}\, X)$$

preserve homotopies.

Proof

a) Let $f : K \to \mathrm{Sing}\ X$ be simplicial. Given $(k,t) \in K_n \times \Delta^n$ in $|K|$, let $(\phi f)(k,t) = f(k)(t) \in X$. This is compatible with the identifications in $|K|$ and so induces $\phi f : |K| \to X$. Conversely, let g map $|K|$ to X. Given $k \in K_n$, let $(\psi g)(k)(t) = g(k,t) \in X$ for $t \in \Delta^n$. Then $(\psi g)(k)$ belongs to $(\mathrm{Sing}\, X)_n$, and ϕ and ψ are inverse bijections.

b) Suppose $f \simeq g : K \to \mathrm{Sing}\, X$. Then $|f| \simeq |g| : |K| \to |\,\mathrm{Sing}\, X|$ by Proposition 8.1.4, so composing with the map $|\,\mathrm{Sing}\, X| \to X$ from the adjunction gives $\phi f \simeq \phi g$. Conversely suppose $H : |K| \times I \to X$ is a homotopy between p and q. The adjoint of $|K \times I| = |K| \times I \xrightarrow{H} X$ is a simplicial map $\tilde H : K \times I \to \mathrm{Sing}\, X$ and $\tilde H : \psi p \simeq \psi q$. $\qquad\square$

Letting K equal S^n yields

Corollary 8.4.3 $\pi_n(\mathrm{Sing}\, X) \cong \pi_n(X)$ *for all spaces* X. $\qquad\square$

5 Simplicial Abelian Groups

Let A be a simplicial abelian group. Define $d : A_n \to A_{n-1}$ by $d = \sum_{j=0}^{n}(-1)^j d_j$. It is easy to check that $d^2 = 0$.

Our next goal is to describe the "normalized" sub-chain-complex NA of the chain complex (A, d). We will show there is a subcomplex DA called the "degenerate" subcomplex of A such that $A \cong NA \oplus DA$ (as chain complexes) and $DA \approx 0$. It will follow that the inclusion $NA \hookrightarrow A$ is a chain homotopy equivalence.

Define a filtration on A by $(F^p A)_n = \{x \in A_n \mid d_i x = 0 \text{ for } i > n - p\}$ and set $(NA)_n = (F^n A)_n$. Let $(DA)_n$ be the subgroup of A_n generated by $\{\mathrm{Im}\, s_j\}_{j=0}^{n-1}$.

Theorem 8.5.1

a) $d_j\big((F^p A)_n\big) \subset (F^p A)_{n-1}$ *for all j and so $(F^p A, d)$ is a subcomplex of (A, d) for all p. Therefore NA is a subcomplex of A.*

b) DA *is a subcomplex of A.*

c) $A \cong NA \oplus DA$ *as chain complexes.*

d) $DA \approx 0$.

e) *The inclusion $NA \hookrightarrow A$ is a chain homotopy equivalence.*

Proof For $x \in (F^p A)_n$, if $j > n - p$ then $d_j x = 0$ while if $j \leq n - 1$ then for $i > n - p - 1$ we have $d_i d_j x = d_j d_{i+1} x = 0$ so that $d_j x$ lies in $(F^p A)_{n-1}$. This proves (a). In the short exact sequence $0 \longrightarrow (F^{p+1} A)_n \xrightarrow{\alpha} (F^p A)_n \xrightarrow{d_{n-p}} (F^p A)_{n-1} \longrightarrow 0$, the map $s_{n-p-1} : (F^p A)_{n-1} \to (F^p A)_n$ provides a right splitting of d_{n-p}. Thus $r_p = 1 - s_{n-p-1} d_{n-p}$ is a left splitting of α and $\mathrm{Ker}\, r_p = \mathrm{Im}\, s_{n-p-1}$. Let r equal $r_{n-1} r_{n-2} \cdots r_1 r_0 : A_n \to (NA)_n$. Then r is a left splitting of the inclusion $(NA)_n \hookrightarrow A$.

Lemma 8.5.2 $DA = \operatorname{Ker} r$.

Proof Let $s_j y$ belong to A_n. For $i < n - j - 1$,

$$
\begin{aligned}
r_i(s_j y) &= s_j y - s_{n-i-1} d_{n-i} s_j y \\
&= s_j y - s_{n-i-1} s_j d_{n-i-1} y \\
&= s_j y - s_j s_{n-i-2} d_{n-i-1} y.
\end{aligned}
$$

Thus $r_i(\operatorname{Im} s_j) \subset \operatorname{Im} s_j$ for $i < n - j - 1$. Let $s_j b \in A_n$ be a generator of $(DA)_n$. Then

$$
r(s_j b) = r_{n-1} \cdots r_{n-j-1}(r_{n-j-2} \cdots r_0 s_j b) = 0
$$

since $r_{n-j-2} \cdots r_0 s_j b$ belongs to $\operatorname{Im} s_j = \operatorname{Ker} r_{n-j-1}$. Conversely if x belongs to $\operatorname{Ker} r$ then $r_{n-1}(r_{n-2} \cdots r_0 x) = rx = 0$ and so $r_{n-2} \cdots r_0 x$ belongs to $\operatorname{Ker} r_{n-1} = \operatorname{Im} s_0 \subset DA$. However $r_i a \equiv a \pmod{DA}$ for all i and a and so $r_i a \in DA$ implies $a \in DA$. Therefore x belongs to DA.

Proof of Theorem 8.5.1 (continued) From the lemma we have that $A \cong NA \oplus DA$ as abelian groups. Let $(F_p DA)_n$ be the subgroup of DA generated by $\{\operatorname{Im} s_j\}_{j \leq p}$. Then $DA = \lim\limits_{\overrightarrow{p}} F_p DA$. Let x be a generator of $(F_p DA)_n$. Then $x = s_j y$ for some $j \leq p$. If $i < j$ then $d_i x = d_i s_j y = s_{j-1} d_i y$ which lies in $F_p DA$ and if $i > j+1$ then $d_i x = d_i s_j y = s_j d_{i-1} y$ which lies in $F_p DA$. Since $d_j x = d_{j+1} x = y$, this implies $dx = \sum\limits_{\substack{i=0 \\ i \neq j, j+1}}^{n} (-1)^i d_i x$ lies in $F_p DA$. Therefore $F_p DA$ is a subcomplex of A and so DA is a subcomplex of A. Thus the splitting $A \cong NA \oplus DA$ is a decomposition as chain complexes. To show that $DA \approx 0$, completing the proof, it suffices to show the following lemma.

Lemma 8.5.3 $F_p DA \approx 0$ *for all* p *with contractions* $J_p : 1_{F_p DA} \simeq 0$ *which are compatible in the sense that* $J_p(x)$ *is independent of* p *for* $p > |x|$.

Proof Suppose by induction that $J_{p-1} : 1_{F_{p-1} DA} \approx 0$ has been constructed. Define $s : (F_p DA)_n \to (F_p DA)_{n+1}$ by letting $s = (-1)^{p+1} s_p$ (setting s equal to 0 if $p > n$). Let $f = ds + sd + 1_{(F_p DA)}$. Thus $s : f \simeq 1_{F_p DA}$. Since $J_{p-1} : F_{p-1} DA \cong 0$, to finish the proof it suffices to show that $\operatorname{Im} f$ is contained in $F_{p-1} DA$ because then $J_p = J_{p-1} f - s$ is a homotopy from $1_{F_p DA}$ to 0 which is compatible with J_{p-1}.

$$
\begin{aligned}
f &= ds + sd + 1 \\[2mm]
&= \sum_{j=0}^{n+1} (-1)^{j+p+1} d_j s_p + \sum_{j=0}^{n} (-1)^{j+p+1} s_p d_j + 1 \\[2mm]
&= \sum_{j<p} (-1)^{j+p+1} s_{p-1} d_j + \sum_{j \leq p} (-1)^{j+p+1} s_p d_j + 1 \\[2mm]
&\equiv \sum_{j \leq p} (-1)^{j+p+1} s_p d_j + 1 \pmod{F_{p-1} DA}.
\end{aligned}
$$

Let $s_i y$ be a generator of $F_p DA$. If $i < p$, $s_p d_j s_i y = s_p s_{i-1} d_j y = s_{i-1} s_{p+1} d_j y$ for $j < i$ while $s_p d_j s_i y = s_p s_i d_{j-1} y = s_i s_{p+1} d_{j-1} y$ for $j > i+1$ and so

$$f(s_i y) \equiv \sum_{j \leq p} (-1)^{j+p+1} s_p d_j (s_i y) + s_i y$$

$$\equiv (-1)^{i+p+1} s_p d_i s_i y + (-1)^{i+1+p+1} s_p d_{i+1} s_i y$$

$$= (-1)^{i+p+1} s_p y + (-1)^{i+1+p+1} s_p y = 0 \quad (\mathrm{mod}\ F_{p-1} DA).$$

If $i = p$ then as above $s_p d_j s_p y$ belongs to $F_{p-1} DA$ for $j < i$ and so

$$f(s_p y) \equiv (-1)^{p+p+1} s_p d_p s_p y + s_p y = 0 \quad (\mathrm{mod}\ F_{p-1} DA). \qquad \square$$

K be a simplicial set. Let $\mathbb{Z}K$ be the simplicial abelian group in which $(\mathbb{Z}K)_n$ is the free abelian group on the set K_n with the face and degeneracy maps induced from those in K. Define $H_*(K) = H_*(\mathbb{Z}K, d)$. From Theorem 8.5.1e and Theorem 8.3.2 it follows that $H_*(K) = \pi_*(\mathbb{Z}K)$. Also examining the definitions shows that $(S_*(X), d) = (\mathbb{Z}(\mathrm{Sing}\, X), d)$ so that $H_*(\mathrm{Sing}\, X) = H_*(X)$ for a topological space X.

Theorem 8.5.4 $K \to \mathrm{Sing}\, |K|$ *induces an isomorphism* $H_*(K) \to H_*(\mathrm{Sing}\, |K|)$ $\cong H_*(|K|)$. *In fact* $N\mathbb{Z}K \to C_*(|K|)$ *is an isomorphism where* C_* *is the cellular chain complex of* $|K|$. $\qquad \square$

Corollary 8.5.5 $|\mathrm{Sing}\, X| \to X$ *induces a homology isomorphism for all spaces* X. $\qquad \square$

6 Simplicial Approximation

This section presents the Simplicial Approximation Theorem, from which the equivalence of the homotopy theories in the categories of CW-complexes and simplicial sets follows.

Theorem 8.6.1 (Simplicial Approximation Theorem) *Let* $f : |K| \to |L|$ *where* L *is a Kan complex. Then there exists a simplicial map* $g : K \to L$ *such that* $|g| \simeq f$.

See Curtis [C, Theorem 12.1]. $\qquad \square$

Theorem 8.6.2 *If* K *and* L *are Kan complexes then* $|K| \approx |L|$ *if and only if* $K \approx L$. $\qquad \square$

Corollary 8.6.3 *Let* K *be a Kan complex. Then* $\pi_n(K) \cong \pi_n(|K|)$. $\qquad \square$

Combining Corollary 8.4.3, Corollary 8.6.3, and the Whitehead theorem gives

Corollary 8.6.4 *If* X *is a* CW-*complex then* $|\mathrm{Sing}\, X| \approx X$. $\qquad \square$

From Corollary 8.6.2 and Corollary 8.6.4 we get

Corollary 8.6.5 *If* K *is a Kan complex then* $K \approx \mathrm{Sing}\, |K|$ *and in particular* $\pi_n(K) \approx \pi_n(\mathrm{Sing}\, |K|) \approx \pi_n(|K|)$. $\qquad \square$

CHAPTER 9

Fibre Bundles and Classifying Spaces

1 Fibre Bundles

Definition 9.1.1 Let B be a pointed topological space. A (*locally trivial*) *fibre bundle* over B consists of a map $p : X \to B$ such that for all $b \in B$ there exists an open neighbourhood U of b for which there is a homeomorphism $\phi : p^{-1}(U) \to p^{-1}(*) \times U$ satisfying $\pi'' \circ \phi = p|_U$, where π'' denotes projection onto the second factor. If there is a numerable open cover of B by open sets with homeomorphisms as above then the bundle is said to be *numerable*.

Covering spaces are the special case of fibre bundles in which the fibre has the discrete topology. As with covering spaces, B is called the *base space* of the fibre bundle, X the *total space*, $p^{-1}(*)$ the *fibre*, and p the *projection*. A *bundle map* of fibre bundles consists of maps between the total spaces and base spaces which form a commutative square with the bundle projections.

It follows from Theorem 7.1.6 that a numerable fibre bundle is always a fibration and, in particular, Theorem 2.5.6 implies that every fibre bundle over a paracompact Hausdorff space is a fibration.

Definition 9.1.2 Let G be a topological group and B a topological space. A *principal G-bundle* over B consists of a fibre bundle $p : X \to B$ together with an action $G \times X \to X$ such that:

1) the map $G \times X \to X \times X$ given by $(g, x) \mapsto (x, g \cdot x)$ maps $G \times X$ homeomorphically to its image;
2) $B = X/G$ and $p : X \to X/G$ is the quotient map;
3) for all $b \in B$ there exists an open neighbourhood U of b such that $p : p^{-1}(U) \to U$ is G-bundle isomorphic to the trivial bundle $\pi'' : G \times U \to U$. That is, there exists a homeomorphism $\phi : p^{-1}(U) \to G \times U$ satisfying $p = \pi'' \circ \phi$ and $\phi(g \cdot x) = g \cdot \phi(x)$, where $g \cdot (g', u) = (gg', u)$.

Note that (1) implies that $g \cdot x = g' \cdot x$ if and only if $g = g'$ which is the definition of a *free* action. This together with (2) implies that the fibre of p is G.

A map of principal G-bundles is a map of fibre bundles which is also a G-map.

Let ξ be a principal G-bundle $p : X \to B$. Given a map $f : B' \to B$, the pullback yields a principal G-bundle over B', written $f^*(\xi)$ or $f^!(\xi)$, and the pullback square becomes a bundle map from $f^*(\xi)$ to ξ.

Let $p : X \to B$ be a principal G-bundle. Then we have a covering of B by open sets $\{U_j\}$ which have G-homeomorphisms $\phi_{U_j} : p^{-1}(U_j) \to G \times U_j$. Choose such a covering and collection of G-homeomorphisms. For every pair of sets U, V in our covering, the homeomorphisms ϕ_U and ϕ_V restrict to homeomorphisms $p^{-1}(U \cap V) \to G \times (U \cap V)$. The composite $G \times (U \cap V) \xrightarrow{\phi_U^{-1}} p^{-1}(U \cap V) \xrightarrow{\phi_V} G \times (U \cap V)$ commutes with the action of G and is thus determined by the images of the elements $(1, b)$. Let $\phi_V \circ \phi_U^{-1}(1, b) = (\tau_{V,U}(b), b)$. The function $\tau_{V,U} : U \cap V \to$

G is continuous by condition (1). The functions τ_{U_j, U_i} are called the *transition functions* of the bundles (with respect to the covering $\{U_j\}$). These functions are compatible in the sense that $\tau_{U,U}(b) = $ identity of G, $\tau_{U_j, U_i}(b) = \left(\tau_{U_i, U_j}(b)\right)^{-1}$, and if b lies in $U_i \cap U_j \cap U_k$ then $\tau_{U_k, U_i}(b) = \tau_{U_k, U_j}(b) \cdot \tau_{U_j, U_i}(b)$. Conversely, given a covering $\{U_j\}_{j \in J}$ of a space B and a compatible collection of continuous functions $\tau_{U_j, U_i} : U_i \cap U_j \to G$, we can construct a principal G-bundle having these as transition functions by setting $X = \coprod_{j \in J}(G \times U_j)/\sim$ where $(g, b) \sim \left(g \cdot \tau_{U_j, U_i}(b), b\right)$ for all $i, j \in J$ and $b \in U_i \cap U_j$. Thus a principal bundle is equivalent to a compatible collection of transition functions.

Let A and B be G-spaces. Then we define a G-action on $A \times B$ by $g \cdot (a, b) = (g \cdot a, g \cdot b)$ for $a \in A$, $b \in B$, and $g \in G$. The quotient space $(A \times B)/G$ is denoted $A \times_G B$.

The action of a group G on a set S is called *effective* if for all $g \in G$ different from the identity of G there exists $s \in S$ such that $g \cdot s \neq s$. Equivalently, G acts effectively if the homomorphism $G \to \mathrm{Aut}(S)$ coming from the action is injective.

Let G be a topological group. Let $p : X \to B$ be a principal G-bundle and let F be a G-space on which the action of G is effective. The *fibre bundle with structure group G* formed from p and F is $q : F \times_G X \to B$ where $q(f, x) = p(x)$. In the special case where $F = \mathbb{R}^n$ and G is the orthogonal group $O(n)$ acting in the standard way, such a fibre bundle is called an *n-dimensional (real) vector bundle*. Similarly if $F = \mathbb{C}^n$ and G is the unitary group $U(n)$ with the standard action we have an *n-dimensional complex vector bundle*.

2 Classifying Principal Bundles and Milnor's Construction

Theorem 9.2.1 *Let G be a topological group and let ξ be a numerable principal G-bundle over a space B. Suppose $f \simeq h : B' \to B$. Then $f^*(\xi) \cong h^*(\xi)$.*

Outline of Proof

1) It suffices to consider the special case where $B = B' \times I$ and f and h are the inclusions i_0 and i_1.

2) A partition of unity argument shows that we may assume that our numerable cover of $B' \times I$ has the form $\{U_j \times I\}_{j \in J}$, where $\{U_j\}_{j \in J}$ forms a numerable cover of B'.

3) Use the partition of unity to construct functions $v_j : B' \to I$ for $j \in J$ having the property that v_j is 0 outside of U_j and $\max_{j \in J} v_j(b) = 1$ for all $b \in B'$.

4) Let ξ be the bundle $p : X \to B' \times I$ and let $\phi_j : U_j \times I \times G \to p^{-1}(U_j \times I)$ be a local trivialization of ξ over $U_j \times I$. For each $j \in J$ define a function
$$h_j : p^{-1}(U_j \times I) \to p^{-1}(U_j \times I) \text{ by } h_j\big(\phi_j(b, t, g)\big) = \phi_j\Big(b, \max\big(v_j(b), t\big), g\Big).$$

5) Pick a total ordering on the index set J and for each $x \in X$ define $h(x) \in X$ by composing, in order, the functions h_j for the (finitely many) j such that the first component of $p(x)$ lies in U_j. Check that $h : X \to X$ is continuous using the local finiteness property of the cover U_j. Then $\mathrm{Im}\, h \subset p^{-1}(X \times 1)$ and the restriction $h|_{p^{-1}(X \times 0)}$ is a G-bundle isomorphism $i_0^*(\xi) \to i_1^*(\xi)$.

$\square$

A numerable principal G-bundle γ over a pointed space $\tilde{B}$ is called a *universal G-bundle* if:

1) for any numerable principal G-bundle ξ there exists a map $f : B \to \tilde{B}$ from the base space B of ξ to the base space $\tilde{B}$ of γ such that $\xi = f^*(\gamma)$;

2) whenever f, h are two pointed maps from some space B into the base space $\tilde{B}$ of γ such that $f^*(\gamma) \cong h^*(\gamma)$ then $f \simeq h$.

In other words, a numerable principal G-bundle γ with base space $\tilde{B}$ is a universal G-bundle if, for any pointed space B, pullback induces a bijection from $[B, \tilde{B}]$ to isomorphism classes of numerable principal bundles over B. The standard categorical argument shows that up to homotopy equivalence there can be at most one space $\tilde{B}$ with this property.

Let G be a topological group. Let EG be the infinite join $EG = \varinjlim_n G^{*n}$.

Explicitly, as a set (not as a topological space),

$$EG = \{(g_0, t_0, g_1, t_1, \ldots, g_n, t_n, \ldots) \in (G \times I)^\infty\}/\sim$$

such that at most finitely many t_i are nonzero, $\sum t_i = 1$, and

$$(g_0, t_0, g_1, t_1, \ldots, g_n, 0, \ldots) \sim (g_0, t_0, g_1, t_1, \ldots, g'_n, 0, \ldots).$$

A G-action on EG is given by

$$g \cdot (g_0, t_0, g_1, t_1, \ldots, g_n, t_n, \ldots) = (gg_0, t_0, gg_1, t_1, \ldots, gg_n, t_n, \ldots).$$

Let $BG = EG/G$. We also write $E_nG = G^{*(n+1)}$ and $B_n(G) = E_nG/G$, referring to the inclusions $B_0G \hookrightarrow B_1G \hookrightarrow B_2G \hookrightarrow \ldots \hookrightarrow B_nG \hookrightarrow \ldots$ as the Milnor filtration on BG.

Theorem 9.2.2 (Milnor) *For every topological group G, the quotient map $EG \to BG$ is a numerable principal G-bundle and this bundle is a universal G-bundle.*

Outline of Proof Let ζ be the bundle $p : EG \to BG$. To show that ζ is a numerable principal G-bundle, define

$$V_i = \{(g_0, t_0, g_1, t_1, \ldots, g_n, t_n, \ldots) \in EG \mid t_i > 0\}$$

and $U_i = V_i/G \subset BG$. Then $\{U_i\}_{i=0}^\infty$ forms an open cover of BG and one checks that it is a numerable cover. The maps $\phi : V_i = p^{-1}(U_i) \to G \times U_i$ defined by

$$\phi(g_0, t_0, g_1, t_1, \ldots, g_n, t_n, \ldots) = (g_i, p(g_0, t_0, g_1, t_1, \ldots, g_n, t_n, \ldots))$$

give a local trivialization relative to the cover $\{U_i\}_{i=0}^\infty$ and the conditions for a principal bundle are easily checked.

Let $q : X \to B$ be any numerable principal G-bundle. To construct a map $f : B \to BG$ inducing q from $EG \to BG$ the first step is to choose an arbitrary numerable cover of B over which q is locally trivial and use its partition of unity to replace it by a countable numerable cover $\{W_i\}_{i=0}^\infty$ of B over which q is trivial. Suppose now that $\phi_i : q^{-1}(W_i) \to G \times W_i$ is a local trivialization over W_i and let $\{h_i\}_{i=0}^\infty$ be a partition of unity relative to $\{W_i\}_{i=0}^\infty$. Define $\tilde{f} : X \to EG$ by

$$\tilde{f}(z) = (\pi'(\phi_0 z), h_0(qz), \pi'(\phi_1 z), h_1(qz), \ldots, \pi'(\phi_i z), h_i(qz), \ldots)$$

where π' denotes projection onto the first factor. This makes sense since for any i for which $\phi_i z$ is not defined, $h_i z = 0$ and so the first element of the ith component is irrelevant. Since $\tilde{f}$ commutes with the G-action it induces a well defined map $f : B \to BG$ whose pullback with $EG \to BG$ produces the bundle q.

Conversely suppose that $f, f' : B \to BG$ satisfy $f^*(\zeta) \cong f'^*(\zeta)$. Let e denote the identity of G. Let $\alpha : EG \to EG$ be the map

$$\alpha(g_0, t_0, g_1, t_1, \ldots, g_{2n}, t_{2n}, \ldots) = (g_0, t_0, e, 0, g_1, t_1, e, 0, \ldots, e, 0, g_{2n}, t_{2n}, e, 0, \ldots)$$

whose image is contained in the even components and let $\alpha' : EG \to EG$ be the map

$$\alpha'(g_0, t_0, g_1, t_1 \ldots, g_{2n+1}, t_{2n+1}, \ldots)$$
$$= (e, 0, g_0, t_0, e, 0, g_1, t_1, \ldots, e, 0, g_{2n+1}, t_{2n+1}, e, 0, \ldots)$$

whose image is contained in the odd components. There exist homotopies $\alpha \simeq 1_{EG}$ and $\alpha' \simeq 1_{EG}$ which commute with the action of G and thus induce homotopies $\bar{\alpha} \simeq 1_{BG}$ and $\bar{\alpha}' \simeq 1_{BG}$, where $\bar{\alpha}$ and $\bar{\alpha}'$ are the maps on BG induced by α and α'. Let $\theta : E\big(f^*(\zeta)\big) \to E\big(f'^*(\zeta)\big)$ be the homeomorphism on total spaces coming from the G-bundle isomorphism and let $\sigma : E\big(f^*(\zeta)\big) \to EG$ and $\sigma' : E\big(f'^*(\zeta)\big) \to EG$ be the maps induced on total spaces by f and f'. Define $H : E\big(f^*(\zeta)\big) \times I \to EG$ by

$$H(z, s) = \big(g_0, st_0, g_0', (1 - s)t_0', g_1, st_1, g_1', (1 - s)t_1', g_2, st_2, g_2', (1 - s)t_2', \ldots\big)$$

where $\sigma(z) = (g_0, t_0, g_1, t_1, g_2, t_2, \ldots)$ and $\sigma'\theta(z) = (g_0', t_0', g_1', t_1', g_2', t_2', \ldots)$. Then H commutes with the G-action and thus induces a homotopy $\bar{\alpha}f \simeq \bar{\alpha}'f'$. Therefore $f \simeq \bar{\alpha}f \simeq \bar{\alpha}'f' \simeq f'$. $\qquad\square$

Example 9.2.3 Let $G = \mathbb{Z}/2\mathbb{Z}$ with the discrete topology. Then $G^{*(n+1)} = S^n$ with the nontrivial element acting via the antipodal map. Therefore by definition, $B_n(G) = \mathbb{R}P^n$, $EG = S^\infty$, and $BG = \mathbb{R}P^\infty$. Similarly if $G = S^1$ with the standard topology then $B_n(G) = \mathbb{C}P^n$, $EG = S^\infty$, and $BG = \mathbb{C}P^\infty$ and if $G = S^3$ with the standard topology then $B_n(G) = \mathbb{H}P^n$, $EG = S^\infty$, and $BG = \mathbb{H}P^\infty$. Although $EG = S^\infty$ in all three cases, we are thinking of S^∞ as the unit ball in $\mathbb{R}^\infty$, $\mathbb{C}^\infty$, and $\mathbb{H}^\infty$ respectively, where these spaces are given the direct limit topology. $\qquad\square$

Since homotopy classes of maps into BG "classify" numerable principal G-bundles, BG is called the *classifying space* of the group G. It follows from Proposition 7.7.1 that the space EG is weak homotopy equivalent to a point, and this gives a weak homotopy equivalence $\Omega BG \to G$ which is an H-map and which is a homotopy equivalence if G is a CW-complex.

Conversely,

Theorem 9.2.4 (Milnor) *For any space X there exists a topological group G such that G is weak homotopy equivalent to ΩX.*

See Milnor[Mi1]. $\qquad\square$

The map $E_n(G) \to B_n(G)$ is also a numerable principal G-bundle.

Proposition 9.2.5 $B_n(G)$ *has Lusternik-Schnirrelmann category at most n for any topological group G.*

Proof Choose ϵ such that $0 < \epsilon < 1/(n+1)$. For $i = 0, \ldots, n$, let

$$V_i = \{(g_0, t_0, g_1, t_1, \ldots, g_n, t_n) \subset E_n(G) \mid t_i \geq \epsilon\}$$

and let $U_i = V_i/G \subset B_n(G)$. Since $\sum t_i = 1$, the sets U_i cover $B_n(G)$. The map $\gamma_i : V_i \times I \to V_i$ given by

$$\gamma_i\big((g_0, t_0, g_1, t_1, \ldots, g_n, t_n), s\big) = \Big(g_0, st_0, g_1, st_1, \ldots, g_i, 1 - s\big(\sum_{k \neq i} t_k\big), \ldots, g_n, st_n\Big)$$

commutes with the action of G and so induces a well defined map $U_i \times I \to U_i$ which is a contraction of U_i. $\square$

Let $F \longrightarrow X \xrightarrow{p} B$ be a homotopy fibration. By repeated application of Ganea's theorem we get a sequence of fibrations

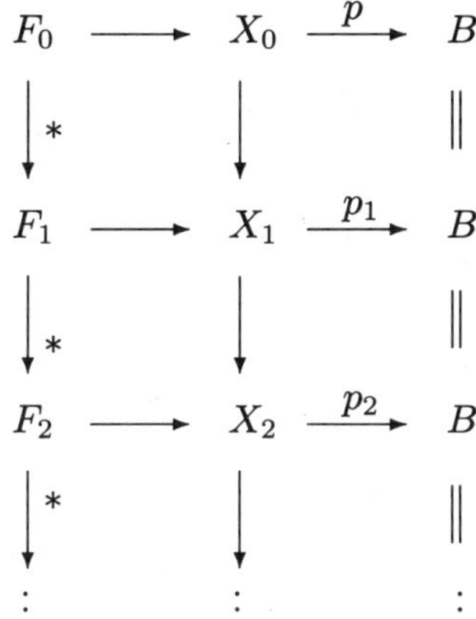

in which $F_0 \to X_0 \to B$ is the original fibration, $F_n \approx (\Omega B)^{*n} * F$ and $X_n \approx X_{n-1}/F_{n-1}$. I will refer to this sequence of fibrations as the iterated Ganea construction. In the case of the path fibration $\Omega B \to PB \to B$, the induced fibration $\Omega B \to F_n \to X_n$ is equivalent up to homotopy to Milnor's bundle fibration sequence $G \to E_n G \to B_n G$.

There are also homotopy constructions due to Milgram, Dold-Lashof, Stascheff, and May which apply in more general situations and produce BG when applied to a topological group G.

CHAPTER 10

Hopf Algebras and Graded Lie Algebras

1 Algebras and Coalgebras

Let R be a commutative ring. R will be regarded as a graded R-module concentrated in degree 0. A nonnegatively graded R-module M is called *connected* if $M_0 \cong R$. Throughout this chapter, $V \otimes W$ will stand for $V \otimes_R W$ unless otherwise indicated. We use $T : V \otimes W \to W \otimes V$ to denote the map given by $v \otimes w \mapsto (-1)^{|v||w|} w \otimes v$.

Definition 10.1.1 A *(graded) R-algebra* consists of a (graded) R-module A together with R-module maps $\mu : A \otimes A \to A$ and $\eta : R \to A$ such that:

1) the composite $A \cong A \otimes R \xrightarrow{A \otimes \eta} A \otimes A \xrightarrow{\mu} A$ equals 1_A;

2) the composite $A \cong R \otimes A \xrightarrow{\eta \otimes A} A \otimes A \xrightarrow{\mu} A$ equals 1_A.

An R-algebra A is called *associative* if $\mu \circ (\mu \otimes 1_A) = \mu \circ (1_A \otimes \mu) : A \otimes A \otimes A \to A$. It is called *commutative* if $\mu \circ T = \mu : A \otimes A \to A$. (As noted earlier, this was once called "graded commutative".)

Observe that R can be made into an R-algebra in a unique way such that $\eta = 1_R$. An *augmentation* of an R-algebra A is an R-algebra morphism $\epsilon : A \to R$. If A is an augmented R-algebra, the kernel $I(A)$ of the augmentation is called the *augmentation ideal*.

If A and B are R-algebras then $A \otimes B$ becomes an R-algebra under the multiplication $A \otimes B \otimes A \otimes B \xrightarrow{A \otimes T \otimes B} A \otimes A \otimes B \otimes B \xrightarrow{\mu_A \otimes \mu_B} A \otimes B$.

Definition 10.1.2 A *(graded) R-coalgebra* consists of a (graded) R-module C together with R-module maps $\psi : C \to C \otimes C$ and $\epsilon : C \to R$ such that:

1) the composite $C \xrightarrow{\psi} C \otimes C \xrightarrow{C \otimes \epsilon} C \otimes R \cong C$ equals 1_C;

2) the composite $C \xrightarrow{\psi} C \otimes C \xrightarrow{\epsilon \otimes C} R \otimes C \cong C$ equals 1_C.

An R-coalgebra C is called *coassociative* if $(\psi \otimes 1_C) \circ \psi = (1_C \otimes \psi) \circ \psi : C \to C \otimes C \otimes C$. It is called *cocommutative* if $T \circ \psi = \psi : C \to C \otimes C$.

R can be made into an R-coalgebra in a unique way such that $\epsilon = 1_R$. An *augmentation* of an R-coalgebra C is an R-coalgebra morphism $\eta : R \to C$. If C is an augmented R-coalgebra, the cokernel of the augmentation is denoted $J(C)$.

The tensor product of R-coalgebras can be given a coalgebra structure in a manner dual to that used above to define the tensor product of R-algebras.

Definition 10.1.3 A *(graded) Hopf algebra* over R consists of a (graded) R-module H together with R-module maps $\mu, \eta, \psi, \epsilon$ such that:

1) (H, μ, η) is an R-algebra;

2) (H, ψ, ϵ) is an R-coalgebra;

3) ψ is a homomorphism of R-algebras.

105

Explicitly, condition (3) requires that

$$
\begin{array}{ccc}
H \otimes H & \xrightarrow{\qquad\qquad \mu \qquad\qquad} & H \\
\downarrow{\scriptstyle \psi \otimes \psi} & & \downarrow{\scriptstyle \psi} \\
H \otimes H \otimes H \otimes H \xrightarrow{H \otimes T \otimes H} H \otimes H \otimes H \otimes H \xrightarrow{\mu \otimes \mu} & & H \otimes H
\end{array}
$$

commutes. Examination of this diagram shows that this condition is equivalent to

3') μ is a homomorphism of R-coalgebras.

A Hopf algebra is called associative, commutative, coassociative, or cocommutative according to whether the underlying algebra or coalgebra structures have these properties.

Remark The basic source for properties of Hopf algebras is the paper by Milnor and Moore [MM]. Their terminology differs slightly in that their definition of a Hopf algebra is an associative coassociative Hopf algebra and they use the term "quasi-Hopf-algebra" for the concept defined above.

Definition 10.1.4 A graded R-module M is said to have *finite type* if M_n is finitely generated for all n.

Given a graded R-module M, denote the dual graded module $\operatorname{Hom}_R(M, R)$ by M^*.

Proposition 10.1.5

a) *If M is a projective R-module of finite type then the canonical map $M \to M^{**}$ is an isomorphism.*

b) *If M and N are projective R-modules with at least one having finite type, then the canonical map $M^* \otimes N^* \to (M \otimes N)^*$ is an isomorphism.*   □

Proposition 10.1.6 *Let H be a Hopf algebra over R such that H is projective and has finite type. Then H^* is also a Hopf algebra.*

a) *H^* is associative $\iff$ H is coassociative.*
b) *H^* is commutative $\iff$ H is cocommutative.*
c) *H^* is coassociative $\iff$ H is associative.*
d) *H^* is cocommutative $\iff$ H is commutative.*   □

Example 10.1.7 If X is an H-space then $H_*(X; R)$ is a Hopf algebra, and if $H_*(X; R)$ is projective of finite type then it has a dual Hopf algebra $H^*(X; R)$.

□

On a polynomial algebra on one generator $R[x]$ with $|x| > 0$, there is a unique comultiplication $\psi : R[x] \to R[x] \otimes R[x]$ which turns $R[x]$ into a coalgebra, namely that determined by $\psi(x) = x \otimes 1 + 1 \otimes x$. The resulting algebra $R[x]^*$ is called a *divided polynomial algebra*, denoted $\Gamma[x]$. It has a basis $\gamma_j(x)$ defined by

$$
\langle \gamma_j(x), x^k \rangle = \begin{cases} 1 & \text{if } j = k; \\ 0 & \text{if } j \neq k, \end{cases}
$$

and multiplication given by $\gamma_i(x)\gamma_j(x) = \binom{i+j}{i}\gamma_{i+j}(x)$. The comultiplication on $R[x]^*$ is given by $\psi\big(\gamma_n(x)\big) = \sum_{i+j=n} \gamma_i(x) \otimes \gamma_j(x)$. From Theorem 7.3.1 we get,

Corollary 10.1.8 $H^*(\Omega S^n) \cong \Gamma[x]$ *where* $|x| = n - 1$. $\qquad\square$

Let A be an R-algebra and let M be an R-module. A map $\theta_M : A \otimes M \to M$ is said to give M the structure of a (*left*) A-*module* if:

1) $\theta_M \circ (\mu \otimes M) = \theta_M \circ (A \otimes \theta_M) : A \otimes A \otimes M \to M$;

2) the composite $M \cong R \otimes M \xrightarrow{\eta \otimes M} A \otimes M \xrightarrow{\theta_M} M$ equals 1_M.

Let C be an R-coalgebra and let M be an R-module. A map $\psi_M : M \to C \otimes M$ is said to give M the structure of a (*left*) C-*comodule* if:

1) $(\psi \otimes M) \circ \psi_M = (C \otimes \psi_M) \circ \psi_M : M \to C \otimes C \otimes M$;

2) the composite $M \xrightarrow{\psi_M} C \otimes M \xrightarrow{\epsilon \otimes M} R \otimes M \cong M$ equals 1_M.

Let H be a Hopf algebra over R and let M and N be (left) H-modules. Then $M \otimes N$ becomes a (left) H-module under the map

$$H \otimes M \otimes N \xrightarrow{\psi \otimes M \otimes N} H \otimes H \otimes M \otimes N \xrightarrow{H \otimes T \otimes N} H \otimes M \otimes H \otimes N \xrightarrow{\theta_M \otimes \theta_N} M \otimes N,$$

where θ_M and θ_N are the module structure maps on M and N. Similarly if M and N are (left) H-comodules then $M \otimes N$ becomes a (left) H-comodule under the map

$$M \otimes N \xrightarrow{\psi_M \otimes \psi_N} H \otimes M \otimes H \otimes N \xrightarrow{H \otimes T \otimes N} H \otimes H \otimes M \otimes N \xrightarrow{\mu \otimes M \otimes N} H \otimes M \otimes N.$$

A (*graded*) *algebra over a Hopf algebra* H consists of a (graded) R-algebra A together with a map $H \otimes A \to A$ turning A into an H-module in such a way that μ_A becomes an H-module homomorphism.

2 Tensor Products and Cotensor Products

Let A be an R-algebra, let M be a right A-module and let N be a left A-module. The tensor product $M \otimes_A N$ can be expressed as the cokernel of the map $(\theta_M \otimes N - M \otimes \theta_N) : M \otimes A \otimes N \to M \otimes N$, where θ_M and θ_N are the module structure maps on M and N.

With the above as motivation, we dualize to form the definition of the cotensor product. Let C be an R-coalgebra, let M be a right C-comodule and let N be a left C-comodule. The *cotensor product* of M and N over C, written $M \square_C N$, is defined as the kernel of the map $(\psi_M \otimes N - M \otimes \psi_N) : M \otimes N \to M \otimes C \otimes N$, where ψ_M and ψ_N are the comodule structure maps on M and N.

Proposition 10.2.1

a) $C \square_C N \cong N$.

b) *If* $0 \to N' \to N \to N'' \to 0$ *is a short exact sequence of left C-comodules which is split when regarded as a sequence of R-modules, then*

$$0 \to M \square_C N' \to M \square_C N \to M \square_C N''$$

is an exact sequence of R-modules for any right C-comodule M. $\qquad\square$

3 Kernels and Cokernels

Observe that if $f : W \to X$ is a morphism of augmented R-algebras then $R \otimes_W X = X/(f(IW)X)$. Dually, if $g : Y \to Z$ is a morphism of augmented R-coalgebras then $Y \square_Z R$ is the kernel of the composite map $Y \xrightarrow{\psi} Y \otimes Y \longrightarrow Y \otimes JY \xrightarrow{Y \otimes g} Y \otimes JZ$.

A sub-R-algebra A of an augmented R-algebra B is called *left normal* if the inclusion $(IA)B \hookrightarrow B$ splits as a map of R-modules (equivalently $B \twoheadrightarrow R \otimes_A B$

splits) and $B(IA) \subset (IA)B$. It is called *right normal* if $B(IA) \hookrightarrow B$ splits and $(IA)B \subset B(IA)$. Dually, a surjection $B \twoheadrightarrow C$ of augmented R-coalgebras is called *right (co)normal* if $R\square_C B \hookrightarrow B$ splits and $B\square_C R \subset R\square_C B$. Left (co)normal can be defined similarly.

Theorem 10.3.1

a) *Let $A \hookrightarrow B$ be a left normal inclusion of augmented R-algebras. Then the multiplication on B induces a well defined multiplication on $R \otimes_A B$ which turns it into an R-algebra and $R \otimes_A B$ forms the cokernel of $A \hookrightarrow B$ in the category of augmented R-algebras.*

b) *Let $B \twoheadrightarrow C$ be a right normal surjection of augmented R-coalgebras. Then the comultiplication on B restricts to a comultiplication on $B\square_C R$ which turns it into an R-coalgebra and $B\square_C R$ forms the kernel of $B \twoheadrightarrow C$ in the category of augmented R-coalgebras.*

c) *Let $A \hookrightarrow B$ be an inclusion of Hopf algebras over R. Then the comultiplication on B induces a well defined comultiplication on $R \otimes_A B$ which turns it into an R-coalgebra. If $A \hookrightarrow B$ is left normal then the multiplication on B induces a well defined multiplication on $R \otimes_A B$ which turns it into a Hopf algebra over R and $R \otimes_A B$ forms the cokernel of $A \hookrightarrow B$ in the category of Hopf algebras over R.*

d) *Let $B \twoheadrightarrow C$ be a surjection of Hopf algebras over R. Then the multiplication on B restricts to a multiplication on $B\square_C R$ which turns it into an R-algebra. If $B \twoheadrightarrow C$ is right normal then the comultiplication on B restricts to a comultiplication on $B\square_C R$ which turns it into a Hopf algebra over R and $B\square_C R$ forms the kernel of $B \twoheadrightarrow C$ in the category of Hopf algebras over R.* $\square$

Of course left and right can be interchanged in this theorem by interchanging the factors in the tensor products or cotensor products.

Theorem 10.3.2 *Let $A \xrightarrow{i} B \xrightarrow{\pi} C$ be maps of Hopf algebras over R in which i is injective, π is surjective and $\pi \circ i$ is the zero map in the category of Hopf algebras over R. (That is, the map whose restriction to the augmentation ideal is trivial.) Then the following are equivalent:*

1) *i is left normal and $\pi = \operatorname{CoKer} i$;*
2) *π is right normal and $i = \operatorname{Ker} \pi$;*
3) *the induced map $R \otimes_A B \to C$ is an isomorphism;*
4) *the induced map $A \to B\square_C R$ is an isomorphism.* $\square$

If these equivalent conditions hold then $A \xrightarrow{i} B \xrightarrow{\pi} C$ is called a *short exact sequence of Hopf algebras* over R.

Theorem 10.3.3 *Let $A \xrightarrow{i} B \xrightarrow{\pi} C$ be maps of nonnegatively graded connected Hopf algebras over R in which i is a injective, π is surjective and $\pi \circ i$ is the zero map in the category of Hopf algebras over R. Then $A \to B \to C$ is a short exact sequence of Hopf algebras if and only if $B \cong A \otimes C$ as left A-modules and as right C-comodules.* $\square$

4 Primitives and Indecomposables

Let A be an augmented R-algebra. The *indecomposable quotient* of A, denoted $Q(A)$, is defined by $Q(A) = R \otimes_A I(A) = I(A)/(I(A))^2$. Elements of $Q(A)$ are sometimes called "indecomposable elements" although this terminology is somewhat misleading in that $Q(A)$ is a quotient of A, not a subspace. Dually, for an R-coalgebra C, we define the primitives of C, denoted $P(C)$ by

$$P(C) = J(C) \square_C R = \mathrm{Ker}\left(J(C) \xrightarrow{\psi} J(C) \otimes J(C)\right) = \{x \in C \mid \psi(x) = x \otimes 1 + 1 \otimes x\}.$$

Proposition 10.4.1 *Let A be an augmented, projective, finite type R-algebra such that $I(A) \twoheadrightarrow Q(A)$ splits. Then $P(A^*) = Q(A)^*$ and $Q(A^*) = P(A)^*$.* $\square$

Proposition 10.4.2

a) *Let $f : A \to B$ be a morphism of nonnegatively graded R-algebras such that A is augmented and B connected. Then f is an epimorphism if and only if $Q(f)$ is an epimorphism.*

b) *Let $f : A \to B$ be a morphism of nonnegatively graded R-coalgebras such that A and B are flat R-modules, A is connected, and B is augmented. Then f is an monomorphism if and only if $P(f)$ is an monomorphism.* $\square$

Proposition 10.4.3

a) *Let $A \hookrightarrow B$ be a left normal morphism of augmented R-algebras. Then*

$$Q(A) \to Q(B) \to Q(R \otimes_A B) \to 0$$

is an exact sequence of R-modules.

b) *Let $B \twoheadrightarrow C$ be a right normal morphism of augmented R-coalgebras. Then*

$$0 \to P(B \square_C R) \to P(B) \to P(C)$$

is an exact sequence of R-modules. $\square$

5 Graded Lie Algebras

Definition 10.5.1 A *graded Lie algebra* over R is a graded R-module L together with an R-bilinear map $[\ ,\ ] : L_n \times L_k \to L_{n+k}$ such that:

1) $[x, y] = (-1)^{|x||y|+1}[y, x]$;
2) $[x, [y, z]] = [[x, y], z] + (-1)^{|x||y|}[y, [x, z]]$ (Jacobi Identity);
3) $[x, [x, x]] = 0$.

Remark The last condition is redundant if 6 is invertible in R.

If H is a Hopf algebra over R then $P(H)$ forms a Lie algebra under the operation $[x, y] = xy - (-1)^{|x||y|}yx$. Conversely, let L be a Lie algebra over R. Define a Hopf algebra called the *universal enveloping algebra* of L, denoted $U(L)$, by $U(L) = T(L)/I$ where $T(L)$ is the tensor algebra on L and I is the two-sided ideal of $T(L)$ generated by $\{x \otimes y - (-1)^{|x||y|}y \otimes x - [x, y] \mid x, y \in L\}$ and the comultiplication is determined by making the images of the elements of L primitive.

Theorem 10.5.2 *Let L be a Lie algebra over R where R is either a field or a subring of $\mathbb{Q}$.*

1) *If $\mathrm{char}\, R = 0$ then $P\left(U(L)\right) = L$.*
2) *If $\mathrm{char}\, R = p$ then $P\left(U(L)\right)$ has the basis $\{x^{p^k} \mid x \in B, k \geq 0\}$ where B is a basis for L.* $\square$

Theorem 10.5.3 *Let $0 \to L' \to L \to L'' \to 0$ be a short exact sequence of augmented Lie algebras over R such that the R-module inclusion $(I(UL'))(UL) \hookrightarrow UL$ splits. Then $UL' \to UL \to UL''$ is a short exact sequence of Hopf algebras over R.* $\square$

Any Hopf algebra H has an associated multiplicative filtration called the *primitive filtration*. (See Section 11.1 for definitions concerning filtrations and Section 11.6 for the definition of a multiplicative filtration.) This filtration is defined by setting $F_p H = 0$ for $p < 0$, $F_0 H = R$, and $F_p H = F_{p-1}H + (P(H))(F_{p-1}H)$ for $p > 0$. Equivalently, $F_p H$ is the submodule of H generated by products of at most p primitives. This filtration is cocomplete if H is generated as an algebra by primitives.

The Hopf algebra $U(L)$ has a similar multiplicative filtration called the *Lie filtration* defined by $F_p U(L) = 0$ for $p < 0$, $F_0 U(L) = R$, and

$$F_p U(L) = F_{p-1}U(L) + L(F_{p-1}U(L))$$

for $p > 0$. Theorem 10.5.2 implies that the Lie filtration and primitive filtration of $U(L)$ coincide when R is a subring of $\mathbb{Q}$ or a field of characteristic 0.

Notice that for $a \in F_p U(L)$ and $b \in F_q U(L)$ the definition of $U(L)$ implies that the commutator $ab - (-1)^{|a||b|}ba$ lies in $F_{p+q-1}U(L)$. Therefore the induced multiplication on the associated graded object, $\mathrm{Gr}(U(L))$, is commutative. Therefore the inclusion $L \cong \mathrm{Gr}_1(U(L)) \hookrightarrow \mathrm{Gr}(U(L))$ induces an algebra homomorphism $\phi : S(L) \to \mathrm{Gr}(U(L))$ from the free commutative algebra $S(L)$ on L. Both sides are Hopf algebras under the comultiplication determined by making the elements of L primitive and ϕ is a morphism of Hopf algebras.

Theorem 10.5.4 (Poincaré-Birkhoff-Witt) *If the Lie algebra L is a free R-module then $\phi : S(L) \to \mathrm{Gr}\, U(L)$ is a Hopf algebra isomorphism.*

The proof is easier in the case where K is a field of characteristic 0. The general case is given in Serre [Se4, Theorem 4.3]. $\square$

Suppose that the Lie algebra L is a free R-module. Choose a basis B for L and a total ordering on B. By repeatedly applying the relation

$$y \otimes x = (-1)^{|x||y|}(x \otimes y - [x, y])$$

whenever $y > x$ in the chosen ordering, every monomial in $U(L)$ can be written uniquely as a sum of monomials with nondecreasing factors. This gives a map $\theta : U(L) \to S(L)$. Notice that θ preserves the coproduct.

Theorem 10.5.5 *If the Lie algebra L is a free R-module then $\theta : U(L) \to S(L)$ is a coalgebra isomorphism.* $\square$

Since the map θ is the key ingredient of the proof of Theorem 10.5.4 (see [Se4, p. 3.7]), Theorem 10.5.5 is also sometimes called the Poincaré-Birkhoff-Witt Theorem.

6 The Canonical Conjugation

Let C be a connected coassociative coalgebra and A a connected associative algebra over R, and let $G = \{f \in \mathrm{Hom}_R(C, A) \mid f_0 = 1_R\}$. Define a binary operation $*$ on G by setting $f * g$ to be the composite $C \xrightarrow{\psi_C} C \otimes C \xrightarrow{f \otimes g} A \otimes A \xrightarrow{\phi_A} A$. It is easy to see that this is an associative operation and that the map which is zero in positive degrees acts as the identity under this operation. Furthermore, every element has an inverse since given $f \in G$, it is possible to solve inductively for the components of the inverse. Explicitly, suppose f^{-1} has already been determined on elements with degrees less than n, and let x belong to C_n. Then if $\psi(x) = x \otimes 1 + \sum_i x'_i \otimes x''_i + 1 \otimes x$, the definition requires that

$$0 = \left(f^{-1} * f\right)(x) = f^{-1}(x) + \sum_{i=1}^{|x|-1} f^{-1}(x'_i)f(x''_i) + f(x),$$

and so $f^{-1}(x) = -f(x) - \sum_{i=1}^{|x|-1} f^{-1}(x'_i)f(x''_i)$. Therefore G forms a group under the operation $*$.

If H is a connected Hopf algebra, the inverse in this group of the identity map of H (which is not the identity element of the group!) is called the *canonical conjugation* or *canonical anti-automorphism* on H and is often denoted χ.

The motivating example of the canonical conjugation occurs when H is the homology of an H-group X of finite type with coefficients in a field. In this case the conjugation is induced by the map $X \to X$ given by $x \mapsto x^{-1}$. The conjugation has the following properties, generalizing obvious properties of the motivating example.

Proposition 10.6.1 $\chi(ab) = (-1)^{|a||b|}\chi(b)\chi(a)$. $\qquad\qquad\square$

Proposition 10.6.2 *If H is either commutative or cocommutative then χ is an involution (i.e., $\chi^2 = 1_H$).* $\qquad\qquad\square$

CHAPTER 11

Spectral Sequences

1 Filtrations

Throughout this chapter let $\underline{C}$ be an abelian category. Since explicit descriptions are often useful for intuitive purposes, we will assume that each object of $\underline{C}$ is also a set (i.e., that $\underline{C}$ comes with a forgetful functor to the category of sets) although there are ways to interpret the statements without this assumption.

Let X be an object of $\underline{C}$. A *filtration* of X consists of objects $\{F_n X\}_{n \in \mathbb{Z}}$ and monomorphisms $\ldots \subset F_n X \subset F_{n+1} X \subset \ldots \subset X$. Decreasing filtrations $X \supset \ldots \supset F^{n+1} X \supset F^n X \supset \ldots$ (customarily written using upper notation) can be handled by converting them to the form above by means of the convention $F^n = F_{-n}$. A *morphism of filtered objects* $f : \mathcal{F}_X \to \mathcal{F}_Y$ consists of a morphism $f = f_\infty : X \to Y$ such that $f(F_n X) \subset F_n Y$. Given a filtration $\mathcal{F}_X = \{F_n X\}_{n \in \mathbb{Z}}$ of X, we form the *associated graded object* of the filtration, $\mathrm{Gr}_*(\mathcal{F}_X)$, by setting $\mathrm{Gr}_q(\mathcal{F}_X) = F_q X / F_{q-1} X$.

The fundamental question in connection with filtrations is, "What can we deduce about X from knowledge of $\mathrm{Gr}_*(\mathcal{F}_X)$?"

A morphism $f : \mathcal{F}_X \to \mathcal{F}_Y$ induces $\mathrm{Gr}_*(f) : \mathrm{Gr}_*(\mathcal{F}_X) \to \mathrm{Gr}_*(\mathcal{F}_Y)$. To have a chance to determine X from $\mathrm{Gr}_*(\mathcal{F}_X)$ we must have some conditions such that given $f : \mathcal{F}_X \to \mathcal{F}_Y$, where $\mathcal{F}_X$ and $\mathcal{F}_Y$ satisfy the conditions, then $\mathrm{Gr}_*(f)$ is an isomorphism if and only if f_∞ is an isomorphism. Intuitively $\mathrm{Gr}_n(\mathcal{F}_X)$ tells us about elements of $F_n X$ which are not in $F_{n-1} X$. Elements in $\bigcap_n F_n X$ or outside of $\bigcup_n F_n X$ do not show up in any $\mathrm{Gr}_n(\mathcal{F}_X)$ so we clearly need $\bigcap_n F_n X = 0$ and $\bigcup_n F_n X = X$. However this is not sufficient as can be seen from the following example.

Example 11.1.1 Let $X = \bigoplus_{k \in \mathbb{N}} \mathbb{Z}$, $Y = \prod_{k \in \mathbb{N}} \mathbb{Z}$ and let $f : X \to Y$ be the canonical inclusion. Set

$$F_n X = \begin{cases} X & \text{if } n \geq 0; \\ \bigoplus_{k \geq -n} \mathbb{Z} & \text{if } n < 0, \end{cases} \qquad F_n Y = \begin{cases} Y & \text{if } n \geq 0; \\ \prod_{k \geq -n} \mathbb{Z} & \text{if } n < 0. \end{cases}$$

Then the conditions $\bigcap_n F_n X = 0$, $\bigcap_n F_n Y = 0$, $\bigcup_n F_n X = X$, and $\bigcup_n F_n Y = Y$ are satisfied, and $\mathrm{Gr}_n(f) : \mathrm{Gr}_n(\mathcal{F}_X) \cong \mathrm{Gr}_n(\mathcal{F}_Y)$ for all n, but f is not an isomorphism. $\qquad\square$

Definition 11.1.2 Let $\mathcal{F}_X = \{F_n X\}_{n \in \mathbb{Z}}$ be a filtration of X. $\mathcal{F}_X$ is called *cocomplete* if the canonical map $\varinjlim_n F_n X \to X$ is an isomorphism. $\mathcal{F}_X$ is called *complete* if the canonical map $X \to \varprojlim_n (X/F_n X)$ is an isomorphism. A filtration which is both complete and cocomplete is called *bicomplete*.

113

The condition $\bigcup_n F_n X = X$ is equivalent to $\{F_n\}$ cocomplete. At the other end we have

Proposition 11.1.3 *Let $\mathcal{F}_X = \{F_n\}_{n \in \mathbb{Z}}$ be a complete filtration of X. Then* $\bigcap_n F_n X = 0$. $\hfill\square$

The failure of the converse is illustrated by X in Example 11.1.1.

Theorem 11.1.4 (Comparison Theorem) *Let $f : \mathcal{F}_X \to \mathcal{F}_Y$ be a morphism between bicomplete filtrations. Then $\mathrm{Gr}_*(f)$ is an isomorphism if and only if f_∞ is an isomorphism.*

Proof Trivially if f_∞ is an isomorphism then $\mathrm{Gr}_*(f)$ is an isomorphism. Conversely, suppose $\mathrm{Gr}_*(f)$ is an isomorphism. Consider first the special case where the filtrations are bounded below. That is, suppose that there exists N such that $F_n X = F_n Y = 0$ for $n < N$. Then $f_n : F_n X \to F_n Y$ is an isomorphism for all n by induction and the 5-Lemma and so taking $\lim_{n \to \infty}$ gives $f_\infty : X \cong Y$. Turning now to the general case, for each integer k define an induced filtration on $X/F_k X$ by

$$F_n(X/F_k X) = \begin{cases} F_n X/F_k X & \text{if } n \ge k; \\ 0 & \text{if } n < k, \end{cases}$$

and similarly filter $Y/F_k Y$. These filtrations are bounded below, so for each k the induced map $X/F_k X \to Y/F_k Y$ is an isomorphism by the special case above. But then $X \cong \varprojlim_k (X/F_k X) \cong \varprojlim_k (Y/F_k Y) \cong Y$. $\hfill\square$

2 Definition of Spectral Sequence

A *differential $\underline{C}$-object* (X, d) consists of an object X of $\mathrm{Obj}\,\underline{C}$ and a $\underline{C}$-morphism $d : X \to X$ such that $d^2 = 0$. If (X, d) is a differential object in $\underline{C}$ then we set $Z(X) = \mathrm{Ker}\, d$, $B(X) = \mathrm{Im}\, d$, and $H(X) = Z(X)/B(X)$, called respectively the *cycles*, *boundaries*, and *homology* of (X, d).

Definition 11.2.1 A *spectral sequence* in $\underline{C}$ consists of a sequence of differential objects $(E^n, d^n)_{n=n_0}^\infty$ such that $E^{n+1} = H(E^n)$.

The integer n_0 which indexes the first object in the sequence plays no significant role in the definition and is almost always taken to be either 1 or 2. Often the differentials are dropped in the notation and we refer to the spectral sequence $\{E^n\}$. A morphism of spectral sequences is a collection of maps $\{E^n \to {}'E^n\}$ commuting with the differentials and we let $SpecS(\underline{C})$ denote the category of spectral sequences in $\underline{C}$.

Let $\{E^n\}$ be a spectral sequence. For all n there is a monomorphism $Z(E^n) \rightarrowtail E^n$ and an epimorphism $Z(E^n) \twoheadrightarrow E^{n+1}$. Set $Z^1(E^n) = Z(E^n)$ and inductively define $Z^k(E^n)$ to be the pullback of $Z^{k-1}(E^{n+1}) \rightarrowtail Z^{k-2}(E^{n+1})$ and $Z^{k-1}(E^n) \twoheadrightarrow Z^{k-2}(E^{n+1})$ as illustrated in the following diagram in which all of the squares are pullbacks.

$$
\begin{array}{ccccc}
\cdots \longrightarrow & E^n & & & \\
& \uparrow & & & \\
\cdots \longrightarrow & Z^1(E^n) \longrightarrow & E^{n+1} & & \\
& \uparrow & \uparrow & & \\
\cdots \longrightarrow & Z^2(E^n) \longrightarrow & Z^1(E^{n+1}) \longrightarrow & E^{n+2} & \\
& \uparrow & \uparrow & \uparrow & \\
\cdots \longrightarrow & Z^3(E^n) \longrightarrow & Z^2(E^{n+1}) \longrightarrow & Z^1(E^{n+2}) \longrightarrow & \cdots \\
& \uparrow & \uparrow & \uparrow & \\
& \vdots & \vdots & \vdots &
\end{array}
$$

Explicitly,

$$
Z^k(E^n) = \{x \in E^n \mid d^n x = 0,\ d^{n+1}[x] = 0,\ \ldots,\ d^{n+k-1}[x] = 0\}.
$$

Usually we write simply $d^r x$ rather than $d^r[x]$, but note however that $d^r x$ is not defined unless $d^k x = 0$ for $n \le k < r$. Set

$$
Z^\infty(E^n) = \varprojlim_k Z^k(E^n) = \bigcap_k Z^k(E^n) = \{x \in E^n \mid d^m x = 0 \text{ for all } m \ge n\}.
$$

From the diagram we get an epimorphism $Z^\infty(E^n) \twoheadrightarrow Z^\infty(E^{n+1})$ which fits into the pullback square

$$
\begin{array}{ccc}
Z^\infty(E^n) & \longrightarrow & Z^\infty(E^{n+1}) \\
\downarrow & & \downarrow \\
Z^1(E^n) & \longrightarrow & E^{n+1}.
\end{array}
$$

Elements of $Z^\infty(E^{n_0})$ are referred to as *infinite cycles*. Set

$$
E^\infty = \varinjlim_n Z^\infty(E^n) = \{x \in E^{n_0} \mid d^k x = 0 \text{ for all } k\}/\sim
$$

where $\sim$ is the equivalence relation generated by $x \sim 0$ if $x = d^k y$ for some y and k. Nonzero elements of E^∞ are said to have "survived" the spectral sequence. An element $x \in E^{n_0}$ disappears from the spectral sequence in the transition from E^k to E^{k+1} either because $d^k x \ne 0$ or $x = d^k y$ for some y. If there exists r such that $d^k = 0$ for all $k \ge r$ then $E^r = E^\infty$. In this case the spectral sequence is said to *collapse* at E^r.

The spectral sequences which arise in algebraic topology are spectral sequences of chain complexes. That is, E^r is graded for each r and d^r lowers degree by 1. In fact, most often they come with a natural bigrading.

3 Exact Couples

Definition 11.3.1 An *exact couple* in $\underline{C}$ consists of a diagram in $\underline{C}$,

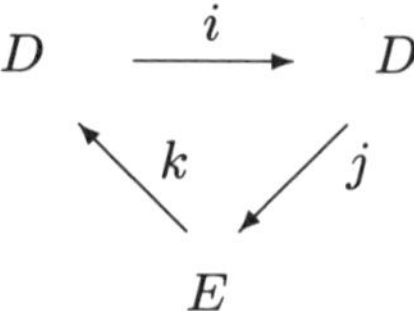

which is exact at each vertex.

We now show how an exact couple gives rise to a spectral sequence. Let

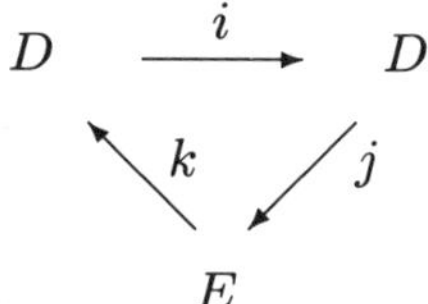

be an exact couple. Set $d = jk : E \to E$. Note that $d^2 = jkjk = 0$ since $kj = 0$. Form the "derived couple"

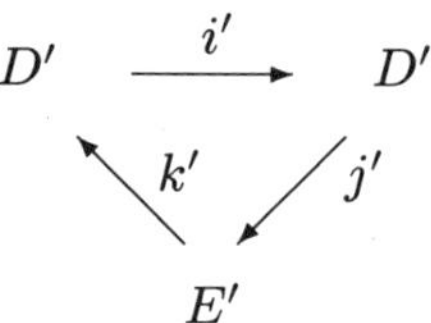

by letting $D' = \operatorname{Im} i$, $E' = H(E, d)$, $i' = i|_{D'}$, $j'(iy) = [j(y)]$ and $k'([x]) = k(x)$. It is easy to check that j' and k' are well defined.

Proposition 11.3.2 *The derived couple of an exact couple is an exact couple.*

$\qquad\qquad\qquad\qquad\qquad\qquad\qquad\qquad\qquad\qquad\qquad\qquad\qquad\qquad\qquad\qquad\quad\square$

Continuing this procedure gives an infinite sequence of exact couples. Setting

$$(E^1, d^1) = (E, d), \qquad (E^2, d^2) = (E', d'), \qquad (E^3, d^3) = (E'', d''), \qquad \dots,$$

where E^n comes from the $(n-1)$st derived couple, yields a spectral sequence since $E^{n+1} = H(E^n)$ by construction.

Next we relate E^∞ from this spectral sequence to quantities obtained from the exact couple.

Let $D \xrightarrow{i} D' = \operatorname{Im} i \xrightarrow{\alpha} D$ be the factorization of $i : D \to D$ as the composition of an epimorphism and a monomorphism. Let $\hat{D}$ be the inverse limit of the system $\dots \xrightarrow{\alpha} D^{n+1} \xrightarrow{\alpha} D^n \xrightarrow{\alpha} \dots \xrightarrow{\alpha} D^1$. Explicitly, $\hat{D} = \bigcap_n i^n D$ equals elements of D infinitely "divisible" by i. If x is infinitely divisible by i then so is ix and so i induces a map $\hat{i} : \hat{D} \to \hat{D}$. At the other end, let $\check{D}$ be the direct limit of $D^1 \xrightarrow{i} \dots \xrightarrow{i} D^n \xrightarrow{i} D^{n+1} \xrightarrow{i} \dots$. Thus $\check{D} = D/\left(\cup_n i^{-n}(0)\right)$ and again i induces a canonical map $\check{i} : \check{D} \to \check{D}$.

For $y \in D$, using the fact that $kj = 0$ and induction we see that $d^n(jy)$ is defined and equal to zero for all n. Thus jy is an infinite cycle, so we get an induced map $\tilde{j} : D \to Z_\infty(E)$. Applying the same considerations to the derived exact couples gives an induced map $\tilde{j} : D^n \to Z_\infty(E^n)$ for each n. The diagram

$$
\begin{array}{ccccccccc}
D & \xrightarrow{i} & \cdots & \xrightarrow{i} & D^n & \xrightarrow{i} & D^{n+1} & \xrightarrow{i} & \cdots \\
\downarrow{\tilde{j}} & & \cdots & & \downarrow{\tilde{j}} & & \downarrow{\tilde{j}} & & \cdots \\
Z^\infty(E) & \twoheadrightarrow & \cdots & \twoheadrightarrow & Z^\infty(E^n) & \twoheadrightarrow & Z^\infty(E^{n+1}) & \twoheadrightarrow & \cdots
\end{array}
$$

commutes since $\tilde{j} : D^n \to Z_\infty(E^n)$ is induced by $j^n = ji^{-(n-1)}$. Therefore there is an induced map of direct limits $\tilde{j} : \check{D} \to E^\infty$.

Turning to the other end, the commutative square on the right of the diagram

$$
\begin{array}{ccccccc}
Z(E^n) & \mapsto & E^n & \xrightarrow{d^n} & E^n \\
\downarrow{\tilde{k}} & & \downarrow{k} & & \| \\
D^{n+1} & \mapsto & D^n & \xrightarrow{j^n} & E^n
\end{array}
$$

yields the induced map $\tilde{k}$ of kernels. Due to the equality on the far right, the left square is a pullback. Taking inverse limits in the diagram of pullbacks

$$
\begin{array}{ccccccccc}
\cdots & \mapsto & Z^{n+1}(E) & \mapsto & Z^n(E) & \mapsto & \cdots & \mapsto & \cdots \quad E \\
\cdots & & \downarrow{\tilde{k}} & & \downarrow{\tilde{k}} & & \cdots & & \cdots \quad \downarrow{k} \\
\cdots & \xrightarrow{\alpha} & D^{n+1} & \xrightarrow{\alpha} & D^n & \xrightarrow{\alpha} & \cdots & \xrightarrow{\alpha} & \cdots \quad D
\end{array}
$$

yields an induced map $\tilde{k} : Z^\infty(E) \to \hat{D}$ which fits into the pullback square

$$
\begin{array}{ccc}
Z^\infty(E) & \mapsto & E \\
\downarrow{\tilde{k}} & & \downarrow{k} \\
\hat{D} & \mapsto & D.
\end{array}
$$

Applying the same considerations to the derived exact couples gives an induced map $\tilde{k} : Z^\infty(E^n) \to \widehat{D^n}$ for each n. However $\widehat{D^n} = \bigcap_m i^m D^n = \bigcap_m i^{m+n} D = \hat{D}$. The maps $\tilde{k} : Z^\infty(E^n) \to \widehat{D^n} = \hat{D}$ are compatible and so induce a map $\tilde{k} : E^\infty \to \hat{D}$ from the direct limit.

Theorem 11.3.3 *The sequence* $0 \longrightarrow \check{D} \xrightarrow{\check{i}} \check{D} \xrightarrow{\tilde{j}} E^\infty \xrightarrow{\tilde{k}} \hat{D} \xrightarrow{\hat{i}} \hat{D}$ *is exact.*

Proof The fact that $\check{\imath}$ is a monomorphism can be seen from the explicit description of $\check{D}$. Consider the diagram

$$
\begin{array}{ccccccccc}
D & \xrightarrow{\,i\,} & D & \xrightarrow{\,\tilde{\jmath}\,} & Z^\infty(E) & \xrightarrow{\,\tilde{k}\,} & \hat{D} & \xrightarrow{\,\hat{\imath}\,} & \hat{D} \\
\big\| & & \big\| & & \downarrow & & \downarrow & & \downarrow \\
D & \xrightarrow{\,i\,} & D & \xrightarrow{\,j\,} & E & \xrightarrow{\,k\,} & D & \xrightarrow{\,i\,} & D
\end{array}
\qquad (*)
$$

in which the bottom row is known to be exact and the second last square is a pullback. In general, given a diagram

$$
\begin{array}{ccccc}
A' & \longrightarrow & B' & \longrightarrow & C' \\
\downarrow & & \downarrow & & \downarrow \\
A & \longrightarrow & B & \longrightarrow & C
\end{array}
$$

in which the bottom row is exact and the left square is a pullback, the top row must be exact. It follows that the top row of $(*)$ is exact at the first $\hat{D}$. Also, $\operatorname{Ker}\tilde{\jmath} = \operatorname{Ker}j = \operatorname{Im}i$, and since the second last square is a pullback, $\operatorname{Ker}\tilde{k} = \operatorname{Ker}k = \operatorname{Im}j = \operatorname{Im}\tilde{\jmath}$. Therefore the top row of $(*)$ is exact everywhere. Applying the same considerations to the derived exact couples and using $\widehat{D^n} = \hat{D}$ gives a commutative diagram

$$
\begin{array}{ccccccccc}
D & \xrightarrow{\,i\,} & \cdots & \xrightarrow{\,i\,} & D^n & \xrightarrow{\,i\,} & D^{n+1} & \xrightarrow{\,i\,} & \cdots \\
\downarrow{\scriptstyle i} & & & & \downarrow{\scriptstyle i} & & \downarrow{\scriptstyle i} & & \\
D & \xrightarrow{\,i\,} & \cdots & \xrightarrow{\,i\,} & D^n & \xrightarrow{\,i\,} & D^{n+1} & \xrightarrow{\,i\,} & \cdots \\
\downarrow{\scriptstyle \tilde{\jmath}} & & & & \downarrow{\scriptstyle \tilde{\jmath}} & & \downarrow{\scriptstyle \tilde{\jmath}} & & \\
Z^\infty(E) & \longrightarrow & \cdots & \longrightarrow & Z^\infty(E^n) & \longrightarrow & Z^\infty(E^{n+1}) & \longrightarrow & \cdots \\
\downarrow{\scriptstyle \tilde{k}} & & & & \downarrow{\scriptstyle \tilde{k}} & & \downarrow{\scriptstyle \tilde{k}} & & \\
\hat{D} & = & \cdots & = & \hat{D} & = & \hat{D} & = & \cdots \\
\downarrow{\scriptstyle \hat{\imath}} & & & & \downarrow{\scriptstyle \hat{\imath}} & & \downarrow{\scriptstyle \hat{\imath}} & & \\
\hat{D} & = & \cdots & = & \hat{D} & = & \hat{D} & = & \cdots
\end{array}
$$

with exact columns. Passing to direct limits, which preserves exactness (Theorem 4.2.4), completes the proof. $\qquad\square$

4 Exact Couple of a Filtered Object

We now describe the most common situation which gives rise to an exact couple and thus to a spectral sequence. Let

$$\ldots \rightarrowtail X_{n-1} \overset{g_{n-1}}{\rightarrowtail} X_n \overset{g_n}{\rightarrowtail} X_{n+1} \rightarrowtail \ldots \rightarrowtail X \qquad (**)$$

be maps of chain complexes. Then for each map or composition of maps in $(**)$, $g : A \to B$, we have an exact triangle of abelian groups

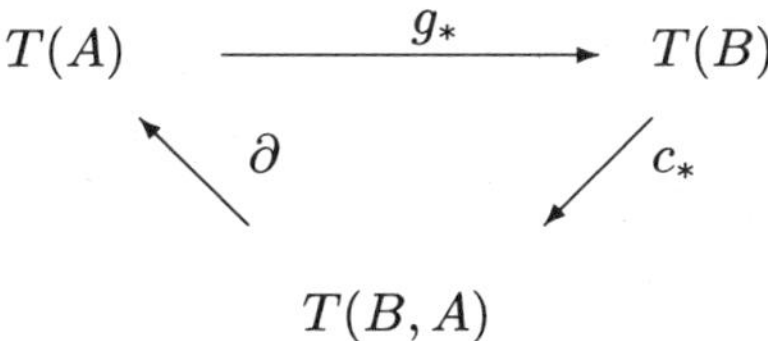

(where ∂ lowers degrees by 1).

We will consider the above situation in greater generality to allow for the various situations to which it applies. Let $\mathcal{S}$ be a category and let $\underline{X}$ be a diagram

$$\ldots \longrightarrow X_{n-1} \overset{g_{n-1}}{\longrightarrow} X_n \overset{g_n}{\longrightarrow} X_{n+1} \longrightarrow \ldots \longrightarrow X \qquad (***)$$

in $\mathcal{S}$ indexed by the ordered set $\mathbb{Z} \cup \{\infty\}$. Typically in the applications these maps are (or can be replaced by) monomorphisms so that $\underline{X}$ can be thought of as a filtration on X. Suppose that for each morphism $g : A \to B$ in $\mathcal{S}$, we are given an exact triangle

$$
\begin{array}{ccc}
T(A) & \overset{g_*}{\longrightarrow} & T(B) \\
 & \searrow \partial \qquad c_* \swarrow & \\
 & T(B, A) &
\end{array}
$$

in $\underline{C}$. (In the applications, $T(\ )$ is usually graded with ∂ lowering degrees by 1.) We assume that the maps g_*, c_*, and ∂ are natural in the sense that they satisfy the analogues of the first three Eilenberg-Steenrod homology axioms.

Remark More precisely, the naturality condition can be expressed by describing T as a pair of functors $T' : \mathcal{S} \to \underline{C}$ and $T'' : \mathcal{M}ap_{\mathcal{S}} \to \underline{C}$, where $\mathcal{M}ap_{\mathcal{S}}$ is the category whose objects are morphisms in $\mathcal{S}$ and whose morphisms are commutative squares in $\mathcal{S}$. The naturality condition can then be expressed as saying that $g_* = T'(\ )$, ∂ is a natural transformation from T'' to the functor $f \mapsto T'(\text{Domain } f)$, and c_* is a natural transformation from the functor $f \mapsto T'(\text{Range } f)$ to T'.

An example other than chain complexes to which our hypotheses apply occurs when $\mathcal{S}$ equals the category of 2-connected topological spaces with $T(X, A) = \pi_*(X, A)$ and ∂ the connecting map in the long exact homotopy sequence, where again $\underline{C}$ is abelian groups. (The restriction to 2-connected spaces guarantees that $T(X, A)$ is in $\underline{C}$.)

Let $\underline{C}$, $\mathcal{S}$, $\underline{X}$, T, and ∂ be as above. We now describe how to associate an exact couple (and thus a spectral sequence) to this situation. As we shall see, the resulting spectral sequence is a tool which provides information about $T(X)$

from a knowledge of $T(X_n, X_{n-1})$ for all n. Define $D = \bigoplus_n T(X_n)$ and $E = \bigoplus_n T(X_n, X_{n-1})$. The map $i : D \to D$ is defined as the one whose restriction to $T(X_n)$ is $g_* : T(X_n) \to T(X_{n+1})$. Similarly $j : D \to E$ is defined as the map whose restriction to $T(X_n)$ is $c_* : T(X_n) \to T(X_n, X_{n-1})$, and $k : E \to D$ is the map whose restriction to $T(X_n, X_{n-1})$ is $\partial : T(X_n, X_{n-1}) \to T(X_{n-1})$. We will sometimes write $\big(E^r(\underline{X})\big)$ for the resulting spectral sequence.

Remark For $\underline{C}$, $\mathcal{S}$, $\underline{X}$, T, and ∂ as above there is in fact a second associated exact couple, however it yields the same spectral sequence. Namely, we could set $\tilde{D} = \bigoplus_n T(X, X_n)$ with again $E = \bigoplus_n T(X_n, X_{n-1})$ and the maps defined in a manner similar to the those in the previous exact couple.

Examining the definitions, $\check{D} = \bigoplus_n \check{D}_n$ where $\check{D}_n = T(X_n)/\big(\cup_m i^{-m}(0)\big)$. Define $f : D \to T(X)$ to be the map whose restriction to $T(X_n)$ is $(f_n)_*$, where $f_n : X_n \to X$ is the map in $(\ast\ast\ast)$. Since $f_n i = f_{n-1}$, the map f induces a map $\check{f} : \check{D} \to T(X)$ such that the composite $D \twoheadrightarrow \check{D} \xrightarrow{\check{f}} T(X)$ is f. Filter $T(X)$ by setting $F_n\big(T(X)\big) = \mathrm{Im}(f_n)_*$. Observe that the restriction of $\check{f} : \check{D}_n \to F_n\big(T(X)\big)$ to $i\check{D}_{n-1}$ is contained in $F_{n-1}\big(T(X)\big)$ and so $\check{f}$ induces a map $\bar{f} : \check{D}/i\check{D} \to \mathrm{Gr}_*\big(T(X)\big)$.

Theorem 11.4.1 *If the filtration $\mathcal{F}_{T(X)} = \big\{F_n\big(T(X)\big)\big\}$ is cocomplete then $\bar{f} : \check{D}/i\check{D} \to \mathrm{Gr}_*\big(T(X)\big)$ is an isomorphism.*

Corollary 11.4.2 *If the filtration $\mathcal{F}_{T(X)}$ is cocomplete then the sequence*

$$0 \longrightarrow \mathrm{Gr}_*\big(T(X)\big) \xrightarrow{\tilde{j}} E^\infty \xrightarrow{\tilde{k}} \hat{D} \xrightarrow{\hat{i}} \hat{D}$$

is exact.

Proof of Theorem 11.4.1 Suppose $y \in F_n\big(T(X)\big)$ represents an element of $\mathrm{Gr}_n\big(T(X)\big)$. Then y belongs to $\mathrm{Im}\,(f_n)_* : T(X_n) \to T(X)$ and so $[y] = \bar{f}[y_n]$ where $(f_n)_*(y_n) = y$. Therefore $\bar{f}$ is onto. Suppose now that x, belonging to $\check{D}_n$, satisfies $\bar{f}(x) = 0$ in $\mathrm{Gr}_n\big(T(X)\big)$. Then $\check{f}_n(x)$ belongs to $F_{n-1}\big(T(X)\big)$ so there exists $y \in \check{D}_{n-1}$ such that $\check{f}_n(x) = \check{f}_{n-1}(y) = \check{f}_n(iy)$. To complete the proof, it suffices to show that $\check{f}$ is a monomorphism, so that $x = iy$ and thus $[x] = 0$ in $\check{D}/i\check{D}$. This follows from the following lemma.

Lemma 11.4.3 $\mathcal{F}_{T(X)}$ *cocomplete implies that $\check{f}$ is a monomorphism.*

Proof Suppose $\check{f}(z) = 0$ for $z \in \check{D}_n$. Since $\mathcal{F}_{T(X)}$ is cocomplete, $T(X) = \varinjlim_k F_k\big(T(X)\big)$ and so by Lemma 4.2.3 there exists m such that $i^m(z) = 0$. But then z belongs to $\cup_m i^{-m}(0)$ and so $[z] = 0$ in $\check{D}$. $\qquad\square$

Corollary 11.4.4 $\mathcal{F}_{T(X)}$ *cocomplete implies that $F_n\big(T(X)\big) = \check{D}_n$.*

Proof $F_n\big(T(X)\big)$ was defined as the image of $\mathrm{Im}\,f : D_n \to T(X)$ which equals $\mathrm{Im}\,\check{f} : \check{D}_n \to T(X)$. However if $\mathcal{F}_{T(X)}$ is cocomplete then by Lemma 11.4.3, $\check{f} : \check{D}_n \to F_n\big(T(X)\big)$ is also injective. $\qquad\square$

In general we say that the spectral sequence (E^r) obtained from the exact couple described above *abuts* to $\mathcal{F}_{T(X)}$ and write $(E^r) \Rightarrow \mathcal{F}_{T(X)}$ or simply $(E^r) \Rightarrow T(X)$. If $\mathcal{F}_{T(X)}$ is bicomplete and the induced injection $\mathrm{Gr}_*\big(T(X)\big) \xrightarrow{\tilde{j}} E^\infty$ is an isomorphism then we say that (E^r) *converges* to $\mathcal{F}_{T(X)}$. If we consider an analogy from calculus, the statement that (E^r) abuts to $\mathcal{F}_{T(X)}$ may be considered to be somewhat analogous to the statement that a certain power series is the Taylor series of some function. One does not normally expect to conclude much about the function from this unless conditions are given under which the Taylor series converges to the function. The question of whether or not a spectral sequence converges can often be difficult. Determining whether or not $\mathcal{F}_{T(X)}$ is cocomplete can usually be handled fairly easily. For example if $\underline{S}$ is a category of chain complexes and $T(\ )$ is homology then in many cases this can be deduced from the fact that homology commutes with direct limits (Theorem 4.2.4). The hard part is determining whether or not the filtration is complete and whether or not the map $\hat{\imath} : \hat{D} \to \hat{D}$ is a monomorphism. A common case in which it is easy to see that the spectral sequence converges occurs when $\mathcal{F}_{T(X)}$ is cocomplete and there exists M such that $T(X_n) = 0$ for $n < M$. In this case $\hat{D} = 0$, so it is clear that the filtration is complete and therefore Theorem 11.4.1 shows convergence.

Theorem 11.4.5 (Spectral Sequence Comparison Theorem) *Let S and T be as above and let $\underline{X}$ and $\underline{Y}$ be diagrams in S indexed by the ordered set $\mathbb{Z} \cup \{\infty\}$. Let η be a morphism from $\underline{X}$ to $\underline{Y}$ such that for some n the induced map of spectral sequences $\eta_* : E^n(\underline{X}) \to E^n(\underline{Y})$ becomes an isomorphism. Suppose that $\big(E^r(\underline{X})\big)$ converges to $\mathcal{F}_{T(X)}$ and that $\big(E^r(\underline{Y})\big)$ converges to $\mathcal{F}_{T(Y)}$. Then $\eta_* : T(X) \to T(Y)$ is an isomorphism.*

Proof The commutative diagram

$$
\begin{array}{ccc}
E^n(\underline{X}) & \xrightarrow{\;\overset{\eta_*}{\cong}\;} & E^n(\underline{Y}) \\
\Big\downarrow{\scriptstyle d^n} & & \Big\downarrow{\scriptstyle d^n} \\
E^n(\underline{X}) & \xrightarrow{\;\overset{\eta_*}{\cong}\;} & E^n(\underline{Y})
\end{array}
$$

shows that $\eta_* : E^{n+1}(\underline{X}) \to E^{n+1}(\underline{Y})$ and $\eta_* : Z\big(E^{n+1}(\underline{X})\big) \to Z\big(E^{n+1}(\underline{Y})\big)$ are isomorphisms. It follows by induction that these statements hold for all $r \geq n$ and so $\eta_* : Z_\infty\big(E^r(\underline{X})\big) \to Z_\infty\big(E^r(\underline{Y})\big)$ is an isomorphism for all $r \geq n$. Therefore $\eta_* : E^\infty(\underline{X}) \to E^\infty(\underline{Y})$ is an isomorphism and so the theorem follows from Theorem 11.1.4 and the definition of convergence. $\qquad\square$

5 Bigraded Spectral Sequences

Let S and T and ∂ be as in the preceding section and suppose that $T(\ )$ is graded and that ∂ has degree -1. For a diagram $\underline{X}$ we bigrade the associated exact couple by setting $D_{p,q} = T_{p+q}(X_p)$ and $E_{p,q} = T_{p+q}(X_p, X_{p-1})$. Terms in the derived exact couples are bigraded by

$$
D^r_{p,q} = i(D^{r-1}_{p,q}) \qquad \text{and} \qquad E^r_{p,q} = \frac{\operatorname{Ker} d^{r-1}_{p,q} : E^{r-1}_{p,q} \to E^{r-1}_{p-r+1,q+r-2}}{\operatorname{Im} d^{r-1}_{p+r-1,q-r+2} : E^{r-1}_{p+r-1,q-r+2} \to E^{r-1}_{p,q}}.
$$

With these bigradings we find that the bidegrees of the maps i^n, j^n, k^n, and d^n are $(1, -1)$, $(-n+1, n-1)$, $(-1, 0)$ and $(-n, n-1)$ respectively. Note that $D^r_{p,q} = 0$ implies that $D^{r+1}_{p,q} = 0$ and that $E^r_{p,q} = 0$ implies that $E^{r+1}_{p,q} = 0$.

We now consider the important special case of diagrams $\underline{X}$ in which $T(X_p) = 0$ for $p < 0$. By exactness, $E_{p,q} = 0$ for $p < 0$ so that we have what is referred to as a right half-plane spectral sequence. Since the conditions imply that $\hat{D} = 0$, the spectral sequence converges to $\mathcal{F}_{T(X)}$ if and only if $\mathcal{F}_{T(X)}$ is cocomplete, which we shall now assume. We will interpret Corollary 11.4.2 in this special case. Inserting our bigradings into Theorem 11.4.1 gives the exact sequence

$$0 \to \check{D}_{p-1,q+1} \to \check{D}_{p,q} \to E^\infty_{p,q} \to 0$$

for each p and q, and by Corollary 11.4.4, $\check{D}_{p,q} = F_p\big(T_{p+q}(X)\big)$. Thus for each n we have a sequence

$$0 = F_{-1}\big(T_n(X)\big) \rightarrowtail F_0\big(T_n(X)\big) \rightarrowtail F_1\big(T_n(X)\big) \rightarrowtail \ldots \rightarrowtail F_k\big(T_n(X)\big) \rightarrowtail \ldots$$

such that

$$F_k\big(T_n(X)\big)/F_{k-1}\big(T_n(X)\big) = E^\infty_{k,n-k}$$

and $T_n(X) = \varinjlim_k F_k\big(T_n(X)\big)$. Of course, by re-indexing any case where the nonzero terms of $T(X_p)$ are bounded below is equivalent to the above.

Let us now specialize further to the case of a first quadrant spectral sequence by adding the hypothesis $E_{p,q} = 0$ for $q < 0$. Notice that these conditions are satisfied in the case where $\underline{S}$ is Top, $T(\ )$ is homology, and X_p/X_{p-1} is $(p-1)$-connected. By exactness, $F_n\big(T_n(X)\big) = F_{n+1}\big(T_n(X)\big) = \ldots = F_\infty\big(T_n(X)\big)$, which equals $T_n(X)$ since we assumed that the filtration is cocomplete. In other words $\{E^\infty_{0,n}, E^\infty_{1,n-1}, \ldots, E^\infty_{n,0}\}$ forms a finite sequence of filtration quotients for $T_n(X)$. Furthermore, in the first quadrant case it is clear from the bidegrees of d_n that for each p, q there exists R such that both the differential entering and the differential leaving $E^R_{p,q}$ are zero, and thus $E^R_{p,q} = E^{R+1}_{p,q} = \cdots = E^\infty_{p,q}$, so that each of these filtration quotients is reached at a finite stage.

6 Multiplicative Spectral Sequences

Definition 11.6.1 A *differential graded algebra* (DGA) is a graded algebra A together with a map $d : A_n \to A_{n-1}$ such that $d^2 = 0$ and

$$d(xy) = d(x)y + (-1)^{|x|}xd(y).$$

It is easy to check that if (A, d) is a differential graded algebra then the multiplication on A induces a well defined multiplication on $H(A)$.

In many spectral sequences of interest, each E^r carries the additional structure of a differential algebra such that the multiplication on E^{r+1} is induced from that on E^r. In this case we say that the spectral sequence is *multiplicative*. If (E^r) is a multiplicative spectral sequence then there is an induced multiplication on E^∞.

If Q is an algebra, we say that a filtration $\mathcal{F}_Q$ on Q is *multiplicative* or an *algebra filtration* if $F_n(Q)F_k(Q) \subset F_{n+k}(Q)$ for all n and k. In this case there is an induced algebra structure on the associated graded object $Gr_*(Q)$. To say that a multiplicative spectral sequence converges multiplicatively to a filtered algebra Q we impose the additional condition that the isomorphism $Gr_*(Q) \cong E^\infty$ be a ring homomorphism.

Massey has given the following necessary and sufficient condition, whose proof is straightforward, for the spectral sequence arising from an exact couple to be multiplicative.

Theorem 11.6.2 (Massey) *Let*

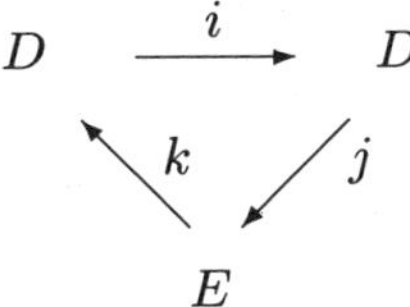

be a graded exact couple in which E is a graded algebra. Suppose whenever elements x, $y \in E$ and elements a, $b \in D$ satisfy $k(x) = i^n(a)$ and $k(y) = i^n(b)$ for some integer n, that there is an element $c \in D$ such that $k(xy) = i^n(c)$ and $j(c) = j(a)y + (-1)^{|x|}xj(b)$. Then the differential on E turns E into a differential graded algebra, and so the multiplication on E induces one on E'. Furthermore the derived couple satisfies the preceding condition also and thus the resulting spectral sequence is multiplicative. $\qquad\square$

It appears that in most cases checking that Massey's conditions are satisfied is no easier than giving a direct proof that a spectral sequence is multiplicative, so perhaps the theorem serves merely to focus attention on what needs to be proved.

The following is an example where Massey's conditions are easy to check.

Proposition 11.6.3 *Let $\mathcal{F}_A$ be a filtered differential graded algebra. Then the associated spectral sequence (abutting to $H_*(A)$) is multiplicative.*

Proof Let x be an element of $E_{p+*} = H_*\big(F_p(A)/F_{p-1}(A)\big)$ and let y be an element of $E_{p'+*} = H_*\big(F_{p'}(A)/F_{p'-1}(A)\big)$. Suppose $k(x) = i^n(a)$ and $k(y) = i^n(b)$ where a belongs to $H_*\big(F_{p-1-n}(A)\big)$ and b belongs to $H_*\big(F_{p'-1-n}(A)\big)$. Pick representatives $\tilde{x} \in F_p(A)$, $\tilde{y} \in F_{p'}(A)$, $\tilde{a} \in F_{p-1-n}(A)$, $\tilde{b} \in F_{p'-1-n}(A)$ in A for x, y, a, and b. Then $d(\tilde{x}) = \tilde{a} + d(w)$ and $d(\tilde{y}) = \tilde{b} + d(w')$ for some $w \in F_{p-1}(A)$ and $w' \in F_{p'-1}(A)$, where d denotes the differential in A. In $F_{p+p'-1}(A)$ we have

$$d(\tilde{x}\tilde{y}) = \tilde{a}\tilde{y} + (-1)^{|x|}\tilde{x}\tilde{b} + d(w)\tilde{y} + (-1)^{|x|}\tilde{x}d(w')$$
$$= \tilde{a}\tilde{y} + (-1)^{|x|}\tilde{x}\tilde{b} + d(w\tilde{y}) + d(\tilde{x}w') - d(ww') + (-1)^{|x|}w\tilde{b} - \tilde{a}w'.$$

Set $c = [\tilde{a}\tilde{y} + (-1)^{|x|}\tilde{x}\tilde{b} + (-1)^{|x|}w\tilde{b} - \tilde{a}w'] \in H_*\big(F_{p+p'-1-n}(A)\big)$. Then $k(xy) = i^n(c)$ and since $(-1)^{|x|}w\tilde{b} - \tilde{a}w'$ lies in filtration $p + p' - 2 - n < p + p' - 1$, $j(c) = j(a)y + (-1)^{|x|}xj(b)$. $\qquad\square$

7 Spectral Sequence of a Double Complex

Definition 11.7.1 A *double complex* in $\underline{C}$ consists of an object $C_{p,q}$ of $\mathrm{Obj}\,\underline{C}$ for each pair of integers, together with morphisms $d' : C_{p,q} \to C_{p-1,q}$ and $d'' : C_{p,q} \to C_{p,q-1}$ such that:

1) $d'^2 = 0 : C_{p,q} \to C_{p-2,q}$;
2) $d''^2 = 0 : C_{p,q} \to C_{p,q-2}$;
3) $d'd'' = d''d' : C_{p,q} \to C_{p-1,q-1}$.

Thus $C_{p,*}$ is a chain complex for each p, $C_{*,q}$ is a chain complex for each q, and $d' : C_{p,*} \to C_{p-1,*}$ and $d'' : C_{*,q} \to C_{*,q-1}$ are chain maps.

Given a double complex C, form the total complex, $\mathrm{Tot}\,C$, by $(\mathrm{Tot}\,C)_n = \bigoplus_{p+q=n} C_{p,q}$ with $d : (\mathrm{Tot}\,C)_n \to (\mathrm{Tot}\,C)_{n-1}$ defined as the map whose restriction to $C_{p,q}$ is

$$\left(d' + (-1)^p d''\right) : C_{p,q} \to C_{p-1,q} \oplus C_{p,q-1} \hookrightarrow (\operatorname{Tot} C)_{n-1}.$$

There are two natural filtrations, $\mathcal{F}'_{\operatorname{Tot} C}$ and $\mathcal{F}''_{\operatorname{Tot} C}$ on $\operatorname{Tot} C$ given by

$$\left(F'_p(\operatorname{Tot} C)\right)_n = \bigoplus_{\substack{s+t=n \\ s \le p}} C_{s,t} \quad \text{and} \quad \left(F''_p(\operatorname{Tot} C)\right)_n = \bigoplus_{\substack{s+t=n \\ t \le p}} C_{s,t}$$

yielding two spectral sequences abutting to $H_*(\operatorname{Tot} C)$. In $E'_{p,q}$ we have

$$\left(F'_p(\operatorname{Tot} C)/F'_{p-1}(\operatorname{Tot} C)\right)_n = C_{p,n-p}$$

with the differential given by $d(x) = (-1)^p d''(x)$. So $F'_p(\operatorname{Tot} C)/F'_{p-1}(\operatorname{Tot} C)$ equals $S^p C_{p,*}$, where S denotes the suspension functor on chain complexes defined by $(SC)_n = C_{n-1}$ (as in Section 4.1). Thus

$${E'}^1_{p,q} = H_{p+q}\left(F_p(\ )/F_{p-1}(\ )\right) = H_{p+q}(S^p C_{p,*}) = H_q(C_{p,*}).$$

The map $d^1 = j^1 k^1$ is induced by d' and so ${E'}^2_{p,q} = H_p\left(H_q(C_{*,*})\right)$. Similarly ${E''}^2_{p,q} = H_q\left(H_p(C_{*,*})\right)$. In general convergence of these spectral sequences is not guaranteed, however by Section 11.5 the first will converge if there exists N such that $C_{p,q} = 0$ for $p < N$ and the second will converge if there exists N such that $C_{p,q} = 0$ for $q < N$.

The following is easy to check.

Proposition 11.7.2 *If one of the differentials in a double complex is trivial, then the spectral sequence collapses at E^2.* $\square$

Example 11.7.3 We will prove the statement made in Section 4.3 that computing $\operatorname{Tor}^R_*(M, N)$ using a projective resolution of M equals the result obtained using a projective resolution of N. Let (P'_*, d') and (P''_*, d'') be projective resolutions of M and N respectively. Let $\operatorname{Tor}'^R_*(M, N) = H_*(P' \otimes N)$ and let $\operatorname{Tor}''^R_*(M, N) = H_*(M \otimes P''_*)$. Define the double complex $C_{p,q} = P'_p \otimes P''_q$ with differentials $d' \otimes 1$ and $1 \otimes d''$. Since the nonzero terms are contained in the first quadrant, both spectral sequences converge. Projectives are flat, so

$$H_q(C_{p,*}) = P'_p \otimes H_q(P''_*) = \begin{cases} P'_p \otimes N & \text{if } q = 0; \\ 0 & \text{if } q \neq 0. \end{cases}$$

Therefore,

$${E'}^2_{p,q} = \begin{cases} \operatorname{Tor}'^R_p(M, N) & \text{if } q = 0; \\ 0 & \text{if } q \neq 0. \end{cases}$$

Since all the nonzero terms of ${E'}^2$ are contained on the line $y = 0$, for degree reasons $d^r = 0$ for $r \geq 2$. Thus ${E'}^2 = {E'}^\infty$ and so all the terms in the filtration quotients for $\mathcal{F}'_{H_n(\operatorname{Tot} C)}$ are zero except for $E^\infty_{n,0} = \operatorname{Tor}'^R_n(M, N)$. Therefore, $H_n(\operatorname{Tot} C) = \operatorname{Tor}'^R_n(M, N)$. Similarly, $H_n(\operatorname{Tot} C) = \operatorname{Tor}''^R_n(M, N)$. $\square$

From a double complex C, we can also form a "product" total complex $\operatorname{Tot}^\pi C$ by $(\operatorname{Tot}^\pi C)_n = \prod_{p+q=n} C_{p,q}$ with $d : (\operatorname{Tot}^\pi C)_{n+1} \to (\operatorname{Tot}^\pi C)_n$ defined as the map whose projection to $C_{p,q}$ is

$$\left(d' + (-1)^p d''\right) : (\operatorname{Tot}^\pi C)_{n+1} \twoheadrightarrow C_{p+1,q} \times C_{p,q+1} \to C_{p,q}.$$

The two spectral sequences resulting from the two filtrations on $\operatorname{Tot}^\pi C$ defined analogously to those on $\operatorname{Tot} C$ are identical to the ones coming from $\operatorname{Tot} C$. However this time the spectral sequence in which one takes homology first vertically and then horizontally will converge if there exists N such that $C_{p,q} = 0$ for $p > N$ and the other will converge if there exists N such that $C_{p,q} = 0$ for $q > N$.

8 Bockstein Spectral Sequence

Let C be a chain complex of free abelian groups and let p be a positive integer. The short exact sequence of groups $0 \to \mathbb{Z}/p\mathbb{Z} \to \mathbb{Z}/p^2\mathbb{Z} \to \mathbb{Z}/p\mathbb{Z} \to 0$ yields a short exact sequence of chain complexes $0 \to C \otimes \mathbb{Z}/p\mathbb{Z} \to C \otimes \mathbb{Z}/p^2\mathbb{Z} \to C \otimes \mathbb{Z}/p\mathbb{Z} \to 0$. The boundary map $\beta_p : H_n(C; \mathbb{Z}/p\mathbb{Z}) \to H_{n-1}(C; \mathbb{Z}/p\mathbb{Z})$ from the corresponding long exact homology sequence is called the mod p Bockstein. When p is clear from the context, it is dropped in the notation. The commutative diagram of short exact sequences

$$
\begin{array}{ccccccccc}
0 & \longrightarrow & \mathbb{Z} & \xrightarrow{p} & \mathbb{Z} & \xrightarrow{\theta_p} & \mathbb{Z}/p\mathbb{Z} & \longrightarrow & 0 \\
 & & \downarrow & & \downarrow & & \| & & \\
0 & \longrightarrow & \mathbb{Z}/p\mathbb{Z} & \xrightarrow{p} & \mathbb{Z}/p^2\mathbb{Z} & \longrightarrow & \mathbb{Z}/p\mathbb{Z} & \longrightarrow & 0
\end{array}
$$

shows that β_p factors as the composition $H_*(C; \mathbb{Z}/p\mathbb{Z}) \xrightarrow{\partial} H_{*-1}(C; \mathbb{Z}) \to H_{*-1}(C; \mathbb{Z}/p\mathbb{Z})$. There is a graded exact couple in which $D_s = H_s(C; \mathbb{Z})$ and $E_s = H_s(C; \mathbb{Z}/p\mathbb{Z})$ with the maps induced from the short exact sequence of groups $0 \to \mathbb{Z} \xrightarrow{p} \mathbb{Z} \longrightarrow \mathbb{Z}/p\mathbb{Z} \to 0$. The corresponding spectral sequence is called the mod p *Bockstein spectral sequence* of C. It is clear from the definitions that β is the d^1-differential of the Bockstein spectral sequence. The d^r-differential is written $\beta^{(r)}$ and is called the rth Bockstein modulo p.

From the definitions in Section 11.3, in the mod p Bockstein spectral sequence $\check{\imath}$ = multiplication by p, $\check{D} = H_*(C)/(p\text{-torsion})$, $\check{D}/(\check{\imath}\check{D}) = (H_*(C)/\text{torsion}) \otimes \mathbb{Z}/p\mathbb{Z}$, and $\hat{D} = \{x \in H_*(C) \mid x \text{ is infinitely divisible by } p\}$. Therefore Theorem 11.3.3 yields,

Proposition 11.8.1 *If $H_*(C)$ has no infinitely p-divisible elements then*

$$E^\infty \cong (H_*(C)/\text{torsion}) \otimes \mathbb{Z}/p\mathbb{Z}$$

in the mod p *Bockstein spectral sequence for $H_*(C)$. In particular this holds if $H_*(C)$ has finite type.* $\square$

From the definition of the differentials we get

Proposition 11.8.2

a) *Let $x \in C_s \otimes \mathbb{Z}/p\mathbb{Z}$ represent a homology class and let $y \in C_s \otimes \mathbb{Z}$ be any element such that $\theta_p(y) = x$, where θ_p denotes reduction modulo p. Then $\beta^{(j)}[x] = 0$ for $j < r$ if and only if y is divisible by p^r. If $\beta^{(j)}[x] = 0$ for $j < r$ then $\beta^{(r)}[x]$ is defined and is given by $[\theta_p(dy/p^r)]$.*

b) *The Bockstein spectral sequence of C collapses at E^r if and only if $H_*(C)$ has no elements of order p^m for $m \geq r$.* $\square$

For the rest of this section we will assume that p is a prime and let $(\ )_{(p)}$ denote localization at p. If $H_*(C)$ is finitely generated then $H_*(C)_{(p)}$ can be reconstructed from its Bockstein spectral sequence. Specifically, by the structure theorem for finitely generated modules over a PID,

$$H_n(C)_{(p)} = Z_{(p)}^s \oplus \mathbb{Z}/p^{t_1}\mathbb{Z} \oplus \mathbb{Z}/p^{t_2}\mathbb{Z} \oplus \ldots \oplus \mathbb{Z}/p^{t_k}\mathbb{Z}$$

for some integers s, t_1, t_2, ..., t_k. By the universal coefficient theorem and the preceding proposition, corresponding to a summand $Z_{(p)}$ there will be a basis element $x \in H_n(C; \mathbb{Z}/p\mathbb{Z}) = E_n^1$ such that $\beta^{(r)}x = 0$ for all r, and corresponding to a summand $\mathbb{Z}/p^t\mathbb{Z}$ there will be a pair of basis elements $x \in E_n^1$, $y \in E_{n+1}^1$ such that $\beta^{(r)}x = 0$ for all r, $\beta^{(r)}y = 0$ for $r < t$ and $\beta^{(t)}y = x$.

If $H_*(X)$ is finitely generated, it is common practice to describe $H_*(X)$ indirectly by exhibiting $H_*(X; \mathbb{Z}/p\mathbb{Z})$ together with a description of the mod p Bockstein spectral sequence of $H_*(X)$ for all p. As seen above, this completely determines $H_*(X)$ but often has the advantage that $H_*(X; \mathbb{Z}/p\mathbb{Z})$ is simpler and/or has more structure than $H_*(X)$. For example, one could determine $H_*(\mathbb{R}P^\infty)$ from the description $H_*(\mathbb{R}P^\infty; \mathbb{Z}/p\mathbb{Z}) = 0$ for $p > 2$, $H_*(\mathbb{R}P^\infty; \mathbb{Z}/2\mathbb{Z}) = \Gamma[x]$ where $|x| = 1$, and $\beta(\gamma_{2n}(x)) = \gamma_{2n-1}(x)$ for all n. (This describes the E^1-term and d^1-differential of the mod 2 Bockstein spectral sequence; normally one would have to go on to describe the higher Bocksteins but in this case $E^2 = H(E^1, d^1) = 0$ so the description is complete.) The advantage of this indirect description of $H_*(\mathbb{R}P^\infty)$ is that the H-space structure on $\mathbb{R}P^\infty$ induces a Hopf algebra structure on $H_*(\mathbb{R}P^\infty; F)$ for any field F so this description actually contains more information than $H_*(\mathbb{R}P^\infty)$. (See Chapter 13 for details on why $\mathbb{R}P^\infty = K(\mathbb{Z}/2\mathbb{Z}, 1)$ is a loop space.) This indirect form of description shows to even better advantage when one can go on to describe the action of the Steenrod operations on $H^*(X; \mathbb{Z}/p\mathbb{Z})$. (See Chapter 14.)

The Bockstein spectral sequence is unusual in two respects:

1) it is the only spectral sequence in common use in algebraic topology which is not naturally bigraded;

2) its primary use is not in computing some object to which it "converges" but rather in studying the p-torsion in $H_*(C)$ by considering the level of the spectral sequence at which various classes disappear.

Proposition 11.8.3 *If C is a differential graded algebra of free abelian groups then the Bockstein spectral sequence of C is multiplicative.*

Proof The proof is similar to that of Proposition 11.6.3 with $c = [\tilde{a}\tilde{y} + (-1)^{|x|}\tilde{x}\tilde{b} + (-1)^{|x|}pw\tilde{b} - p\tilde{a}w']$ where $d\tilde{x} = p^n\tilde{a} + dw$ and $d\tilde{y} = p^n\tilde{b} + dw'$, for $[\tilde{x}]$, $[\tilde{y}] \in H_*(C)$. $\qquad\square$

9 Serre Spectral Sequence

Let $F \longrightarrow X \xrightarrow{\pi} B$ be a fibration in which B is a connected CW-complex. For $b \in B$ set $F_b = \pi^{-1}(b)$. Define a filtration on X by $F_n(X) = \pi^{-1}(B^{(n)})$. This is a nonnegative filtration whose image under homology is cocomplete, so it yields a spectral sequence converging to $H_*(X)$, called the *Serre spectral sequence of the fibration*, in which $E_{p,q}^1 = H_{p+q}(F_p(X), F_{p-1}(X))$. We wish to determine the E^2-term of this spectral sequence.

Definition 11.9.1 A *local coefficient system* on a space Y is a functor from the fundamental groupoid of Y to abelian groups.

By Proposition 7.1.3, the assignment

$$L(b) = H_k(F_b) \qquad \text{and} \qquad L([\gamma]) = ([\gamma]_!)_* : H_k(F_{\gamma(0)}) \cong H_k(F_{\gamma(1)})$$

is a local coefficient system on B for all k.

Let L be a local coefficient system on a space Y. Define a chain complex $S_*(Y; {}^tL)$, called the "singular chain complex of Y with *local* (or *twisted*) coefficients in L" as follows. Set $\big(S_n(Y); {}^tL\big) = \{\text{finite sums } \sum_{i=1}^m a_i\sigma_i\}$ where $\sigma_i : \Delta^n \to Y$ and $a_i \in L\big(\sigma_i(\epsilon_0)\big)$. $\big(S_n(Y); {}^tL\big)$ can be regarded as a subspace of $S_n(Y) \otimes \bigoplus_{y\in Y} L(y)$, so we will use the notation $\sigma \otimes a$ for the element $a\sigma$. Let $\epsilon_{i,j}$ denote the line from ϵ_i to ϵ_j in Δ^n. The differential in $S_*(Y; {}^tL)$ is defined by

$$\partial(\sigma \otimes a) = (\sigma \circ \delta^0) \otimes L(\sigma|_{\epsilon_{0,1}})(a) + \sum_{i=1}^n (-1)^i (\sigma \circ \delta^i) \otimes a.$$

Notice that for $i > 0$ the "first" vertex of $\delta^i(\Delta^n)$ is ϵ_0 while the first vertex of $\delta^0(\Delta^n)$ is ϵ_1. However $L(\sigma|_{\epsilon_{0,1}})(a)$ lies in $L(\epsilon_1)$, so all the terms make sense. The standard computation to show that $\partial^2 = 0$ works although it requires use of the fact that Δ^n is simply connected so that $[\sigma|_{\epsilon_{0,2}}] = [\sigma|_{\epsilon_{0,1}}][\sigma|_{\epsilon_{1,2}}]$. The complex $S_*(Y; {}^tL)$ reduces to $S_*(Y; A)$ when all the morphisms $L([\gamma])$ are the identity map on an abelian group A.

Theorem 11.9.2 (Serre) *Let $F \longrightarrow X \xrightarrow{\pi} B$ be a fibration sequence where B is a connected CW-complex. Then in the fibration's Serre spectral sequence converging to $H_{p+q}(X)$ the $E^2_{p,q}$-term is given by $E^2_{p,q} \cong H_p\big(B; {}^tH_q(F)\big)$.*

Proof Let $\{e_j\}_{j\in J_p}$ be the set of (open) p-cells of B and let $f_j : \big(D^p, \mathrm{Int}(D^p)\big) \to (\overline{e_j}, e_j)$ be the characteristic map for e_j. Since π is a fibration, the inverse image of a homotopy equivalence between subsets of B is a homotopy equivalence, and in general the inverse image of an excision is an excision. Therefore the argument (given in proof of Theorem 5.4.2) that shows the maps $\big(D^p, \partial(D^p)\big) \xrightarrow{f_j} (\overline{e_j}, \partial\overline{e_j}) \hookrightarrow \big(B^{(p)}, B^{(p-1)}\big)$ induce isomorphisms

$$\bigoplus_{j\in J_p} H_*(D^p, \partial(D^p)) \cong \bigoplus_{j\in J_p} H_*(\overline{e_j}, \partial\overline{e_j}) \cong H_*\left(B^{(p)}, B^{(p-1)}\right)$$

shows that their pullbacks induce isomorphisms

$$\bigoplus_{j\in J_p} H_{p+q}(Q_j, Q'_j) \cong \bigoplus_{j\in J_p} H_{p+q}\big(\pi^{-1}(\overline{e_j}), \pi^{-1}(\partial\overline{e_j})\big)$$

$$\cong H_{p+q}\big(F_p(X), F_{p-1}(X)\big) = E^1_{p,q}$$

where $Q_j \to D^p$ is the pullback of π under f_j and $Q'_j \to \partial(D^p)$ is its restriction to $\partial(D^p)$. Since D^p is contractible, by Theorem 7.1.2 $Q_j \to D^p$ is fibre homotopy equivalent to the projection $D^p \times F \to D^p$ and the pullback to D^p of the local coefficient system on B is trivial. Therefore

$$H_{p+q}(Q_j, Q'_j) \cong H_{p+q}\big(D^p \times F, \partial(D^p) \times F\big) \cong H_p\big(D^p, \partial(D^p); {}^tH_q(F)\big).$$

In particular, we see that this is a first quadrant spectral sequence. In order to calculate the differential on the E^1-term, we give a direct description of the isomorphism ψ making the diagram

$$\bigoplus_{j\in J_p} H_p\big(D^p,\partial D^p;{}^tH_q(F)\big) \xrightarrow{\;\cong\;} H_p\big(B^{(p)},B^{(p-1)};{}^tH_q(F)\big)$$

$$\downarrow{\cong} \qquad\qquad\qquad\qquad\qquad\qquad\qquad \downarrow{\psi}$$

$$\bigoplus_{j\in J_p} H_{p+q}(Q_j,Q_{j'}) \xrightarrow{\cong} \bigoplus_{j\in J_p} H_{p+q}\big(\pi^{-1}(\overline{e_j}),\pi^{-1}(\partial\overline{e_j})\big) \xrightarrow{\cong} H_{p+q}\big(F_p(X),F_{p-1}(X)\big)$$

commute.

By an argument similar to that in the proof of Theorem 7.4.1 we get

Lemma 11.9.3 *Let Y be a CW-complex and let $S'_*(Y)$ be the subcomplex of $S_*(Y)$ generated by cellular maps. Then the inclusion $S'_*(Y)\hookrightarrow S_*(Y)$ is a chain homotopy equivalence. The chain homotopies can be chosen to preserve the image of the vertex ϵ_0 so the corresponding inclusion $S'_*(Y;{}^tL)\hookrightarrow S_*(Y;{}^tL)$ is also a chain homotopy equivalence for a local coefficient system L.*

Proof of Theorem 11.9.2 (continued) Let $\sigma:\Delta_p\to B^p$ be cellular and let $Q_\sigma\to\Delta^p$ be the pullback of π and σ, with $\alpha_\sigma:Q_\sigma\to F_pX$ denoting the induced map. Since Δ^s is contractible, it is possible to choose a fibre homotopy equivalence $\phi_\sigma:\Delta^s\times F_{\sigma(\epsilon_0)}\to Q_\sigma$ having the property that for any $z\in\Delta^s$, the composite $F_{\sigma(\epsilon_0)}=\{z\}\times F_{\sigma(\epsilon_0)}\xrightarrow{\phi_\sigma}F_z$ is $\gamma_!$, where γ is any path from $\sigma(\epsilon_0)$ to $\sigma(z)$. Let $\xi_s\in H_s\big(\Delta^s,\partial(\Delta^s)\big)$ be the generator corresponding to the identity map of Δ^s. Define $\psi:S'_p\big(B^{(p)},B^{(p-1)};{}^tH_q(F)\big)\to H_{p+q}\big(F_p(X),F_{p-1}(X)\big)$ on generators by $\psi(\sigma\otimes w)=(\alpha_\sigma)_*(\phi_\sigma)_*(\xi_p\otimes w)$. For a generator $\tau\in S'_{p+1}\big(B^{(p)},B^{(p-1)}\big)$,

$$\psi\big(\partial(\tau\otimes w)\big)=\psi\big(\tau\delta^0\otimes([\epsilon_{0,1}]_!)_*w\big)+\sum_{j=1}^{p+1}(-1)^j\psi(\tau\delta^j\otimes w)$$

$$=(\alpha_{\tau\circ\delta^0})_*(\phi_{\tau\circ\delta^0})_*\big(\xi_p\otimes([\epsilon_{0,1}]_!)_*w\big)$$

$$+\sum_{j=1}^{p+1}(-1)^j(\alpha_{\tau\circ\delta^j})_*(\phi_{\tau\circ\delta^j})_*(\xi_p\otimes w).$$

For $j>0$ we can choose $\phi_{\tau\circ\delta^j}$ to be $\phi_\tau\circ\big(\delta^j\times F_{\tau(\epsilon_0)}\big)$ and we can choose $\phi_{\tau\circ\delta^0}$ to be $\phi_\tau\circ\big(\delta^0\times[\epsilon_{1,0}]_!\big)$. Therefore,

$$(\phi_{\tau\circ\delta^j})_*(\xi_p\otimes w)=\big(\phi_\tau\circ(\delta^j\times F_{\tau(\epsilon_0)})\big)_*(\xi_p\otimes w)=(\phi_\tau)_*\big((\delta^j)_*\xi_p\otimes w\big)$$

for $j>0$, while

$$(\phi_{\tau\circ\delta^0})_*\big(\xi_p\otimes([\epsilon_{0,1}]_!)_*w\big)=(\phi_\tau)_*\big((\delta^0)_*\xi_p\otimes([\epsilon_{1,0}]_!)_*([\epsilon_{0,1}]_!)_*w\big)=(\phi_\tau)_*\big((\delta^0)_*\xi_p\otimes w\big).$$

Hence

$$\psi\big(\partial(\tau\otimes w)\big)=(\alpha_\tau)_*(\phi_\tau)_*\left(\sum_{j=0}^{p+1}(-1)^j(\delta^j)_*\xi_p\otimes w\right).$$

However, $\sum_{j=0}^{p+1}(-1)^j(\delta^j)_*\xi_p=0$ in $H_p\big(\Delta^{p+1},(\Delta^{p+1})^{(p-1)}\big)$ since it is the image of ξ^{p+1} under the trivial composition

$$H_{s+1}\big(\Delta^{s+1},\partial(\Delta^{s+1})\big)\to H_s\big(\partial(\Delta^{s+1}),(\Delta^{s+1})^{(s-1)}\big)\to H_s\big(\Delta^{s+1},(\Delta^{s+1})^{(s-1)}\big).$$

Therefore ψ induces a well defined map

$$\psi : H_p\big(B^{(p)}, B^{(p-1)}; {}^t H_q(F)\big) \to H_{p+q}\big(F_p(X), F_{p-1}(X)\big).$$

From the construction, the diagram commutes, as can be seen by checking its restriction to each factor. Therefore ψ is an isomorphism.

To conclude the proof we now show that ψ is a chain map when the differential on the domain is that from cellular chain complex for $H_*\big(B; {}^t H_q(F)\big)$ and the differential on the range is the d^1-differential of the spectral sequence. Note that the definitions imply that the latter differential is the boundary map in the long exact homology sequence of the triple $\big(F_p(X), F_{p-1}(X), F_{p-2}(X)\big)$. Let $[\sigma] \otimes x$ represent a generator of $H_p\big(B^{(p)}, B^{(p-1)}; {}^t H_q(F)\big)$ where $\sigma : \Delta^p \to B$ is cellular. As above $\psi d([\sigma] \otimes w) = (\alpha_\sigma)_*(\phi_\sigma)_* \big(\sum_{j=0}^{p}(-1)^j(\delta^j)_*\xi_{p-1} \otimes w\big)$. In

$$H_{p+q-1}\big(\partial\Delta^p \times F_{\sigma(\epsilon_0)}, (\Delta^p)^{(p-2)} \times F_{\sigma(\epsilon_0)}\big)$$

we have $\sum_{j=0}^{p}(-1)^j(\delta^j)_*\xi_{p-1} \otimes w = d(\xi_p \otimes w)$ where $d : H_{p+q}\big(\Delta^p \times F_{\sigma(\epsilon_0)}, \partial\Delta^p \times F_{\sigma(\epsilon_0)}\big) \to H_{p+q-1}\big(F_{p-1}(X), F_{p-2}(X)\big)$ is the boundary map in the long exact homology sequence of the triple. Therefore the commutative square

$$
\begin{array}{ccc}
H_{p+q}\big(\Delta^p \times F_{\sigma(\epsilon_0)}, \partial\Delta^p \times F_{\sigma(\epsilon_0)}\big) & \xrightarrow{(\phi_\sigma)_*} & H_{p+q}\big(F_p(X), F_{p-1}(X)\big) \\
\downarrow{\scriptstyle d} & & \downarrow{\scriptstyle d^1} \\
H_{p+q-1}\big(\partial\Delta^p \times F_{\sigma(\epsilon_0)}, (\Delta^p)^{(p-2)} \times F_{\sigma(\epsilon_0)}\big) & \xrightarrow{(\phi_\sigma)_*} & H_{p+q-1}\big(F_{p-1}(X), F_{p-2}(X)\big)
\end{array}
$$

shows that

$$\psi d([\sigma] \otimes w) = (\alpha_\sigma)_*(\phi_\sigma)_* d(\xi_p \otimes w)$$

$$= d^1(\alpha_\sigma)_*(\phi_\sigma)_*(\xi_p \otimes w)$$

$$= d^1\psi([\sigma] \otimes w).$$

$$\square$$

The differential $d^r : E^r_{r,0} \to E^r_{0,r-1}$ in the Serre spectral sequence is called the *transgression*, often denoted τ; we will describe it in more detail.

Theorem 11.9.4 *Let $F \xrightarrow{\ j\ } X \xrightarrow{\ \pi\ } B$ be a fibration sequence where B is a connected CW-complex. Let $\tilde\pi : X/F \to B$ be the induced map. Then for $r \geq 2$, in the Serre spectral sequence for the fibration*

$$E^r_{r,0} = \operatorname{Im} \tilde\pi_* : H_r(X/F) \to H_r(B), \qquad E^r_{0,r-1} = H_{r-1}(F)/\partial(\operatorname{Ker} \tilde\pi_*),$$

and the transgression $d^r : E^r_{r,0} \to E^r_{0,r-1}$ is determined by the commutative diagram

$$
\begin{array}{ccc}
H_r(X/F) & \xrightarrow{\ \tilde\pi_*\ } & E^r_{r,0} \\
\downarrow{\scriptstyle \partial} & & \downarrow{\scriptstyle d^r} \\
H_{r-1}(F) & \longrightarrow & E^r_{0,r-1}
\end{array}
$$

where ∂ comes from the long exact homology sequence.

Proof From the E^2-term on, the spectral sequence is independent of the CW-structure on B, so we will assume that we have chosen one such that $B^{(0)}$ is a single point and thus $F_0(X) = F$. Filter X/F by inverse images of skeletons under $\tilde{\pi}$ and let $('E^r)$ denote the resulting spectral sequence converging to $H_*(X/F)$. The filtrations are compatible so the collapsing map $c : X \to X/F$ induces a morphism of spectral sequences.

Lemma 11.9.5 *For each r, $c_* : E^r_{m,*} \to {}'E^r_{m,*}$ is a monomorphism for all $m > 0$ and an isomorphism for $m \geq r$.*

Proof Proceed by induction on r. To begin the induction note that

$$F_m(X/F)\big/F_{m-1}(X/F) = F_m X\big/F_{m-1}X \quad \text{for } m > 0,$$

so $E^1_{m,*} = {}'E^1_{m,*}$ for $m > 0$. For all $m > 0$, the induction hypothesis that $E^{r-1}_{m,*} \subset {}'E^{r-1}_{m,*}$ implies that

$$\operatorname{Ker} d^{r-1} : E^{r-1}_{m,*} \to E^{r-1}_{m-r+1,*+r-2} \subset \operatorname{Ker} {}'d^{r-1} : {}'E^{r-1}_{m,*} \to {}'E^{r-1}_{m-r-1,*+r-2},$$

with equality for $m \geq r$ since in that case $E^{r-1}_{m-r+1,*} \subset {}'E^{r-1}_{m-r+1,*}$ while $E^{r-1}_{m,*} = {}'E^{r-1}_{m,*}$. Also, for $m > 0$,

$$\operatorname{Im} d^{r-1} : E^{r-1}_{m+r-1,*-r+2} \to E^{r-1}_{m,*} = \operatorname{Im} {}'d^{r-1} : {}'E^{r-1}_{m+r-1,*-r+2} \to {}'E^{r-1}_{m,*}$$

since $m + r - 1 > r - 1$. Therefore $E^r_{m,*} \subset {}'E^r_{m,*}$ for all $m > 0$ with equality when $m \geq r$. $\qquad\square$

Proof of Theorem 11.9.4 (continued) $F_0(X/F) = \{*\}$ and so $'E^r_{0,r-1} = 0$ and therefore the exact sequence $0 \longrightarrow {}'E^\infty_{r,0} \longrightarrow {}'E^r_{r,0} \xrightarrow{\ 'd^r\ } {}'E^r_{0,r-1}$ implies that $'E^\infty_{r,0} = {}'E^r_{r,0}$. From the definition of the filtration, the composition

$$H_r(X/F) = F_r\big(H_r(X/F)\big) \twoheadrightarrow F_r\big(H_r(X/F)\big)\big/F_{r-1}\big(H_r(X/F)\big)$$
$$=' E^\infty_{r,0} =' E^r_{r,0} \subset E^2_{r,0} = H_r(B)$$

is $\tilde{\pi}_*$. By the lemma, $'E^r_{r,0}$ can be replaced in this composition by $E^r_{r,0}$ which gives $E^r_{r,0} = \operatorname{Im} \tilde{\pi}_*$. Since $H_r(F_r X/F) \to H_r(X/F)$ is surjective for connectivity reasons, $E^r_{r,0}$ could equally well be considered as $\operatorname{Im} \tilde{\pi}_* : H_r(F_r X/F) \to H_r(B)$. The differential d^r is induced by the map $\partial : H_r(F_r X/F_{r-1}X) \to H_{r-1}(F_{r-1}X)$ from the exact couple and

$$
\begin{array}{ccc}
H_r(F_r X/F) & \longrightarrow & H_r(F_r X/F_{r-1}X) \\
\Big\downarrow{\partial} & & \Big\downarrow{\partial} \\
H_{r-1}(F) & \longrightarrow & H_{r-1}(F_{r-1}X)
\end{array}
$$

commutes, yielding the commutative square in the theorem. The identification of $E^r_{0,r-1}$ is obtained by applying the 5-Lemma to

$$0 \longrightarrow \operatorname{Im}\partial/\partial(\operatorname{Ker}\tilde{\pi}_*) \longrightarrow H_{r-1}(F)/\partial(\operatorname{Ker}\tilde{\pi}_*) \longrightarrow H_{r-1}(F)/\operatorname{Im}\partial \longrightarrow 0$$

$$0 \longrightarrow \operatorname{Im}d^r \longrightarrow E^r_{0,r-1} \longrightarrow E^\infty_{0,r-1} \longrightarrow 0$$

using the isomorphisms

$$\operatorname{Im}d^r = d^r(E^r_{r,0}) = \partial(\operatorname{Im}\tilde{\pi}_*) \cong \partial\big(H_r(X/F)/\operatorname{Ker}\tilde{\pi}_*\big) = \operatorname{Im}\partial/\partial(\operatorname{Ker}\pi_*)$$

and

$$E^\infty_{0,r-1} \cong F_0\big(H_{r-1}(X)\big) = \operatorname{Im}j_* \cong H_{r-1}(F)/\operatorname{Ker}j_* = H_{r-1}(F)/\operatorname{Im}\partial.$$

$\square$

The elements of the subset $\operatorname{Im}\tilde{\pi}_* \subset H_*(B)$ (which constitutes the domain of transgression) are called the *transgressive elements* of $H_*(B)$.

Applying cohomology to the Serre filtration of X yields a spectral sequence with $E^1_{-p,-q} = H^{p+q}\big(F_p(X), F_{p-1}(X)\big)$, for $p, q \geq 0$, converging to $H^p(X)$. The natural home of this spectral sequence is the third quadrant, however it is customary to rotate it $180°$ into the first quadrant with the differentials running "backwards" by setting $E^{p,q}_1 = E^1_{-p,-q}$. Dualizing the ideas in the proof of Theorem 11.9.2 gives a chain isomorphism $\psi^* : E^{p,q}_1 \cong H^{p+q}\big(B^{(p)}, B^{(p-1)}; {}^tH^q(F)\big)$ which identifies $E^{p,q}_2$ as $H^p\big(B; {}^tH^q(F)\big)$. Since the construction is natural with respect of maps of fibrations, the diagonal map from the fibration p over B to the fibration $p \times p$ over $B \times B$ induces a multiplication on this spectral sequence which corresponds to the cup product on B under our identification of the E_2-term. Theorem 11.9.4 dualizes to the following description of the cohomology transgression.

Theorem 11.9.6 *Let $F \xrightarrow{j} X \xrightarrow{\pi} B$ be a fibration sequence where B is a connected CW-complex. Let $\tilde{\pi} : X/F \to B$ be the induced map. Then for $r \geq 2$, in the cohomological Serre spectral sequence for the fibration*

$$E^{r,0}_r = H^r(B)/\operatorname{Ker}\tilde{\pi}^* \cong \operatorname{Im}\tilde{\pi}^*, \qquad E^{0,r-1}_r = \delta^{-1}(\operatorname{Im}\tilde{\pi}^*) \subset H^{r-1}(F),$$

and the transgression $d_r : E^{r,0}_r \to E^{0,r-1}_r$ is induced by the map $\delta : H^{r-1}(F) \to H^r(X/F)$ from the long exact cohomology sequence. $\square$

The following routine example, which experienced spectral sequence users would do in their head, will illustrate to the beginner how spectral sequences are used to do calculations.

Example 11.9.7 We will calculate $H^*(\mathbb{C}P^\infty)$ from $H^*(S^1)$. Consider the fibre bundle $S^1 \to S^\infty \to \mathbb{C}P^\infty$ (Example 9.2.3) and its associated Serre spectral sequence. Using that $\mathbb{C}P^\infty$ is connected, we initially know that $E^{0,*}_2 = \mathbb{Z} \otimes H^*(S^1) \cong H^*(S^1)$. Therefore

$$E^{p,q}_2 = E^{p,0}_2 \otimes E^{0,q}_2 = H^p(\mathbb{C}P^\infty) \otimes H^q(S^1) = \begin{cases} H^p(\mathbb{C}P^\infty) & \text{if } q = 0 \text{ or } 1; \\ 0 & \text{otherwise,} \end{cases}$$

using the universal coefficient theorem and the fact that $H^*(S^1)$ is torsion-free. In particular, from the E_2-term on, the nonzero terms all lie on the lines $y = 0$ and $y = 1$. Since the spectral sequence converges to $H^*(S^\infty)$ which is acyclic,

$E_\infty^{p,q}$ must be zero unless $p = q = 0$. For each $r \geq 2$ the differential d_r out of $E_r^{1,0}$ goes to a zero group and the differential d_r into $E_r^{1,0}$ comes from a zero group and thus $E_2^{1,0} = E_\infty^{1,0} = 0$. The only possible nonzero differential on $E_2^{0,1}$ is $d_2 : E_2^{0,1} \to E_2^{2,0}$, so $\operatorname{Ker} d_2 = E_\infty^{0,1} = 0$. Similarly, all differentials on $E_2^{2,0}$ are zero and the only possible nonzero differential entering $E_2^{2,0}$ is $d_2 : E_2^{0,1} \to E_2^{2,0}$, and therefore $\operatorname{CoKer} d_2 = E_\infty^{2,0} = 0$. Thus $d_2 : E_2^{0,1} \to E_2^{2,0}$ is an isomorphism and so $E_2^{2,0} \cong E_2^{0,1} \cong \mathbb{Z}$. Our knowledge of $E_2^{1,0}$ and $E_2^{2,0}$ (together with our knowledge of $E_2^{0,*}$) now determine the lines $x = 1$ and $x = 2$ of the E_2-term. In particular, $E_2^{1,1}$, the only term which could have produced a nonzero differential into $E_2^{3,0}$, is now known to be zero and so $E_2^{3,0} = 0$ by the same argument used to show $E_2^{1,0} = 0$. Similarly the argument used to calculate $E_2^{2,0}$ can be repeated to show that $E_2^{4,0} \cong E_2^{2,1} \cong \mathbb{Z}$. Continuing by induction gives $H^{2n-1}(\mathbb{C}P^\infty) = E_2^{2n-1,0} = 0$ and $H^{2n}(\mathbb{C}P^\infty) = E_2^{2n,0} = \mathbb{Z}$ for all n. Let v be a generator of $E_2^{0,1}$ and set $x = d_2(v) \in E_2^{2,0}$. Then x is a generator of $E_2^{2,0}$ and xv is a generator of $E_2^{2,1}$. Therefore $d_2(xv)$ is a generator of $E_2^{4,0}$. However since the spectral sequence is multiplicative, $d_2(xv) = (d_2 x)v + x(d_2 v) = 0 + x^2 = x^2$. Continuing by induction shows that x^n is a generator for the group $H^{2n}(\mathbb{C}P^\infty) = E_2^{2n,0}$ for each n. Therefore $H^*(\mathbb{C}P^\infty) = \mathbb{Z}[x]$ where $|x| = 2$. $\qquad\square$

Exercise 11.9.8 Use the fibration $\Omega S^n \to P S^n \to S^n$ to calculate $H^*(\Omega S^n)$ including its ring structure. Compare the result with that obtained by dualizing the Hopf algebra $H_*(\Omega S^n)$ described by the Bott-Samelson theorem.

Exercise 11.9.9 Show that

$$H_*(\Omega^2 S^{2n+1}; \mathbb{Z}/p\mathbb{Z}) \cong \bigotimes_{k=0}^{\infty} E[u_{2np^k-1}] \otimes \bigotimes_{k=1}^{\infty} \mathbb{Z}/p\mathbb{Z}[v_{2np^k-2}]$$

where $E[x_m]$ and $(\mathbb{Z}/p\mathbb{Z})[x_m]$ denote respectively exterior and polynomial algebras on an element in degree m.

Exercise 11.9.10 The space $\mathbb{C}P^\infty$ can be made into a topological group and thus deloops to give an "Eilenberg-MacLane space" $K(\mathbb{Z}, 3)$ (see Chapter 13) described by the property that

$$\pi_n\big(K(\mathbb{Z}, 3)\big) = \begin{cases} \mathbb{Z} & \text{if } n = 3; \\ 0 & \text{otherwise.} \end{cases}$$

Let $j : S^3 \to K(\mathbb{Z}, 3)$ be a generator of $\pi_3\big(K(\mathbb{Z}, 3)\big)$ and let $S^3\langle 3\rangle$ denote the homotopy-theoretic fibre of j. How do the homotopy groups of $S^3\langle 3\rangle$ compare to those of S^3? Use the Serre spectral sequence for the induced fibration $\mathbb{C}P^\infty \to S^3\langle 3\rangle \to S^3$ to calculate $H_*(S^3\langle 3\rangle)$ and use the result to show that $\pi_4(S^3) \cong \mathbb{Z}/2\mathbb{Z}$.

There is an algebraic analogue of the Serre spectral sequence. Let A be an augmented differential graded R-algebra. A *differential graded A-module* (M, d_M) is a graded A-module M together with an R-module differential d_M on M such that the module structure map $A \otimes M \to M$ is a chain map.

Theorem 11.9.11 (Algebraic Serre Spectral Sequence) *Let A be an augmented differential nonnegatively graded R-algebra and let C be a flat differential nonnegatively graded R-module. Let $\tilde{d}$ be a differential on $A \otimes C$ such that $(A \otimes C, \tilde{d})$ becomes a differential A-module and such that $A \otimes C \xrightarrow{\epsilon \otimes C} R \otimes C \cong C$ is a chain* *map. Then there is a spectral sequence with $E^2_{p,q} = H_p\big(C; H_q(A)\big)$ converging to $H_{p+q}(A \otimes C, \tilde{d})$.*

Proof Filter C by skeletons and $A \otimes C$ by inverse images of skeletons. That is, set $F_p(C) = \bigoplus_{j \le p} C_j$ and $F_p(A \otimes C) = A \otimes F_p(C)$. Then $E^1_{p,q} = H_{p+q}(A \otimes C_p) = C_p \otimes H_q(A)$, where we have used that $F_{p-1}C \hookrightarrow F_pC$ splits to get the first equality and that C is flat for the second. Thus $E^2_{p,q} = H_p\big(C; H_q(A)\big)$. The modules are nonnegatively graded, so the resulting spectral sequence is a first quadrant spectral sequence converging to $H_*(A \otimes C, \tilde{d})$. $\qquad\square$

Note that Theorem 10.3.3 implies that the above conditions are satisfied for any short exact sequence $A \to B \to C$ of nonnegatively graded differential Hopf algebras.

Remark $\tilde{d}$ is sometimes called a "twisted" differential on $A \otimes C$.

In the same manner, one can obtain a third quadrant spectral sequence, or equivalently, a first quadrant "cohomological" spectral sequence (i.e., differentials running backwards) in the case where the differentials on A and C are of degree 1 instead of -1.

Example 11.9.12 Let p be a prime. Let X be a graded set of finite type, and let $S[X]$ denote the free commutative graded $(\mathbb{Z}/p\mathbb{Z})$-algebra on the vector space generated by X. That is, if $p = 2$ then $S[X] = (\mathbb{Z}/2\mathbb{Z})[X]$ while if $p > 2$ then $S[X] = (\mathbb{Z}/p\mathbb{Z})[X_e] \otimes E[X_o]$ where X_e is the subset of even degree elements in X, X_o is the subset of odd degree elements, and $E[Y]$ denotes the exterior algebra over $\mathbb{Z}/p\mathbb{Z}$ on the vector space by the set Y. Let A be the dual space $S[X]^*$ and let $C = \operatorname{Tor}^A(\mathbb{Z}/p\mathbb{Z}, \mathbb{Z}/p\mathbb{Z})$, where $\mathbb{Z}/p\mathbb{Z}$ is considered as a graded vector space concentrated in degree 0 on which A acts trivially (i.e., through its augmentation). Then

$$A \cong \begin{cases} \Gamma[X_e] \otimes E[X_o] & \text{if } p > 2; \\ \Gamma[X] & \text{if } p = 2. \end{cases}$$

We will regard A and C as differential graded algebras with zero differential. We begin by computing C. Since $A = \bigotimes_{x \in X} A(x)$ where $A(x) = S[x]^*$, a projective resolution of $\mathbb{Z}/p\mathbb{Z}$ over A will be given by $\bigotimes_{x \in X} P_*(x)$, where $P_*(x)$ is a projective resolution of $\mathbb{Z}/p\mathbb{Z}$ over $S[x]^*$. Therefore $C = \bigotimes_{x \in X} C(x)$ where $C(x) = P(x) \otimes_{A(x)} \mathbb{Z}/p\mathbb{Z}$. Let x belong to X with $|x| = n$. Define $P(x)$ as a bigraded differential algebra as follows. Set

$$P_{*,*}(x) = \begin{cases} \Gamma[x] \otimes \bigotimes_{k=0}^{\infty} \Gamma[y_k] \text{ with} \\[4pt] \quad |x| = (n,0),\ |y_k| = (2^k n, 1) & \text{if } p = 2; \\[10pt] E[x] \otimes \Gamma[y] \text{ with} \\[4pt] \quad |x| = (n,0),\ |y| = (n,1), & \text{if } p > 2 \text{ and } n \text{ is odd}; \\[10pt] \Gamma[x] \otimes \bigotimes_{k=0}^{\infty} E[y_k] \otimes \bigotimes_{k=1}^{\infty} \Gamma[z_k] \text{ with} \\[4pt] \quad |x| = (n,0),\ |y_k| = (p^k n, 1),\ |z_k| = (p^k n, 2), & \text{if } p > 2 \text{ and } n \text{ is even} \end{cases}$$

where the differential is given by:

$$d\big(\gamma_{2^t}(y_k)\big) = \gamma_{2^k}(x)\gamma_{2^t-1}(y_k) \qquad \text{if } p = 2;$$
$$d\big(\gamma_{p^t}(y)\big) = x\gamma_{p^t-1}(y) \qquad \text{if } p > 2 \text{ and } n \text{ is odd;}$$
$$d(y_k) = \gamma_{p^k}(x) \quad \text{and}$$
$$d\big(\gamma_{p^t}(z_k)\big) = \gamma_{(p-1)p^{k-1}}(x)y_{k-1}\gamma_{p^t-1}(z_k) \qquad \text{if } p > 2 \text{ and } n \text{ is even.}$$

In all cases set $dx = 0$ which makes the inclusion $A(x) \to P(x)$ a chain map. Direct calculation shows that $d^2 = 0$ and that $P_*(x)$ is acyclic with this differential, so that (noting that d has degree -1 in the second gradation) $\cdots \to \oplus_m P_{m,q}(x) \to \oplus_m P_{m,q-1}(x) \to \dots$ becomes a projective $A(x)$-module resolution of $\mathbb{Z}/p\mathbb{Z}$. Therefore

$$C(x) = \mathrm{Tor}^{A(x)}(\mathbb{Z}/p\mathbb{Z}, \mathbb{Z}/p\mathbb{Z}) = H_*\big(P_*(x) \otimes_{A(x)} \mathbb{Z}/p\mathbb{Z}\big)$$

$$= \begin{cases} \displaystyle\bigotimes_{k=0}^{\infty} \Gamma[y_k] & \text{if } p = 2; \\[2em] \Gamma[y] & \text{if } p > 2 \text{ and } |x| \text{ is odd;} \\[2em] \displaystyle\bigotimes_{k=0}^{\infty} E[y_k] \otimes \bigotimes_{k=1}^{\infty} \Gamma[z_k] & \text{if } p > 2 \text{ and } |x| \text{ is even.} \end{cases}$$

With our trivial differentials on $A(x)$ and $C(x)$, the hypotheses of the algebraic Serre spectral sequence hold. In this Serre spectral sequence, $E^2 = H\big(A(x)\big) \otimes H\big(C(x)\big) = A(x) \otimes C(x) \cong P(x)$ although the bigrading of $A(x) \otimes C(x)$ in the spectral sequence is different from that in $P(x)$. Direct calculation shows that the differentials in the spectral sequence are the same as those in $P(x)$, although for each $w \in A(x) \otimes C(x)$, the r for which $d^r(w) = d(w)$ will vary with w. $\qquad\square$

The algebraic Serre spectral sequence for $P = A \otimes C$ is obtained by tensoring the corresponding spectral sequences over $x \in X$. The spectral sequence converges to $H_k(P) = H_k(*; \mathbb{Z}/p\mathbb{Z})$.

There is a dual, "cohomological" spectral sequence with $E_2 = S[X] \otimes C^*$, converging to $H^*(*; \mathbb{Z}/p\mathbb{Z})$, in which $d_{|x^{p^k}|+1} x^{p^k} = y_k$ and, in the case $p > 2$, $|x|$ even, then additionally $d_{|x^{(p-1)p^{k-1}}|+1} x^{(p-1)p^{k-1}} y_{k-1} = z_k$. Notice that although this spectral sequence is multiplicative it is not possible, for example, to compute $d_{|x^p|+1} x^p$ from $d_*(x)$ since x has already disappeared from the spectral sequence before reaching $E_{|x^p|+1}$. In fact, the elements which become indecomposable at some stage of the spectral sequence are $\{x^{p^t}, y_t\}_{t\geq 0}$ plus, in the case $p > 2$, $|x|$ even, then additionally the elements $\{x^{(p-1)p^t} y_t\}_{t\geq 0}$.

Notational Remark Rational homotopy theorists tend to use $\Lambda(V)$ to denote the free commutative algebra on the vector space V. Other topologists and algebraists usually use $\Lambda(V)$ to denote the exterior algebra on the set V.

Theorem 11.9.13 (Comparison Theorem for Serre-type Spectral Sequences) *Let (E) and $('E)$ be first quadrant spectral sequences such that $E^2_{p,q} = E^2_{p,0} \otimes E^2_{0,q}$ and $'E^2_{p,q} = {}'E^2_{p,0} \otimes {}'E^2_{0,q}$. Let $f : E \to {}'E$ be a homomorphism of spectral sequences such that $f^2_{p,q} = f^2_{p,0} \otimes f^2_{0,q}$. Suppose that $f^\infty_{p,q} : E^\infty_{p,q} \to {}'E^\infty_{p,q}$ is an isomorphism for all p and q. Then the following are equivalent:*

1) $f^2_{p,0} : E^2_{p,0} \to {}'E^2_{p,0}$ is an isomorphism for $p \leq n-1$;

2) $f^2_{0,q} : E^2_{0,q} \to {}'E^2_{0,q}$ is an isomorphism for $q \leq n$.

Proof Suppose that $f_{p,0}^2$ is an isomorphism for $p \leq n-1$. Assume by induction on m that $f_{0,q}^2$ is an isomorphism for $q \leq m$. Then $f_{p,q}^2$ is an isomorphism for p, q such that $p \leq n-1$ and $q \leq m$. For all $p \leq n-1$ suppose also by a subsidiary induction on r that $f_{p,q}^r$ is a monomorphism for $q \leq m$ and an isomorphism for $q \leq m-r+2$. This condition holds for $r = 2$. It follows that for $p \leq n-1$, $f_{p,q}^r : Z_{p,q}^r \to {'Z_{p,q}^r}$ is a monomorphism for $q \leq m$ and an isomorphism for $q \leq m-r+1$, and that $f_{p,q}^r : B_{p,q}^r \to {'B_{p,q}^r}$ is an epimorphism for $q \leq m$, where $Z_{s,t}^r$ and $B_{s,t}^r$ denote respectively the kernel of $d^r : E_{p,q}^r \to E_{p-r,q+r-1}^r$ and the image of $d^r : E_{p+r,q-r+1}^r \to E_{p,q}^r$. Applying the Snake Lemma to the diagram

$$
\begin{array}{ccccccccc}
0 & \longrightarrow & B_{p,q}^r & \longrightarrow & Z_{p,q}^r & \overset{d^r}{\longrightarrow} & E_{p,q}^{r+1} & \longrightarrow & 0 \\
 & & \downarrow & & \downarrow & & \downarrow {\scriptstyle f_{p,q}^{r+1}} & & \qquad (*)\\
0 & \longrightarrow & {'B_{p,q}^r} & \longrightarrow & {'Z_{p,q}^r} & \overset{d^r}{\longrightarrow} & {'E_{p,q}^{r+1}} & \longrightarrow & 0
\end{array}
$$

shows that $f_{p,q}^{r+1}$ is a monomorphism for $q \leq m$ and an isomorphism for $q \leq m-(r+1)+2$ (for $p \leq n-1$) completing the subsidiary induction.

Assume that $m+1 \leq n$. Suppose now by downward induction on r that $f_{0,m+1}^{r+1}$ is an isomorphism when $2 \leq r \leq m+1$, observing to start the induction that $f_{0,m+1}^{m+1}$ is the isomorphism

$$
E_{0,m+1}^{m+1} \cong E_{0,m+1}^{\infty} \cong {'E_{0,m+1}^{\infty}} \cong {'E_{0,m+1}^{m+1}}.
$$

Completing this induction will show that $f_{0,m+1}^2$ is an isomorphism, thus also completing the induction on m.

Since $r \leq m \leq n-1$, in the diagram

$$
\begin{array}{ccccccccccc}
0 & \longrightarrow & Z_{r,m-r+2}^r & \longrightarrow & E_{r,m-r+2}^r & \overset{d^r}{\longrightarrow} & E_{0,m+1}^r & \longrightarrow & E_{0,m+1}^{r+1} & \longrightarrow & 0 \\
 & & \downarrow & & \downarrow & & \downarrow & & \downarrow & & \\
0 & \longrightarrow & {'Z_{r,m-r+2}^r} & \longrightarrow & {'E_{r,m-r+2}^r} & \overset{'d^r}{\longrightarrow} & {'E_{0,m+1}^r} & \longrightarrow & {'E_{0,m+1}^{r+1}} & \longrightarrow & 0
\end{array}
$$

the second vertical map was shown earlier to be an isomorphism and the fourth is an isomorphism by the induction hypothesis. Therefore to complete the downward induction on r it suffices to show that $Z_{p,m-p+2}^p \to {'Z_{p,m-p+2}^p}$ is an epimorphism for $2 \leq p \leq m$. To do this, we will show by downward induction on k that $Z_{p,m-p+2}^k \to {'Z_{p,m-p+2}^k}$ is an epimorphism for all $k \geq p$. Certainly this holds for large k since for large k,

$$
Z_{p,m-p+2}^k \cong E_{p,m-p+2}^k \cong E_{p,m-p+2}^{\infty} \cong {'E_{p,m-p+2}^{\infty}} \cong {'E_{p,m-p+2}^k} \cong {'Z_{p,m-p+2}^k}.
$$

Since $m - p + 2 \leq m$, $B_{p,m-p+2}^p \to {'B_{p,m-p+2}^p}$ is an epimorphism as observed earlier. Therefore considering diagram $(*)$ with $r = k$ and $q = m-p+2$ shows that it is suffices to check that $f_{p,m-p+2}^{k+1} : E_{p,m-p+2}^{k+1} \to {'E_{p,m-p+2}^{k+1}}$ is an epimorphism. However $k \geq p$ implies that $E_{p,m-p+2}^{k+1} \cong Z_{p,m-p+2}^{k+1}$ and ${'E_{p,m-p+2}^{k+1}} \cong {'Z_{p,m-p+2}^{k+1}}$ so this follows from the induction hypothesis.

The proof of the converse is similar. $\qquad\square$

Theorem 11.9.14 (Borel) *Let p be a prime. Let G be a topological group such that $H^*(G; \mathbb{Z}/p\mathbb{Z})$ is a free commutative algebra $S[X]$ of finite type over $\mathbb{Z}/p\mathbb{Z}$. Suppose that x^{p^k} is transgressive for all $k \geq 0$ and all $x \in X$. Then*

1) *If $p = 2$, $H^*(BG; \mathbb{Z}/2\mathbb{Z}) = (\mathbb{Z}/2\mathbb{Z})[\{\tau(x^{2^k})\}_{x \in X}]$.*
2) *If $p > 2$, $H^*(BG; \mathbb{Z}/p\mathbb{Z}) = S[\{\tau(x^{p^k})\}_{x \in X} \cup \{\beta(\tau(x^{p^k}))\}_{x \in X_e}]$ where β is the Bockstein and X_e is the subset of even degree elements of X.*

Notational Remark Recall that the notation $S[Y]$ stands for the free commutative algebra on the vector space generated by Y and so it is not affected by the presence of the element 0 in Y. In particular, $x^{p^k} = 0$ for $|x|$ odd, $k > 0$, $p > 2$ and so does not contribute to the answer.

Proof Let $('E_*)$ be the (cohomological) algebraic Serre spectral sequence of Example 11.9.12 for $S[X]$ and let (E_*) denote the mod p (cohomological) Serre spectral sequence of the fibration $G \to EG \to BG$. We use the notation of Example 11.9.12. In particular $C^* = S[Y]$, where $Y = \bigcup_{x \in X} Y(x)$ and $Y(x)$ is the set described in Example 11.9.12 whose nature depends on the parity of p and x. Define $f_2^{p,q} : {}'E_2^{p,q} \to E_2^{p,q}$ by $f_2^{p,q} = f_2^{p,0} \otimes f_2^{0,q}$ where $f_2^{0,*}$ is the isomorphism ${}'E_2^{0,*} = S[X] \cong H^*(G; \mathbb{Z}/p\mathbb{Z}) = E_2^{0,*}$ and $f_2^{*,0} : {}'E^{*,0} = C^* = S[Y] \to H^*(BG; \mathbb{Z}/p\mathbb{Z}) = E^{*,0}$ is the map defined on generators by $f_2^{*,0}(y_k) = \tau(x^{p^k})$ for $y_k \in Y(x)$, and $f_2^{*,0}(z_k) = \beta(\tau(x^{p^k}))$ in the case where $z_k \in Y(x)$ exists. Then $d_2 f_2 = f_2' d_2$ and so f_2 induces a map $f_3 : {}'E_2 \to E_2$. Suppose that f_r induced by f_2 has been defined. If x^{p^t} is a generator of ${}'E_r$, then

$$
{}'d_r\left(x^{p^t}\right) = \begin{cases} 0 & \text{if } r \leq |x^{p^t}|; \\ y_t & \text{if } r = |x^{p^t}| + 1. \end{cases}
$$

The fact that x^{p^t} is transgressive in (E_*) says that

$$
d_r\left(x^{p^t}\right) = 0 \text{ if } r \leq |x^{p^t}| \qquad \text{while} \qquad d_{|x^{p^t}|+1}\left(x^{p^t}\right) = \tau\left(x^{p^t}\right)
$$

and so in either case $f_r' d_r\left(x^{p^t}\right) = d_r f_r\left(x^{p^t}\right)$ since f_r is induced by f_2. In the case where $p > 2$ and $|x|$ is odd, checking that the differentials are given on generators

$$
d_r\left(x^{(p-1)p^{t-1}}y_{t-1}\right) = \begin{cases} 0 & \text{if } r \leq |x^{(p-1)p^{t-1}}y_{t-1}|; \\ y_t & \text{if } r = |x^{(p-1)p^{t-1}}y_{t-1}| + 1 \end{cases}
$$

is messy by this approach and is most easily handled by means of the Eilenberg-Moore spectral sequence (Theorem 11.11.3) where one is in a position to take advantage of the fact that the Bockstein commutes with maps of spaces. See the discussion following Theorem 11.11.3 for plugging this gap in this proof. Granting this we have that $f_r' d_r = d_r f_r$ so it follows that f_r induces f_{r+1}, and continuing inductively a map $f : ('E) \to (E)$ of spectral sequences is produced. Both spectral sequences converge to the cohomology of a point and $f_2^{0,*}$ is an isomorphism. Therefore Theorem 11.9.13 implies that $f_2^{*,0}$ is an isomorphism as claimed. $\qquad \square$

10 Adams-Hilton Models

Let R be a commutative ring. Although it will not appear in the notation, throughout this section the coefficient ring will be assumed to be R. We begin with a brief introduction to the homotopy theory of nonnegatively graded differential algebras.

Given a differential graded algebra A, a morphism $\theta : TV \to A$ which induces an isomorphism on homology is called a *free model* for A. According to the following lemma, every nonnegatively graded differential algebra has a free model.

Lemma 11.10.1 *Let (A, d) be a nonnegatively graded differential algebra. Then there exists a differential graded tensor algebra $(TV, \tilde{d})$ together with a homomorphism $\theta : (TV, \tilde{d}) \to (A, d)$ of differential graded algebras such that θ induces an isomorphism on homology.*

Proof Let $V(-1)$ be a generating set for the cycles of A and make $T\big(V(-1)\big)$ into a differential graded algebra by setting $\tilde{d}v = 0$ for $v \in V(-1)$. The inclusion $V(-1) \hookrightarrow A$ induces a homomorphism of DGA's $\theta(-1) : T\big(V(-1)\big) \to A$ which is surjective on homology. Let $V(0)$ be the graded set obtained by adjoining to $V(-1)$ a generator v_w in degree $|w| + 1$ for each element w of a set of cycles whose homology classes form a generating set for $\mathrm{Ker}\,\big(\theta(-1)_*\big)$. Extend the DGA structure on $T\big(V(-1)\big)$ to $T\big(V(0)\big)$ by setting $\tilde{d}(v_w) = w$. Extend $\theta(-1)$ to a DGA homomorphism $\theta(0) : T\big(V(0)\big) \to A$ by setting $\theta(0)(v_w) = \hat{w}$ where $\hat{w} \in A$ is any element satisfying $d\hat{w} = w$. Then the map induced on homology by $\theta(0)$ is surjective in all degrees and is an isomorphism in degree 0. Iterating this procedure gives, for each integer n, a set $V(n)$ together with a DGA structure on $T\big(V(n)\big)$ and a DGA homomorphism $\theta(n) : T\big(V(n)\big) \to A$ which induces a surjection in homology in all degrees and is an isomorphism in degrees less than $n + 1$. Let $V = \bigcup_{n=-1}^{\infty} V(n)$. Then the map from the colimit $\theta : TV \to A$ induces an isomorphism on homology. $\square$

The same method yields the following generalization.

Lemma 11.10.2 *Let $f : (A', d') \to (A, d)$ be a homomorphism of nonnegatively graded differential algebras. Then there exist a graded set V, a differential $\tilde{d}$ on $A' \amalg TV$ extending d', together with a homomorphism $\theta : (A' \amalg TV, \tilde{d}) \to (A, d)$ of differential graded algebras extending f such that θ induces an isomorphism on homology.* $\square$

The following lemma is also proved by induction.

Lemma 11.10.3 *Let $(TV, \tilde{d})$, (A, d), and (B, e) be nonnegatively graded differential algebras. Let $\theta : \big(TV, \tilde{d}\big) \to (B, e)$ be a morphism of DGA's and let $f : (A, d) \to (B, e)$ be a morphism of DGA's which induces an isomorphism on homology. Then θ "lifts" to a DGA homomorphism $\theta' : (TV, \tilde{d}) \to (A, d)$ satisfying $\theta = f\theta'$.* $\square$

Let (A, d) be a nonnegatively graded differential algebra. A nonnegatively graded differential algebra (C, d_C) together with a DGA homomorphism $p : (C, d_c) \to (A, d)$ and an injective DGA homomorphism $i_1 \bot i_2 : (A, d) \amalg (A, d) \to (C, d_c)$ is called a *cylinder object* for (A, d) in the category of differential graded algebras if p induces an isomorphism on homology and $pi_1 = pi_2 = 1_A$. Applying Lemma 11.10.2 to $1_A \bot 1_A : A \amalg A \to A$ shows that every nonnegatively graded differential algebra has a cylinder object. Two homomorphisms $f, g : (A, d) \to (B, e)$ are called *homotopic* in the category of DGA's (written $f \simeq g$ as usual) if there exists a factorization $(A, d) \amalg (A, d) \xrightarrow{i_1 \bot i_2} (C, d_C) \to (B, e)$ of $f \bot g$ through some cylinder object (C, d_C).

Making use of Lemma 11.10.3 one can show

Proposition 11.10.4 *If f, $g : A \to B$ are homotopic DGA homomorphisms and $i : A' \to A$, $j : B \to B'$ are any DGA homomorphisms then $jfi \simeq jgi : A' \to B'$.* $\qquad\square$

Proposition 11.10.5 *DGA homotopy is an equivalence relation.*

Proof Let f, g, $h : A \to B$ be morphisms of nonnegatively graded differential algebras. Let $\phi_1 : C_1 \to B$ and $\phi_2 : C_2 \to B$ be homotopies $\phi_1 : f \simeq g$, $\phi_2 : g \simeq h$, where $A \coprod A \xrightarrow{i_1 \perp i_2} C_1 \xrightarrow{p_1} A$ and $A \coprod A \xrightarrow{j_1 \perp j_2} C_2 \xrightarrow{p_2} A$ are cylinder objects for A. Let C_3 be the pushout of $i_2 : A \to C_1$ and $j_1 : A \to C_2$, and let $k_1 : C_1 \to C_3$ and $k_2 : C_2 \to C_3$ be the induced maps. Then p_1 and p_2 induce a map $p_3 : C_3 \to A$ which is a homology isomorphism. It follows that $A \coprod A \xrightarrow{k_1 i_1 \perp k_2 j_2} C_3 \xrightarrow{p_3} A$ is a cylinder object for A. The maps $\phi_1 i_2$ and $\phi_2 j_1$ are both $g : A \to B$, and so there is an induced map $\phi_3 : C_3 \to B$ which is a homotopy from f to h. $\qquad\square$

Proposition 11.10.6 *Let f, $g : (TW, d) \to (B, e)$ be homomorphisms of DGA's. Then $f \simeq g$ if and only if there exists a degree 1 map $s : TW \to B$ satisfying $s(xy) = s(x)g(y) + (-1)^{|x|}f(x)s(y)$ and $es + sd = f - g$.*

This is proved by constructing a "universal cylinder" $\left(T(W \oplus W \oplus SW), \tilde{d}\right)$ for TW which comes with a degree 1 map $s : TW \to T(W \oplus W \oplus SW)$ which satisfies $s(xy) = s(x)i_2(y) + (-1)^{|x|}i_1(x)s(y)$ and $\tilde{d}s + sd = i_1 - i_2$, where i_1 and i_2 are the inclusions of TW into the first and second factors of $T(W \oplus W \oplus SW)$. (See [BL].) $\qquad\square$

It is immediate from Proposition 11.10.6 that if $f \simeq g$ as DGA homomorphisms then $f \simeq g$ as chain maps.

Remark The concept of DGA homotopy is intended as an algebraic analogue of the concept of a homotopy $H : G \times I \to K$ between continuous group homomorphisms of topological groups which has the property that H_t is a homomorphism for all t, however we shall not pursue the details of this analogy.

For a topological monoid M, the singular chain complex $S_*(M)$ forms a differential nonnegatively graded algebra under the Eilenberg-Zilber map EZ. In the special case where M is the Moore loops $\Omega'X$ on a pointed space X, we will use $\mathcal{A}(X)$ to indicate a free model for $S_*(\Omega'X)$. For $f : X \to Y$, given any free models $\theta_X : \mathcal{A}(X) \to S_*(\Omega'X)$, $\theta_Y : \mathcal{A}(Y) \to S_*(\Omega'Y)$ there exists a DGA homomorphism $\mathcal{A}(f)$ (unique up to DGA homotopy) such that

$$
\begin{array}{ccc}
\mathcal{A}(X) & \xrightarrow{\mathcal{A}(f)} & \mathcal{A}(Y) \\
\downarrow{\scriptstyle \theta_X} & & \downarrow{\scriptstyle \theta_Y} \\
S_*(\Omega'X) & \xrightarrow{(\Omega'f)_*} & S_*(\Omega'X)
\end{array}
$$

commutes up to DGA homotopy. Since (even given a specific choice of $\mathcal{A}(X)$ and $\mathcal{A}(Y)$) only the DGA homotopy class of $\mathcal{A}(f)$, not $\mathcal{A}(f)$ itself, is determined by f we refer to it as a free model for f. Given a composition $X \xrightarrow{f} Y \xrightarrow{g} Z$ and choices $\mathcal{A}(X)$, $\mathcal{A}(Y)$, $\mathcal{A}(Z)$, $\mathcal{A}(f)$, and $\mathcal{A}(g)$ then $\mathcal{A}(g) \circ \mathcal{A}(f)$ is a free model for $g \circ f$, but it is important to realize that there may be other choices for $\mathcal{A}(g \circ f)$ which do not factor as a composition $\mathcal{A}(g) \circ \mathcal{A}(f)$ for any choices of $\mathcal{A}(f)$ and $\mathcal{A}(g)$.

For a CW-complex X whose 1-skeleton is a single point, Adams and Hilton gave a construction of a free model $\mathcal{A}(X)$ whose algebra generators are in one-to-one correspondence with the cells of X. A free model satisfying the first two conditions of the following theorem will be called an *Adams-Hilton model* for X.

Remark Every simply connected space has a CW-approximation whose 1-skeleton is a single point.

Theorem 11.10.7 (Adams-Hilton) *Let X be a CW-complex such that the 1-skeleton of X is a single point. Then there exists a free model $\theta_X : \mathcal{A}(X) \to S_*(\Omega'X)$ for X such that:*

1) *$\mathcal{A}(X) = TV$ where V is a graded set with one element in degree $m - 1$ for each m-cell of X for each $m > 0$.*

2) *The suspension of the indecomposable quotient $SQ\mathcal{A}(X)$ (sometimes called the linearized chain complex of $\mathcal{A}(X)$) is isomorphic to the reduced cellular chain complex $\tilde{C}_*(X)$ of X.*

3) *$\mathrm{Im}\,\theta_X$ is contained in the 0th Eilenberg subcomplex $S_*^{(0,*)}(\Omega'X)$ for X (Section 7.4).*

4) *If $f \simeq g : X \to Y$ then $f_* \simeq g_* : \mathcal{A}(X) \to \mathcal{A}(Y)$ as DGA homomorphisms.*

5) *$1_{\mathcal{A}(X)}$ is a possible choice for $\mathcal{A}(1_X)$.*

6) *Given a composition $X \xrightarrow{f} Y \xrightarrow{g} Z$ and choices $\mathcal{A}(f)$ and $\mathcal{A}(g)$ then $\mathcal{A}(g) \circ \mathcal{A}(f)$ is a possible choice for $\mathcal{A}(g \circ f)$.*

7) *Let W be a subcomplex of X. Then any Adams-Hilton model $\theta_W : \mathcal{A}(W) \to S_*(\Omega'W)$ has an extension to an Adams-Hilton model $\theta_X : \mathcal{A}(X) \to S_*(\Omega'X)$, and given a map $f : X \to Y$ any choice of $\mathcal{A}(f|_W)$ has an extension to a possible choice for $\mathcal{A}(f)$. Furthermore, if $X = \bigcup_{\alpha \in J} X_\alpha$ and coherent Adams-Hilton models $\mathcal{A}(X_\alpha)$ have been chosen then $\mathrm{colim}_{\alpha \in J} \mathcal{A}(X_\alpha)$ is an Adams-Hilton model for X. Similarly given $f : X \to Y$ and a set of coherent choices $\mathcal{A}(f|_{X_\alpha})$, then $\mathrm{colim}_{\alpha \in J} \mathcal{A}(f|_{X_\alpha})$ is a possible choice for $\mathcal{A}(f)$.*

8) *The differential graded algebra $\mathcal{A}(X)$ satisfying the preceding conditions is uniquely determined up to isomorphism of DGA's and the map θ_X is uniquely determined up to DGA homotopy.*

Outline of Proof We shall concentrate our attention on constructing $\mathcal{A}(X)$ such that the first property is satisfied, the others being fairly straightforward once one has the construction.

Suppose by induction that $\mathcal{A}(X^{(n)})$ and $\theta_{X^{(n)}} : \mathcal{A}(X^{(n)}) \to S_* (\Omega'X^{(n)})$ have been constructed. Property (1) describes the algebra $\mathcal{A}(X^{(n+1)})$ as TV where

$$V = \{\text{cells } v \text{ of } X^{(n+1)} \mid |v| > 0\}$$

with degrees shifted downward by 1; we take the differential on the subalgebra $\mathcal{A}(X^{(n)})$ of $\mathcal{A}(X^{(n+1)})$ to be the same as that on $\mathcal{A}(X^{(n)})$. We must define the differential on the algebra generators of $\mathcal{A}(X^{(n+1)})$ corresponding to $(n+1)$-cells in X.

Pick a cycle $\iota_{n-1} \in S_{n-1}(\Omega'S^n)$ such as the homology class $[\iota_{n-1}]$ is the generator of $\tilde{H}_{n-1}(\Omega'S^n)$ corresponding under the isomorphism $\tilde{H}_{n-1}(S^{n-1}) \cong \tilde{H}_{n-1}(\Omega'S^n)$ to the generator which was denoted by ι_{n-1} in Chapter 7.4. We may choose ι_{n-1} to be in the 0th Eilenberg subcomplex $S_{n-1}^{(0,*)}(\Omega'S^n)$ for $\Omega'S^n$. Let i denote the inclusion $i : S^n \hookrightarrow D^{n+1}$ and let j denote the inclusion $j : X^{(n)} \hookrightarrow X^{(n+1)}$. Since $\Omega'D^{n+1}$ is contractible, there exists $\eta \in S_n(\Omega'D^{n+1})$ such that $i_*(\iota_{n-1}) = d\eta$. Again, η can be chosen to be in the 0th Eilenberg subcomplex.

Let $f_v : (D^{n+1}, S^n) \to (X^{(n+1)}, X^{(n)})$ be the characteristic map of an $(n+1)$-cell v in X. Pick a cycle $z_v \in \mathcal{A}(X^{(n)}) \subset \mathcal{A}(X^{(n+1)})$ representing the homology class $(\theta_{X^{(n)}})_*^{-1}((\Omega'f_v)_*([\iota_{n-1}]))$ and set $dv = z_v$.

Since $[\theta_{X^{(n)}}(z_v)] = [(f_v)_*i_*(\iota_{n-1})]$ in $H_n(\Omega'X^{(n)})$ there exists $w_v \in S_n(\Omega'X^{(n)})$ such that $dw_v = \theta_{X^{(n)}}(z_v) - (f_v)_*i_*(\iota_{n-1})$. Extend $\theta_{X^{(n)}}$ to v by setting $\theta_{X^{(n+1)}}(v) = j_*w_v + (f_v)_*(\eta) \in S_n(\Omega'X^{(n+1)})$. Notice that

$$d\theta_{X^{(n+1)}}(v) = j_*(\theta_{X^{(n)}}(z_v) - (f_v)_*i_*(\iota_{n-1})) + (f_v)_*i_*(\iota_{n-1})$$

$$= j_*\theta_{X^{(n)}}(dv) = \theta_{X^{(n+1)}}(dv).$$

Therefore $\theta_{X^{(n+1)}}$ is a chain map.

We wish to show that $\theta_{X^{(n+1)}} : \mathcal{A}(X^{(n+1)}) \to S_*(\Omega'X^{(n+1)})$ induces an isomorphism on homology. By Theorem 8.5.1 it suffices to show that the composite

$$\mathcal{A}(X^{(n+1)}) \xrightarrow{\theta_{X^{(n+1)}}} S_*(\Omega'X^{(n+1)}) \twoheadrightarrow S'_*(\Omega'X^{(n+1)})$$

induces a homology isomorphism, where $S'_*(Y) = S_*(Y)/DS_*(Y)$ is the singular chain complex of Y modulo the degenerate subcomplex defined in Section 8.5.

Write simply A for $\mathcal{A}(X^{(n+1)})$ and let C be the free graded R-module which has one basis element in degree m for each m-cell of $X^{(n+1)}$. Denote the basis element of C corresponding to $v \in V_{m-1}$ as $u_v \in C_m$ and let 1 denote the basis element of C_0. Regard $A \otimes C$ as a left A-module by letting A act on the left on the first factor. Let $\epsilon : A \otimes C \to R$ be the augmentation. Define self-maps $s : A \otimes C \to A \otimes C$ of degree 1 and $d : A \otimes C \to A \otimes C$ of degree -1 as follows. Set $s(1 \otimes c) = 0$, $s(v \otimes 1) = u_v$ and inductively over tensor length define $s(ax) = \epsilon(x)s(a) + (-1)^{|a|}as(x)$ for $a \in A$ and $x \in A \otimes C$. On $A = A \otimes R \subset A \otimes C$, define d to be the differential on A. Extend d to $A \otimes C$ by setting $d(1 \otimes u_v) = (1-sd)(v \otimes 1)$ and $d(a \otimes c) = da \otimes c + (-1)^{|a|}ad(1 \otimes c)$. Direct computation shows that $ds + sd = 1 - \epsilon$ and that $d^2 = 0$. Therefore $A \otimes C$ forms an acyclic chain complex which satisfies the conditions of Theorem 11.9.11.

Let

$$k : (\Omega'D^{n+1}, \Omega'S^n) \hookrightarrow (P'D^{n+1}, P'S^n)$$

and

$$\tilde{k} : (\Omega'X^{(n+1)}, \Omega'X^{(n)}) \hookrightarrow (P'X^{(n+1)}, P'X^{(n)})$$

denote the inclusion of the Moore loops into the Moore path space and let

$$p : (P'D^{n+1}, P'S^n) \to (D^{n+1}, S^n)$$

be the evaluation map. We now extend $\tilde{k}_*\theta_{X^{(n+1)}}$ to a map $\tilde{\theta} : A \otimes C \to S_*(P'X^{(n+1)})$ of differential A-modules.

Since $P'S^n$ is contractible, there exists $\xi \in S_n(P'S^n)$ such that $k_*(\iota_{n-1}) = d\xi$. In $S_n(P'D^{n+1})$, since $d((P'i)_*\xi - k_*\eta) = (P'i)_*k_*\iota_{n-1} - i_*k_*\iota_{n-1} = 0$, there exists $\gamma \in S_{n+1}(P'D^{n+1})$ such that $d\gamma = (P'i)_*\xi - k_*\eta$. Given an $(n+1)$-cell $v \in V$

choose $z_v \in (TV)_{n-1}$, then

$$d\big((P'f_v)_*(P'i)_*\xi - \tilde{k}_*(\theta_{X^{(n+1)}}sz_v - j_*w_v)\big)$$

$$= (P'f_v)_*k_*\iota_{n-1} - \tilde{k}_*\theta_{X^{(n+1)}}dsz_v + \tilde{k}_*j_*\big(\theta_{X^{(n)}}(z_v) - (f_v)_*i_*(\iota_{n-1})\big)$$

$$= -\tilde{k}_*\theta_{X^{(n+1)}}(1 - \epsilon - sd)z_v + \tilde{k}_*\theta_{X^{(n+1)}}z_v$$

$$= 0$$

in $S_{n-1}\big(P'X^{(n+1)}\big)$. Therefore there exists $y_v \in S_{n+1}\big(P'X^{(n+1)}\big)$ such that

$$dy_v = (P'f_v)_*(P'i)_*\xi - \tilde{k}_*(\theta_{X^{(n+1)}}sz_v - j_*w_v).$$

Set $\tilde{\theta}(1 \otimes u_v) = y_v - (P'f_v)_*(\gamma)$ and extend to an A-module map. Then

$$\tilde{\theta}d(1 \otimes u_v) = \tilde{\theta}(1 - \epsilon - sd)v = \tilde{k}_*\big(j_*w_v + (f_v)_*(\eta)\big) - \tilde{\theta}sz_v$$

and so

$$d\tilde{\theta}(1 \otimes u_v) = d\big(y_v - (P'f_v)_*(\gamma)\big)$$

$$= (P'f_v)_*(P'i)_*\xi - \tilde{k}_*(\theta_{X^{(n+1)}}sz_v - j_*w_v) - (P'f_v)_*\big((P'i)_*\xi - k_*\eta\big)$$

$$= \tilde{\theta}d(1 \otimes u_v).$$

Therefore $\tilde{\theta}$ is a morphism of differential A-modules.

As in Theorem 11.9.11, filter C by skeletons and $A \otimes C$ by inverse images of skeletons. Give $S_*\big(X^{(n+1)}\big)$ and $S_*\big(P'X^{(n+1)}\big)$ the Serre filtration coming from the CW-structure on $X^{(n+1)}$ and give $S'_*\big(X^{(n+1)}\big)$ and $S'_*\big(P'X^{(n+1)}\big)$ the induced filtrations. The spectral sequences associated to the filtrations on $S_*\big(P'X^{(n+1)}\big)$ and $S'_*\big(P'X^{(n+1)}\big)$ are each the Serre spectral sequence for the fibration $\Omega'X^{(n+1)}$ $\xrightarrow{i} P'X^{(n+1)} \xrightarrow{p} X^{(n+1)}$. The composite

$$A \otimes C \xrightarrow{\tilde{\theta}} S_*\big(P'X^{(n+1)}\big) \twoheadrightarrow S'_*\big(P'X^{(n+1)}\big)$$

is filtration preserving. (This depends on the fact that $\tilde{S}'_*(*) = 0$ and therefore $pi = *$ implies that $p_*i_* = 0$ at the chain level when using $S'_*(\ )$, statements which would not be true for $S_*(\ )$.) Therefore there is an induced map from the algebraic Serre spectral sequence of $A \otimes C$ to the Serre spectral sequence of $\Omega'X^{(n+1)} \xrightarrow{i} P'X^{(n+1)} \xrightarrow{p} X^{(n+1)}$. The map induced on $E^2_{0,*}$-terms is an isomorphism $H_*(C) \cong H_*\big(X^{(n+1)}\big)$ and the map induced on E^∞ is also an isomorphism, both sides being acyclic. Therefore Theorem 11.9.13 implies that $\theta_{X^{(n+1)}} : A\big(X^{(n+1)}\big) \to H_*\big(\Omega'X^{(n+1)}\big)$ is an isomorphism, completing the induction. Taking the colimit of the models $A\big(X^{(n)}\big)$ gives a free model $\mathcal{A}(X)$ with the desired properties. $\qquad\square$

11 Eilenberg-Moore Spectral Sequence

Let C be a differential graded R-coalgebra. Let $\mathcal{C}$ be the category of differential graded C-comodules and let CC be the category of chain complexes over R. There is a pair of adjoint functors $F \dashv G$ where $F : \mathcal{C} \to CC$ is the forgetful functor, and $G : CC \to \mathcal{C}$ is given by $G(X) = C \otimes X$. The objects $C \otimes X$ are called *extended comodules.* Let $\mathcal{I}$ be the morphisms of CC which are split monomorphisms. Let $\mathcal{J} = F^{-1}(\mathcal{I})$. Then $\mathcal{J}$ consists of the morphisms of $\mathcal{C}$ which are split monomorphisms as chain maps over R (although possibly not split as morphisms of C-comodules). Since $\mathcal{I}$ forms a closed class of monomorphisms in CC, by the dual of Proposition 4.6.1 $\mathcal{J}$ forms an injective class of monomorphisms in $\mathcal{C}$. Since every object of CC is injective with respect to $\mathcal{I}$, the injectives with respect to $\mathcal{J}$ are the C-comodules which are summands of $C \otimes X$ for some X. Note that in the special case where R is a field and C has trivial differential, $\mathcal{I}$ consists of all monomorphisms of CC and $\mathcal{J}$ becomes all monomorphisms of $\mathcal{C}$. Thus the distinction between "real" injectives and relative injectives disappears in this special case.

Let C be a graded differential R-coalgebra and let M be a differential graded C-comodule. The relative derived functors $R^{\mathcal{J}}(M \square_C \_)$ will be denoted $\mathrm{CoTor}^{C,\mathcal{J}}(M, \_)$, however we will abuse notation and write simply $\mathrm{CoTor}^C(M, \_)$ for $\mathrm{CoTor}^{C,\mathcal{J}}(M, \_)$, even though this notation really ought to be reserved for the absolute derived functor. In the case where the differential on C is 0 and M is a flat R-module, the distinction between the relative and absolute derived functor disappears. For each p, $\mathrm{CoTor}^C_p(M, N)$ is a graded R-module and it is customary to write $\mathrm{CoTor}^C_{pq}(M, N)$ for $\big(\mathrm{CoTor}^C_p(M, N)\big)_q$.

Theorem 11.11.1 (Eilenberg-Moore Spectral Sequence) *Let*

$$\begin{array}{ccc} W & \longrightarrow & Y \\ \downarrow & & \downarrow \pi \\ X & \stackrel{f}{\longrightarrow} & B \end{array}$$

be a pullback square in which π is a fibration and X and B are simply connected. Suppose that $H_*(X)$, $H_*(Y)$, and $H_*(B)$ are flat R-modules, where $H_*(\ )$ denotes homology with coefficients in the commutative ring R. Then there is a (second quadrant) spectral sequence with $E^2_{p,q} = \mathrm{CoTor}^{H_*(B)}_{pq}\big(H_*(X), H_*(Y)\big)$ converging to $H_{p+q}(W)$.

Remark Although we shall use relative derived functors in the proof, notice that $H_*(B)$ has zero differential, so under our flatness hypotheses the CoTor appearing in theorem is absolute, not relative.

Proof We may suppose that X and B are CW-complexes and that f is a cellular map. Let F be the fibre of π. For short we write $L = S_*(X; R)$, $M = S_*(Y; R)$, $C = S_*(B; R)$, and $V = S_*(W; R)$. As in the Serre spectral sequences, filter the base spaces X and B by skeletons, and filter the total spaces W and Y by inverse images of skeletons, resulting in filtrations of each of the chain complexes. If we define a comultiplication on C by means of the Alexander-Whitney map then C is a differential coalgebra and L, M are differential C-comodules. Let $I_0(M) \to I_{-1}(M) \to I_{-2}(M) \to \ldots \to I_p(M) \to \ldots$ be the canonical relative injective C-comodule resolution of M, relative to the class of extended C-comodules. (See Section 4.6.) Let T be the second quadrant double complex with $T_{p,q} = \big(L \square_C I_p(M)\big)_q$. Since C is flat over R, $I_p(F_n M)$ forms a filtration of $I_p(M)$ for each p. This yields a filtration $F_n T_{p,q} = \bigoplus_{i+j} \big(F_i L \square_C I_p(F_j M)\big)_q$ on each $T_{p,q}$. This gives a third filtration on $\mathrm{Tot}^\pi T$ in addition to the horizontal and vertical filtrations. Since T is second quadrant, the spectral sequence in which $E^2_{p,q} = H_p H_q(T_{**})$ converges to $H_{p+q}(\mathrm{Tot}^\pi T)$. In this spectral sequence using our flatness hypotheses and the Künneth formula yields

$$E^2_{p,q} = H_p\Big(H_*(X)\,\square_{H_*(B)}\, I_p\big(H_*(Y)\big)\Big)_q = \mathrm{CoTor}^{H_*(B)}_{pq}\big(H_*(X), H_*(Y)\big).$$

Since B is simply connected, $I_p\big(H_*(Y)\big)$ is p-connected so

$$\mathrm{CoTor}^{H_*(B)}_{pq}\big(H_*(X), H_*(Y)\big) = 0 \quad \text{if } p + q < 0.$$

Therefore $H_n(\mathrm{Tot}^\pi T) = 0$ for $n < 0$. To finish the proof, we now use the third spectral sequence to show that $H_*(\mathrm{Tot}^\pi T) = H_*(W)$.

We will write $('E^r)$ for the third spectral sequence and also write $\big('E^r(L)\big)$ and $\big('E^r(M)\big)$ for the spectral sequences coming from the filtrations on L and M which were used to produce the filtration on $\mathrm{Tot}^\pi T$. The latter two are the Serre spectral sequences for the fibrations $1_X : X \to X$ and $\pi : Y \to B$ respectively. From the definition of the filtration, for each p and q

$$H_q\big(F_n(T_{p,*}), F_{n-1}(T_{p,*})\big) = \bigoplus_{i+j=n} \Big('E^1_i(L)\,\square_{E^1_*(B)}\, I_p\big('E^1_j(M)\big)\Big)_q$$

and so $'E^1_{*q} = (\mathrm{Tot}^\pi U_{p,q})_q$ where

$$U_{p,q} = \Big('E^1_*(L)\,\square_{E^1_*(B)}\, I_p\big('E^1_*(M)\big)\Big)_q.$$

This $'E^1$-term is itself the total complex of a (second quadrant) double complex, so we compute its homology, $'E^2$, by yet another spectral sequence. That is, we have a spectral sequence with $''E^2_{s,t} = H_s\big(H_t('E^1_{**})\big)$ converging to $'E^2_{**}$. In this spectral sequence

$$''E^2_{s,t} = H_s\Big('E^2_*(L)\,\square_{H_*(B)}\, I_s\big(E^2_*(M)\big)\Big)_t$$

$$= H_s\Big(H_*(X)\,\square_{H_*(B)}\, I_s\big(H_*(B)\otimes H_*(F)\big)\Big)_t$$

$$= \mathrm{CoTor}^{H_*(B)}_{st}\big(H_*(X), H_*(B)\otimes H_*(F)\big)$$

$$= \begin{cases} 0 & \text{if } s \neq 0; \\ \big(H_*(X)\otimes H_*(F)\big)_t & \text{if } s = 0. \end{cases}$$

Thus the spectral sequence $(''E^r)$ collapses at $''E^2$ and so we conclude that $'E^2 = H_*(X)\otimes H_*(F)$. In particular, we have just shown that $('E^r)$ is a first quadrant spectral sequence for $r \geq 2$, and so, combined with $H_n(\mathrm{Tot}^\pi T) = 0$ for $n < 0$, this guarantees that $'E^r$ converges to $H_*(\mathrm{Tot}^\pi T)$.

To show that $H_*(\mathrm{Tot}^\pi T) = H_*(W)$ we will use the spectral sequence comparison theorem. For this, consider the fibre square consisting of the identity map from $W \to X$ to itself. Let $\tilde{T}_{p,q} = \big(L\square_L I_p(V)\big)_q$ be the corresponding double complex, which by naturality comes with an induced map $f : \tilde{T} \to T$ which preserves all of the filtrations. The induced map $f_* : '\tilde{E}^2 \to 'E^2$ is the identity, so by the spectral sequence comparison theorem, $f_* : H_*(\mathrm{Tot}^\pi \tilde{T}) \to H_*(\mathrm{Tot}^\pi T)$ is an isomorphism. However it is clear that $H_*(\mathrm{Tot}^\pi \tilde{T}) \cong H_*(W)$ since in its first spectral sequence

$$\tilde{E}^2_{pq} = \mathrm{CoTor}^{H_*(X)}_{pq}\big(H_*(X), H_*(W)\big) = \begin{cases} 0 & \text{if } p \neq 0; \\ H_q(W) & \text{if } p = 0. \end{cases} \qquad \square$$

Looking back over the preceding proof we see that it is divided into two parts. The first part constructs a double complex T from $S_*(B)$, $S_*(X)$, $S_*(Y)$, and identifies the E^2-term in the spectral spectral sequence converging to $\mathrm{Tot}^{\pi} T$. The second part shows that $H_*(\mathrm{Tot}^{\pi} T) = H_*(W)$. The first part is a general construction with its associated spectral sequence and can be made for any pair of differential comodules over a differential coalgebra. The real content of the Eilenberg-Moore spectral sequence is the equality $H_*(\mathrm{Tot}^{\pi} T) = H_*(W)$, which shows how the differential graded algebra $S_*(B)$ together with its differential graded comodules $S_*(X)$, $S_*(Y)$ determine $H_*(W)$. In some contexts, this equality is more useful than the spectral sequence itself which has thrown away data (in that its input contains only $H_*(B)$, $H_*(X)$, $H_*(Y)$ rather than the chain complexes giving rise to these homologies) and does not, in general, retain sufficient information to determine $H_*(W)$.

By a similar method one can prove the following cohomological version of the spectral sequence.

Theorem 11.11.2 (Cohomological Eilenberg-Moore Spectral Sequence) *Let*

$$
\begin{array}{ccc}
W & \longrightarrow & Y \\
\downarrow & & \downarrow {\scriptstyle \pi} \\
X & \xrightarrow{\ f\ } & B
\end{array}
$$

be a pullback square in which π is a fibration and X and B are simply connected. Suppose that $H^*(X)$, $H^*(Y)$, and $H^*(B)$ are flat R-modules of finite type, where $H^*(\ )$ denotes cohomology with coefficients in the Noetherian ring R. Then there is a (second quadrant) spectral sequence with $E_2^{p,q} = \mathrm{Tor}_{pq}^{H^*(B)}\big(H^*(X), H^*(Y)\big)$ converging to $H^{p+q}(W)$.

Remark The Noetherian hypothesis is used to identify the "CoExt" which appears with Tor.

For a simply connected pointed space B, the preceding gives spectral sequences going from the homology and cohomology of B to that of ΩB. There are also versions of the Eilenberg-Moore spectral sequence which go from the homology and cohomology of ΩB to that of B.

Theorem 11.11.3 *Let B be a space such that $H_*(\Omega B)$ is a flat R-module, where $H_*(\ )$ denotes homology with coefficients in the commutative ring R. Then there is (first quadrant) spectral sequence with $E_{p,q}^2 = \mathrm{Tor}_{pq}^{H_*(\Omega B)}(R, R)$ converging to $H_{p+q}(B)$.*

Proof Applying the iterated Ganea construction (end of Section 9.2) to the fibration $\Omega B \to PB \to B$ gives a sequence of fibrations $F_n \xrightarrow{j_n} Y_n \longrightarrow B$ in which $F_n \approx \Omega B^{*(n+1)}$ and $Y_n \approx Y_{n-1}/F_{n-1}$. Since the connectivity of the map $Y_n \to B$ grows with n, B is homotopy equivalent to the direct limit (union) of the spaces obtained by successively turning each map $Y_{n-1} \to Y_n$ into a cofibration. (See the "infinite mapping telescope construction" in Section 13.1.) This filtration is cocomplete, yielding a first quadrant spectral sequence converging to $H_*(B)$. By construction, the homotopy cofibre of $Y_{n-1} \to Y_n$ is $SF_{n-1} \approx S(\Omega B)^{*n} \approx S^n(\Omega B)^{(n)}$ (where A^{*n} denotes the n-fold join of A and $A^{(n)}$ denotes the n-fold smash product). By Ganea's theorem, the composition $S(\Omega B)^{*n} \approx SF_{n-1} \xrightarrow{j_n}$

$Y_{n-1} \longrightarrow S^2(\Omega B)^{*(n-1)}$ (corresponding to $Y_n/Y_{n-1} \to SY_{n-1} \to SY_{n-1}/Y_{n-2}$) is induced by the multiplication of ΩB on the first two factors. Therefore $E_{p,q}^1 = \tilde{H}_{p+q}\big(S^p(\Omega B)^{(p)}\big) = \big(\tilde{H}_*(\Omega B)^{\otimes p}\big)_q$ and the d^1-differential is multiplication of the two leftmost factors. If $P_{*,*}$ is the canonical free resolution of R as a $H_*(B)$-module, then $P_{*,*} \otimes_{H_*(B)} R$ equals $E_{p,q}^1$ and the differential is the same. Therefore $E_{p,q}^2 = \mathrm{Tor}_{pq}^{H_*(\Omega B)}(R,R)$. $\qquad\square$

The same method also yields the cohomological version with $E_2^{p,q} = \mathrm{Ext}_{H^*(\Omega B)}^{pq}(R,R)$ converging to $H^{p+q}(B)$ provided $H^*(\Omega B)$ is flat and of finite type and R is Noetherian. The diagonal map $B \to B \times B$ turns this into a multiplicative spectral sequence.

Borel's Theorem (11.9.14) Revisited Under the conditions of Borel's Theorem, the E^2-term of the Eilenberg-Moore spectral sequence (11.11.3) is precisely the claimed value of $H^*(BG; \mathbb{Z}/p\mathbb{Z})$, so the theorem is equivalent to saying that the Eilenberg-Moore spectral sequence collapses in this case. The map $q_1 : Y_1 = SG \approx S\Omega BG \to BG$ induces the transgression, so the hypotheses concerning the transgression say that certain elements are in the image of q_1^*, and therefore they survive the spectral sequence; i.e., all differentials on them are 0. The gap in the proof of Theorem 11.9.14 when $p > 2$ and $|x|$ even is equivalent to the claim that certain elements are in the image of q_2^* where $q_2 : Y_2 \to B$. Specifically, to handle $x^{(p-1)p^{k-1}} y_{k-1}$ one needs that $q_2^*\big(\beta(\tau(x^{p^k}))\big) \neq 0$, or equivalently that $q_2^*\big(\tau(x^{p^k})\big)$ has order p in $H^*(Y_2; \mathbb{Z}_{(p)})$. Computing integrally in the sequence

$$P[x] \xrightarrow{\psi} P[x] \otimes P[x] \longrightarrow \dots$$

(whose reduction mod p gives the E_1-term of the Eilenberg-Moore spectral sequence) that order is found to be the coefficient of $x^{(p-1)p^{k-1}} \otimes x^{p^{k-1}}$ in $\psi(x^{p^k})$, which is $\binom{p^k}{p^{k-1}}$ and this is divisible by p but not by p^2.

Exercise 11.11.4 Redo the calculations of Example 11.9.7 and Exercises 11.9.8–10 using the Eilenberg-Moore spectral sequence instead of the Serre spectral sequence.

12 Grothendieck Spectral Sequence

Theorem 11.12.1 *Let $\underline{C} \xrightarrow{F} \underline{B} \xrightarrow{G} \underline{A}$ be a composition of additive functors, where $\underline{C}$, $\underline{B}$, and $\underline{A}$ are abelian categories and $\underline{C}$ and $\underline{B}$ have enough projectives. Suppose that*

$$(L_n G)(FP) = \begin{cases} 0 & \text{if } n > 0; \\ G(FP) & \text{if } n = 0 \end{cases}$$

for any projective P of $\underline{C}$. Then for all objects C of $\underline{C}$ there exists a (first quadrant) spectral sequence with $E_{p,q}^2 = (L_p G)\big((L_q F)(C)\big)$ converging to $\big(L_{p+q}(GF)\big)(C)$.

Proof Let P_* be a projective resolution of C in $\underline{C}$. Define a double complex T in $\underline{B}$ by letting $T_{*,q}$ be a projective resolution in $\underline{B}$ of $F(P_q)$, with the d'-differential coming from this resolution. By Theorem 4.3.3 the maps $F(P_q) \to F(P_{q-1})$ yield maps $T_{0,q} \to T_{0,q-1}$ and any choices can be extended to chain maps $T_{*,q} \to T_{*,q-1}$ yielding the d''-differential. In fact, by Proposition 4.3.4 it is always possible to construct the resolutions and maps in such a way that for each p the maps in the vertical direction are each the composition of a split epimorphism and a split monomorphism and also the homology in the vertical direction, $H_q(T_{p,*})$, forms a projective resolution of $H_q\big(F(P_*)\big) = (L_qF)(C)$; we will assume that such a choice has been made. In particular, any additive functor commutes with homology in the vertical direction. The double complex $G(T)_{**}$ in $\underline{A}$ is contained in the first quadrant, so both of its spectral sequences converge to $H_*\big(\operatorname{Tot} G(T)_{**}\big)$. In one spectral sequence we have

$$E^2_{p,q} = H_q\big(H_p(GT_{**})\big) = H_q\big((L_qG)T_{**}\big)$$

$$= H_q\left(\left\{\begin{array}{ll} 0 & \text{if } p > 0; \\ G(T)_{*q} & \text{if } p = 0 \end{array}\right\}\right) = \left\{\begin{array}{ll} 0 & \text{if } p > 0; \\ L_q(FG)(C) & \text{if } p = 0. \end{array}\right.$$

Therefore $H_*\big(\operatorname{Tot} G(T)_{**}\big) = L_*(FG)(C)$. In the other spectral sequence, our choice of the maps $T_{*,q} \to T_{*,q-1}$ gives

$$E^2_{p,q} = H_p\big(H_q(GT_{**})\big) = H_p\big(G(H_qT_{**})\big) = (L_pG)\big((L_qF)(C)\big).$$

□

Notice that the hypothesis on $(L_nG)(FP)$ is automatically satisfied if F takes projectives to projectives.

Example 11.12.2 (Change of Rings Spectral Sequence) Let $f : R \to S$ be a ring homomorphism. Let M be a right S-module and let N be a left R-module. M can also be regarded as an R-module via f. Applying the Grothendieck spectral sequence to the composition

$$\text{Left } R\text{–Modules} \xrightarrow{F} \text{Left } S\text{–Modules} \xrightarrow{G} \text{Abelian Groups}$$

where $F(A) = S \otimes_R A$ and $G(B) = M \otimes_S A$, and noting that $GF(A) = M \otimes_R B$ and that F preserves projectives, yields a "Change of Rings" spectral sequence with $E^2_{p,q} = \operatorname{Tor}^S_p\big(M, \operatorname{Tor}^R_q(S, N)\big)$ converging to $\operatorname{Tor}^R_{p+q}(M, N)$. □

CHAPTER 12

Localization and Completion

1 Triples and Cotriples

Let $\underline{C}$ and $\underline{D}$ be categories and let $F : \underline{C} \to \underline{D}$ and $G : \underline{D} \to \underline{C}$ be functors such that $F \dashv G$. Set $S = G \circ F : \underline{C} \to \underline{C}$ and let $T = F \circ G : \underline{D} \to \underline{D}$. Define $\mu : S \circ S \to S$ be the natural transformation $\mu_X = G(\beta_{F(X)})$, where $\beta : FG \to I_{\underline{D}}$ is the natural transformation defined in Section 4.5. Then letting $\alpha : I_{\underline{C}} \to G\overline{F}$ be as in Section 4.5, $\mu_X \circ \mu_{S(X)} = \mu_X \circ S(\mu_X) : S^3(X) \to S(X)$, $\alpha_{S(X)} \circ \alpha_X = S(\alpha_X) \circ \alpha_X : X \to S^2(X)$, $\mu_X \circ \alpha_{S(X)} = 1_{S(X)}$, and $\mu_X \circ S\alpha_X = 1_{S(X)}$, for every object X of $\underline{C}$.

Definition 12.1.1 Let $\underline{A}$ be a category. A *triple* or *monad* (S, μ, η) on $\underline{A}$ consists of a functor $S : \underline{A} \to \underline{A}$ together with natural transformations $\mu : S^2 \to S$ and $\eta : I_{\underline{A}} \to S$ such that $\mu \circ \mu S = \mu \circ S\mu : S^3 \to S$, $\eta S \circ \eta = S\eta \circ \eta : I_{\underline{A}} \to S^2$, and $\mu \circ \eta S = \mu \circ S\eta = 1_S : S \to S$. A *cotriple* (T, ψ, ϵ) on $\underline{A}$ consists of a functor $T : \underline{A} \to \underline{A}$ together with natural transformations $\psi : T \to T^2$ and $\epsilon : T \to I_{\underline{A}}$, such that $\psi T \circ \psi = T\psi \circ \psi : T \to T^3$, $\epsilon \circ \epsilon T = \epsilon \circ T\epsilon : T^2 \to I_{\underline{A}}$, and $T\epsilon \circ \psi = \epsilon T \circ \psi = 1_T : T \to T$.

An adjoint pair gives rise to a triple on $\underline{C}$ as above and similarly gives a cotriple on $\underline{D}$.

Let T be a cotriple on a category $\underline{D}$. For each object X of $\underline{D}$ we construct a simplicial set $T_{\bullet}X$ as follows. Set $(T_{\bullet}X)_n = T^{n+1}X$ and write T_nX for $(T_{\bullet}X)_n$. For $j = 0, \ldots, n$, define maps $d_j : T_nX \to T_{n-1}X$ by $d_j = T^{n-j}(\epsilon_{T^j X})$ and $s_j : T_nX \to T_{n+1}X$ by $s_j = T^{n-j}\psi_{T^j X}$. These maps satisfy the simplicial identities.

Let $T : \underline{D} \to \underline{D}$ be a cotriple coming from an adjunction $F \dashv G$ as above. Suppose that $X = F(A)$ for some object A of $\underline{C}$. Then we have a map $\gamma = F\alpha_A : X = F(A) \to FGF(A) = T(X)$. Set $s_{-1} = T^{n+1}\gamma : T_nX \to T_{n+1}X$. The simplicial identities extend to this s_{-1}. That is, $s_i s_j = s_{j+1}s_i$ holds for $-1 \leq i \leq j \leq n$, $d_0 s_{-1} = 1$, and $d_j s_{-1} = s_{-1}d_{j-1}$ for $j > 0$.

Theorem 12.1.2 *Let K be a Kan complex with an additional degeneracy s_{-1} (called a "conical contraction") satisfying the simplicial identities. Then $K \approx *$.*

Proof Let x belong to K_n such that $d_j x = *$ for $j = 0, \ldots, n$. Then $d_0 s_{-1}x = x$ and $d_j s_{-1}x = s_{-1}d_{j-1}x = *$ for $j > 0$, so $x \sim *$. Thus $\pi_n(K) = 0$ for all n and so $K \approx *$. $\qquad\qquad\square$

2 The Bousfield-Kan Construction

Given a simply connected space X and a prime p (or more generally any subring of $\mathbb{Q}$), we wish to construct a space $X_{(p)}$ and a map $i : X \to X_{(p)}$ such that the effect of the maps induced by i on homotopy and homology groups is the canonical map from the group into its localization at p. There are many constructions of such a localization. For example, if X equals S^n, noting as motivation that $Z_{(p)}$ is the direct limit of the system $\mathbb{Z} \xrightarrow{2} \mathbb{Z} \xrightarrow{6} \mathbb{Z} \xrightarrow{30} \ldots \longrightarrow \mathbb{Z} \xrightarrow{f(k)} \ldots$, where $f(k)$ is the product of the first k primes excluding p, we can take $S^n_{(p)}$ to be the infinite mapping telescope (see Section 13.1) of the system of spaces formed from the sequence of maps $S^n \xrightarrow{2} S^n \xrightarrow{6} S^n \xrightarrow{30} \ldots \longrightarrow S^n \xrightarrow{f(k)} \ldots$, where $\underline{m}$ denotes the degree m map on S^n. (See Proposition 13.1.2 and Theorem 13.1.3 for the demonstration of the fact that $S^n_{(p)}$ is indeed a localization of S^n.) Then, having constructed $S^n_{(p)}$, we can use these as building blocks in constructing the localization of other CW-complexes.

The most popular localization construction nowadays is that given by Bousfield and Kan [BK] who, for every space X and commutative ring R, define a space $R_\infty X$ and a map $X \to R_\infty X$ which becomes a localization when X is a simply connected CW-complex and R is a subring of $\mathbb{Q}$. One of the reasons the Bousfield-Kan construction is used is that the construction makes sense for any commutative ring. This has the effect of providing a common framework for the discussion of both localization and the theory of "mod p completion", which turn out to be the special cases $R = \mathbb{Z}_{(p)}$ and $R = \mathbb{Z}/p\mathbb{Z}$ respectively. Bousfield has generalized these notions further to define "localization with respect to a generalized homology theory" so that $R_\infty(\ )$ becomes the special case of localization with respect to $H_*(\ ; R)$. Furthermore, the theory has turned out to be useful when applied to non-simply-connected spaces provided they satisfy a certain weaker condition called "nilpotent".

In what follows we will give the description of $R_\infty X$ and an outline of the proof that it has the properties of a localization in the case where X is simply connected and $R \subset \mathbb{Q}$. The construction is given in the category of pointed simplicial sets.

Let R be a commutative ring. Let $\mathcal{S}et_*$ denote the category of pointed sets. Let $A : \mathcal{S}et_* \to R$-mods be given by $A(S) = \left(\bigoplus_{s \in S} R\right)/R$ where the quotient is by the factor corresponding to $s = *$. Let $F : R$-mods $\to \mathcal{S}et_*$ be the forgetful functor with 0 taken as the basepoint of $F(M)$. Then $A \dashv F$. Let $R : \mathcal{S}et_* \to \mathcal{S}et_*$ be the triple resulting from this adjunction. Explicitly, for a pointed set S, elements of $R(S)$ are finite formal sums $\sum r_i \cdot x_i$ with $r \cdot * = 0$ and $\mu_S : R^2(S) \to R(S)$ is induced by $r_1 \cdot r_2 \cdot x \mapsto r_1 r_2 \cdot x$, while $\eta : S \to R(S)$ is induced by $x \mapsto 1 \cdot x$. Note that even if S happens to be an R-module, η is not an R-module homomorphism. For a pointed simplicial set X, let RX denote the simplicial R-module obtained by applying this construction to each degree of X; that is, $(RX)_n = R(X_n)$. Observe that $\pi_*(RX) = H_*(X; R)$ by the definition of homology of a simplicial set.

We saw earlier that for every object X of a category $\underline{C}$ a cotriple on $\underline{C}$ gives rise to a simplicial $\underline{C}$-object, or equivalently, a contravariant functor from Δ^* to $\underline{C}$. Dually, a triple on $\underline{C}$ gives rise to a covariant functor from Δ^* to $\underline{C}$ for every object X. A covariant functor from Δ^* to $\underline{C}$ defines what is called a *cosimplicial* $\underline{C}$-object. That is, a cosimplicial $\underline{C}$-object $K^\bullet$ consists of an object K^n of $\underline{C}$ for each n, together with $\underline{C}$-morphisms $d^j : K^n \to K^{n+1}$ for $j = 0, \ldots, n+1$ and $s^j : K^n \to K^{n-1}$ for $j = 0, \ldots, n-1$ satisfying the obvious identities dual to the simplicial identities. In the case of a cosimplicial object coming from a triple (T, μ, η) we have explicitly $K^n = T^{n+1} X$, $d^j = T^j \eta T^{n-j} : K^{n-1} \to K^n$, and $s^j = T^j \mu T^{n-j} : K^{n+1} \to K^n$. Applying this in the case above, where $\underline{C}$ is the category of simplicial sets, we have for each space X a cosimplicial simplicial set $(RX)^\bullet$

in which $(RX)^n$ is the simplicial R-module $R^{n+1}X$. Note that even though $(RX)^n$ is a simplicial R-module for all n, $(RX)^\bullet$ is not a cosimplicial simplicial R-module since d^0 is not an R-module homomorphism. A cosimplicial simplicial set K is called *grouplike* if K^q is a simplicial group for all q and all coface and codegeneracy maps except possibly d^0 are group homomorphisms. Thus $(RX)^\bullet$ is grouplike.

Observe that a cosimplicial simplicial set is equivalent to a simplicial cosimplicial set. There is a cosimplicial simplicial set written Δ^* whose nth codegree is the simplicial set $\Delta[n]$. Let K be any cosimplicial simplicial set. Define a simplicial set $\operatorname{Tot} K$ by

$$(\operatorname{Tot} K)_n = \operatorname{Hom}_{\text{cosimplicial simplicial}}(\Delta[n] \times \Delta^*, K)$$

where the cosimplicial structure on $\Delta[n] \times \Delta^*$ comes from the second variable, the simplicial structure on $\Delta[n] \times \Delta^*$ in cosimplicial codegree j is the product simplicial set $\Delta[n] \times \Delta[j]$, and the boundary and degeneracy maps giving the simplicial structure on $\operatorname{Tot} K$ are induced from those on $\Delta[n]$ as in the definition of a function complex (given immediately preceding Theorem 8.1.7). Thus, explicitly, an element $f \in (\operatorname{Tot} K)_n$ is a family of maps $f_k^j : (\Delta[n] \times \Delta[j])_k \to K_k^j$ such that

$$
\begin{array}{ccc}
(\Delta[n] \times \Delta[j])_k & \xrightarrow{f_k^j} & K_k^j \\
\downarrow{\scriptstyle d^i} & & \downarrow{\scriptstyle d^i} \\
(\Delta[n] \times \Delta[j+1])_k & \xrightarrow{f_k^{j+1}} & K_k^{j+1}
\end{array}
\qquad \text{and} \qquad
\begin{array}{ccc}
(\Delta[n] \times \Delta[j])_k & \xrightarrow{f_k^j} & K_k^j \\
\downarrow{\scriptstyle d_i} & & \downarrow{\scriptstyle d_i} \\
(\Delta[n] \times \Delta[j])_{k-1} & \xrightarrow{f_{k-1}^j} & K_{k-1}^j
\end{array}
$$

commute along with similar commutative diagrams for s^i and s_i. Notice that an element $\{f_k^j\} \subset (\operatorname{Tot} K)_n$ is determined by its values on

$$\Delta[n] \times (0, 1, \ldots, s) \in (\Delta[n] \times \Delta[s])_s.$$

Define $R_\infty X$ to be the simplicial set $\operatorname{Tot}(RX)^\bullet$ and let $i_X : X \to R_\infty X$ be induced by $x \mapsto 1 \cdot x$.

Theorem 12.2.1 (Bousfield-Kan) *Let X be a simply connected CW-complex and let R be a subring of $\mathbb{Q}$. Then $X \to R_\infty X$ is a localization of the space X.*

Outline of Proof Applying homotopy or homology to $(RX)^\bullet$ gives a cosimplicial abelian group. Let G be a cosimplicial abelian group. Define $d : G^q \to G^{q+1}$ by $d = \sum_{i=0}^{q+1} (-1)^i d^i$ thus producing a cochain complex (G, d). Define the *cohomotopy* of G by $\pi^q(G) = H^q(G, d)$. For each q, define an abelian group NG^q by $NG^q = \operatorname{Ker} s^0 \cap \operatorname{Ker} s^1 \cap \ldots \cap \operatorname{Ker} s^{q-1}$. Then d restricts to NG and the method of Theorem 8.5.1e shows that the inclusion $(NG, d) \hookrightarrow (G, d)$ induces a chain homotopy equivalence.

Given a cosimplicial simplicial set K define simplicial sets $\operatorname{Tot}^m K$ by

$$(\operatorname{Tot}^m K)_n = \operatorname{Hom}_{\text{cosimplicial simplicial}}(\Delta[n] \times \Delta^{(m)}, K),$$

where $\Delta^{(m)}$ is the cosimplicial simplicial set formed by taking the m-skeleton in each codegree of Δ^*. Inclusion of the $(m-1)$-skeleton into the m-skeleton induces a map $\operatorname{Tot}^m K \to \operatorname{Tot}^{m-1} K$ and $\operatorname{Tot} K = \varprojlim_m \operatorname{Tot}^m K$. In the case where K is a grouplike cosimplicial simplicial set this map is a fibration with fibre $\operatorname{Hom}_{SS}(\Delta[m]/\Delta[m]^{(m-1)}, NK^m) \approx \Omega^m(NK^m)$.

Let $\phi : K \to L$ be a map of cosimplicial simplicial sets such that $\phi : K^n \to L^n$ is a homotopy equivalence for all n. If K and L are grouplike then by induction we see that $\phi : \mathrm{Tot}^m K \to \mathrm{Tot}^m L$ is a homotopy equivalence for all m. The homotopy groups of an inverse limit $W = \varprojlim_m W_m$ of fibrations fit into a natural short exact sequence $0 \to \varprojlim_m{}^1 \pi_{q+1}(W_m) \to \pi_q(W) \to \varprojlim_m \pi_q(W_m) \to 0$ analogous to that of Theorem 13.1.3b. (See [BK, p.255] for a proof.) It follows that $\phi : \mathrm{Tot}\, K \to \mathrm{Tot}\, L$ is a homotopy equivalence when K and L are grouplike.

If $\phi : X \to Y$ induces an isomorphism on $H_*(\ ; R)$ then the induced map $\phi : RX \to RY$ is a homotopy isomorphism and so $\phi : (RX)^\bullet \to (RY)^\bullet$ satisfies the conditions in the preceding paragraph. Therefore $\phi : R_\infty(X) \to R_\infty(Y)$ is a homotopy equivalence.

For a grouplike cosimplicial simplicial set, the long exact homotopy sequences from the fibrations $\Omega^m(NK^m) \to \mathrm{Tot}^m K \to \mathrm{Tot}^{m-1} K$ yield an exact couple resulting in a second quadrant spectral sequence with

$$E^1_{p,t} = \pi_{p+t}(\Omega^{-p} NK^{-p}) = \pi_t(NK^{-p}) \Rightarrow \pi_{p+t}(\mathrm{Tot}\, K).$$

It is customary to reflect this spectral sequence into the first quadrant by setting $s = -p$ to obtain $E^1_{s,t} = \pi_t(NK^s) \Rightarrow \pi_{t-s}(\mathrm{Tot}\, K)$, with the differentials in nonstandard directions. The calculation of the fibre of $\mathrm{Tot}^s K \to \mathrm{Tot}^{s-1} K$ also allows one to identify the d^1-differential of the spectral sequence which turns out to be induced by the map $d : NK^{s-1} \to NK^s$. Therefore $E^2_{s,t} = \pi^s \pi_t(K)$. This spectral sequence does not always converge.

If Y is a simplicial R-module then the map $RY \to Y$ induced by $r \cdot y \mapsto ry$ creates an extra codegeneracy on $R_\infty Y$ which shows that $\big(\pi_t(R_\infty Y), d\big)$ is an acyclic cochain complex for each t. Therefore

$$E^2_{s,t} = \pi^s \pi_t\big((RY)^\bullet\big) = \begin{cases} 0 & \text{if } s > 0; \\ \pi_t(Y) & \text{if } s = 0, \end{cases}$$

by the dual of Theorem 12.1.2 when $s > 0$ and by direct calculation when $s = 0$. Because most of the terms are zero, one can show that the spectral sequence converges and conclude that the map $i_Y : Y \to R_\infty Y$ induces an isomorphism on homotopy groups and is thus a homotopy equivalence.

If X is a space $K(G, n)$ having only one nonzero homotopy group G in degree n (Eilenberg-MacLane space, see Section 13.4) then it has a localization which is given by $\phi : K(G, n) \to K(G \otimes R, n)$. Since ϕ induces an isomorphism on $H_*(\ ; R)$ and $K(G \otimes R, n)$ is a simplicial R-module, we have $R_\infty\big(K(G, n)\big) \xrightarrow{\approx} R_\infty\big(K(G \otimes R, n)\big) \xleftarrow{\approx} K(G \otimes R, n)$ and so $X \to R_\infty(X)$ is a localization in this case. An arbitrary space can be made higher and higher connected by repeatedly mapping to Eilenberg-MacLane spaces and taking the fibre (iterating the procedure of Exercise 11.9.10) and so the theorem follows from the facts that R_∞ preserves fibrations ([BK, pp. 54–57]) and connectivity ([BK, p. 30]). $\square$

Henceforth we will write $X_{(p)}$ for $(\mathbb{Z}_{(p)})_\infty X$, where by convention $\mathbb{Z}_{(0)} = \mathbb{Q}$. Its properties are summarized as follows.

Theorem 12.2.2 *Let p be a prime or 0. Then there is a functor $(\ )_{(p)}$ from simply connected CW-complexes to simply connected CW-complexes such that:*

1) $\pi_*(X_{(p)}) = \big(\pi_*(X)\big)_{(p)}$;

2) $H_*(X_{(p)}) = \big(H_*(X)\big)_{(p)}$;

3) *if $F \to E \to B$ a homotopy fibration then $F_{(p)} \to E_{(p)} \to B_{(p)}$ is a homotopy fibration;*

4) *if $A \to X \to C$ a homotopy cofibration then $A_{(p)} \to X_{(p)} \to C_{(p)}$ is a homotopy cofibration.* $\square$

The other important special case of the construction occurs when $R = \mathbb{Z}/p\mathbb{Z}$ in which case $R_\infty X$ will be denoted $\hat{X}_p$ or simply $\hat{X}$. Let $\hat{\mathbb{Z}}_p$ denote the p-adics defined as the inverse limit of the system $\ldots \longrightarrow \mathbb{Z}/p^r\mathbb{Z} \xrightarrow{\theta_{p^{r-1}}} \mathbb{Z}/p^{r-1}\mathbb{Z} \longrightarrow \ldots \xrightarrow{\theta_p} \mathbb{Z}/p\mathbb{Z}$ where θ_{p^m} denotes reduction modulo p^m.

Theorem 12.2.3 *If X is a simply connected CW-complex of finite type, then* $\pi_*(\hat{X}_p) = \pi_*(X) \otimes \hat{\mathbb{Z}}_p$.

See [BK]. $\square$

CHAPTER 13

Generalized Homology and Stable Homotopy

1 Generalized Homology

In Chapter 5 we defined a generalized homology theory to be a sequence of functors h_n satisfying the first six conditions of Definition 5.1.1. Given a generalized homology theory h on the category of pointed spaces, we can define the corresponding reduced theory $\tilde{h}$ by setting $\tilde{h}(X) = \mathrm{Ker}\,(h(X) \to h(*))$. The two concepts are equivalent since conversely given a reduced theory $\tilde{h}$, setting $h(X) = \tilde{h}(X \amalg *)$ yields an unreduced theory.

Using the results of Chapter 7, on the category of pointed CW-complexes we can eliminate the use of pairs and express a (reduced) homology theory in the following way:

Definition 13.1.1 A *reduced homology theory* on the category of pointed CW-complexes consists of a sequence of functors Y_n : pointed CW-complexes $\to \mathcal{AB}$ (the category of abelian groups) together with natural transformations $\sigma : Y_n \to Y_{n+1} \circ S$ such that:

1) Homotopy: $f \simeq g \Rightarrow f_* = g_*$;
2) Suspension: $\sigma_X : Y_n(X) \to Y_{n+1}(SX)$ is an isomorphism for all X;
3) Exactness: For every cellular inclusion $A \to X$ the induced sequence $Y_n(A) \to Y_n(X) \to Y_n(X/A)$ is exact for all n.

From these axioms, the infinite sequence of homotopy cofibrations

$$A \xrightarrow{j} X \xrightarrow{q} C \xrightarrow{\delta} SA \xrightarrow{(Sj)^{-1}} SX \longrightarrow \ldots \longrightarrow S^{k-1}C$$
$$\xrightarrow{\delta} S^k A \xrightarrow{(S^k j)^{(-1)^k}} S^k X \xrightarrow{(S^k q)^{(-1)^k}} S^k C \longrightarrow \ldots$$

associated to a cofibration $A \to X$ allows the construction of a long exact sequence and Mayer-Vietoris follows from Lemma 4.1.8.

We will also assume that our homology theory satisfies the following additional condition:

Milnor Wedge Axiom For any spaces $\{X_j\}_{j \in J}$, the maps $X_j \to \bigvee_{j \in J} X_j$ induce an isomorphism $\bigoplus_{j \in J} Y_n(X_j) \to Y_n(\bigvee_{j \in J} X_j)$ for each n.

Remark 1 If J is finite, this condition follows from the other axioms.

Remark 2 The corresponding statement for an unreduced theory has $\amalg$ replacing $\bigvee$.

These conditions are readily dualized to obtain the definition of a (reduced) cohomology theory (where dualizing includes replacing $\bigoplus$ by $\prod$ in the wedge axiom).

Let X be a direct system $X_1 \xrightarrow{j_1} X_2 \xrightarrow{j_2} \ldots \longrightarrow X_n \xrightarrow{j_n} \ldots$ of topological spaces. The *infinite mapping telescope* of X is the space T which results from repeatedly applying the mapping cylinder construction of Chapter 7. Precisely, $T = \left(\coprod_{n=1}^{\infty}(X_n \times [n-1, n])\right)/\sim$ with $X_n \times \{n\}$ identified to $j_n(X_n) \times \{n\}$.

Proposition 13.1.2 *Let T be the infinite mapping telescope of*

$$X_1 \xrightarrow{j_1} X_2 \xrightarrow{j_2} \ldots \longrightarrow X_n \xrightarrow{j_n} \ldots .$$

Then $\pi_q(T) = \varinjlim_n \pi_q(X_n)$.

Proof Consideration of the obvious map $T \to [0, \infty)$ shows that any compact subset of T is contained in a finite portion of the telescope. Therefore the result follows from the fact that images of maps from spheres and homotopies between such maps are compact. $\qquad\qquad\square$

Theorem 13.1.3 (Milnor) *Suppose $X_1 \subset X_2 \subset \ldots \subset X_n \subset \ldots$ and let $X = \bigcup_{n=1}^{\infty} X_n$.*

    a) *If Y is a reduced homology theory satisfying the wedge axiom then there is a natural isomorphism $Y_q(X) \cong \varinjlim_n Y_q(X_n)$.*

    b) *If Y is a reduced cohomology theory satisfying the wedge axiom then there is a natural short exact sequence*

$$0 \to \varprojlim_n{}^1 Y^{q-1}(X_n) \to Y^q(X) \to \varprojlim_n Y^q(X_n) \to 0.$$

Proof Let $j_n : X_n \to X_{n+1}$ denote the inclusion map. $X \approx T$ where T is the infinite mapping telescope of the system of inclusions. Let

$$T_1 = \left(\bigcup_{n=1}^{\infty} X_{2n-1} \times [2n-2, 2n-1]\right) \bigcup (* \times [0, \infty)) \subset T$$

and let

$$T_2 = \left(\bigcup_{n=1}^{\infty} X_{2n} \times [2n-1, 2n]\right) \bigcup (* \times [0, \infty)) \subset T.$$

Then $T = T_1 \cup T_2$. Let $T_0 = T_1 \cap T_2 = \bigcup_{n=1}^{\infty}(X_n \times \{n\}) \bigcup (* \times [0, \infty))$.

    a) By the wedge axiom, $Y_*(T_1) = \bigoplus_{n=1}^{\infty} Y_*(X_{2n-1})$, $Y_*(T_2) = \bigoplus_{n=1}^{\infty} Y_*(X_{2n})$, and $Y_*(T_0) = \bigoplus_{n=1}^{\infty} Y_*(X_n)$. In the resulting Mayer-Vietoris sequence, the map $Y_q(T_0) \to Y_q(T_1) \oplus Y_q(T_2)$, which we will write as

$$1 + j\langle q \rangle : \bigoplus_{n=1}^{\infty} Y_q(X_n) \to \bigoplus_{n=1}^{\infty} Y_q(X_n),$$

is the map whose restriction to $Y_q(X_n)$ is $1 + (j_n)_*$. So there is a short exact sequence

$$0 \to \operatorname{CoKer}(1 + j\langle q \rangle) \to Y_q(T) \to \operatorname{Ker}(1 + j\langle q-1 \rangle) \to 0.$$

Since $\operatorname{Ker}(1 + j\langle q-1 \rangle) = 0$ and $\operatorname{CoKer}(1 + j\langle q \rangle) \cong \varinjlim_n Y_q(X_n)$, the result follows.

    b) By the wedge axiom, $Y^*(T_1) = \prod_{n=1}^{\infty} Y^*(X_{2n-1})$, $Y^*(T_2) = \prod_{n=1}^{\infty} Y^*(X_{2n})$, and $Y^*(T_0) = \prod_{n=1}^{\infty} Y^*(X_n)$. In the resulting Mayer-Vietoris sequence, the map $Y^q(T_0) \to Y^q(T_1) \oplus Y^q(T_2)$, which we will write as

$$1 + j\langle q \rangle : \prod_{n=1}^{\infty} Y^q(X_n) \to \prod_{n=1}^{\infty} Y^q(X_n),$$

is the map whose projection to $Y^q(X_n)$ is $1 + (j_n)^*$. So there is a short exact sequence

$$0 \to \mathrm{CoKer}\,(1 + j\langle q - 1\rangle) \to Y^q(T) \to \mathrm{Ker}\,(1 + j\langle q\rangle) \to 0.$$

Since $\mathrm{Ker}\,(1 + j\langle q\rangle) \cong \varprojlim_n Y^q(X_n)$ and $\mathrm{CoKer}\,(1 + j\langle q - 1\rangle) \cong \varprojlim_n^1 Y^{q-1}(X_n)$, the result follows. $\qquad\square$

Given a fibration and a generalized homology theory, the Serre filtration on the total space yields an exact couple. If the wedge axiom is satisfied, the proof of Theorem 11.9.2 carries over to give the following theorem which is stated most easily for unreduced theories.

Theorem 13.1.4 (Generalized Serre Spectral Sequence) *Let $F \to X \to B$ be a fibration and let Y be an unreduced homology theory satisfying the wedge axiom. Then there is a (right half-plane) spectral sequence with $E^2_{p,q} \cong H_p\big(B; {}^tY_q(F)\big)$ converging to $Y_{p+q}(X)$.* $\qquad\square$

In the cohomology spectral sequence, the identification of the E_2-term proceeds as before, but in general there are convergence problems due to the appearance of the $\varprojlim^1$ "error" term in Theorem 13.1.3b, although conditions which imply convergence have been studied. Note that for convergence it is not sufficient to know that $\varprojlim_n^1 F_n\big(Y(X)\big) = 0$, since this shows only that the filtration is bicomplete, but not that the spectral sequence converges to it. In the case where B is a finite dimensional complex the filtration is finite so there is no problem and we get:

Theorem 13.1.5 (Generalized Cohomological Serre Spectral Sequence) *Let $F \to X \to B$ be a fibration over a finite dimensional complex B and let Y be an unreduced cohomology theory satisfying the wedge axiom. Then there is a spectral sequence with $E_2^{p,q} \cong H^p\big(B; {}^tY^q(F)\big)$ converging to $Y^{p+q}(X)$.* $\qquad\square$

In the special case of the fibration $* \to X \to X$ we get the "Atiyah-Hirzebruch" Spectral Sequence.

Theorem 13.1.6 (Atiyah-Hirzebruch Spectral Sequence)

a) *Let X be a CW-complex and let Y be an unreduced homology theory satisfying the wedge axiom. Then there is a (right half-plane) spectral sequence with $E^2_{p,q} \cong H_p\big(X; Y_q(*)\big)$ converging to $Y_{p+q}(X)$.*

b) *Let X be a finite dimensional CW-complex and let Y be an unreduced homology theory satisfying the wedge axiom. Then there is a spectral sequence with $E_2^{p,q} \cong H^p\big(X; Y^q(*)\big)$ converging to $Y^{p+q}(X)$.* $\qquad\square$

Theorem 13.1.7 (Equivalence of Theories)

a) *Let Y and Y' be unreduced homology theories satisfying the wedge axiom. Let $\eta : Y \to Y'$ be a natural transformation of homology theories. (That is, a collection of natural transformations $\eta_n : Y_n \to Y'_n$ commuting with the suspension isomorphisms.) Suppose that $\eta_* : Y_*(*) \to Y'_*(*)$ is an isomorphism. Then $\eta_X : Y_*(X) \to Y'_*(X)$ is an isomorphism for all CW-complexes X.*

b) *Let Y and Y' be unreduced cohomology theories satisfying the wedge axiom. Let $\eta : Y \to Y'$ be a natural transformation of cohomology theories. Suppose that $\eta_* : Y^*(*) \to Y'^*(*)$ is an isomorphism. Then $\eta_X : Y^*(X) \to Y'^*(X)$ is an isomorphism for all CW-complexes X.*

Proof Part (a) follows from the spectral sequence comparison theorem, as does (b) in the case where X is a finite dimensional complex. However the naturality of Theorem 13.1.3b then gives (b) for all CW-complexes. $\qquad\square$

2 Stable Homotopy Groups

For pointed spaces W and X, the direct limit of the system

$$[W, X] \xrightarrow{S(\ )} [SW, SX] \xrightarrow{S(\ )} \ldots \longrightarrow [S^n W, S^n X] \longrightarrow \ldots$$

(which is a system of abelian groups for $n \geq 2$) is called the group of *stable homotopy classes of maps* from W to X and denoted $\{W, X\}$.

By adjointness the maps in the above sequence are induced by the canonical maps $S^n X \to \Omega S^{n+1} X$ whose connectivity grows with n by the Bott-Samelson Theorem. Precisely, this yields

Theorem 13.2.1 (Freudenthal Suspension Theorem) *Let Y be $(n-1)$-connected with $n \geq 2$. Then $S : [W, Y] \to [SW, SY]$ is a bijection for CW-complexes W of dimension less than $2n - 1$ and a surjection for CW-complexes W of dimension $2n - 1$.* $\qquad\square$

Thus if W is a finite CW-complex then the above system stabilizes. In the special case where $W = S^n$, $\{S^n, X\}$ is called the nth *stable homotopy group* of X and written $\pi_n^S(X)$.

Proposition 13.2.2 *Let $A \xrightarrow{j} X \xrightarrow{c} C$ be a cofibration. Then $\{W, A\} \xrightarrow{j_\#} \{W, X\} \xrightarrow{c_\#} \{W, C\}$ is exact for all W.*

Proof It is trivial that the composition $c_\# j_\# = 0$. Conversely, let $f \in \{W, X\}$ satisfy $c_\#(f) = 0$. By Lemma 4.2.3 there exists $f_n : S^n W \to S^n X$ such that $(S^n c)(f_n) \simeq *$. This gives rise to an extension $h : C S^n W \to S^n C$ and an induced map of homotopy cofibres $q : S^{n+1} W \to S^{n+1} A$ fitting into the following homotopy commutative diagram

$$
\begin{array}{ccccccc}
S^n W & \longrightarrow & C S^n W & \longrightarrow & S^{n+1} W & \xrightarrow{\pm(1_{S^{n+1}W})} & S^{n+1} W \\
\downarrow{\scriptstyle f_n} & & \downarrow{\scriptstyle h} & & \downarrow{\scriptstyle q} & & \downarrow{\scriptstyle f_{n+1}} \\
S^n X & \xrightarrow{S^n c} & S^n C & \longrightarrow & S^{n+1} A & \xrightarrow{\pm(S^{n+1}j)} & S^{n+1} X
\end{array}
$$

which shows that f_{n+1} belongs to $\operatorname{Im} S^{n+1} j_\#$. $\qquad\square$

3 Spectra

To establish the connection between generalized homology theory and stable homotopy we take a glimpse at "spectra". We begin with some historical notes. Spectra as described below were originally defined by Lima. To every space is associated a spectrum called its "suspension spectrum", but there are other spectra which do not come from spaces. To every spectrum one can associate a special type of spectrum called an Ω-spectrum. George Whitehead showed that every spectrum gives rise through its Ω-spectrum to both a cohomology theory and a homology theory which satisfy the wedge axiom. Brown then showed that conversely every such theory comes from a Ω-spectrum. Later generations found it desirable to create a category of spectra and a corresponding homotopy category by defining morphisms of spectra in such a way that if one applies the definition to the special case of suspension spectra one gets that the group of morphisms in the homotopy category of spectra from the suspension spectrum of W to that of X equals $\{W, X\}$. Furthermore, other standard constructions from the category of spaces, such as smash products, were introduced. Adams [A2] wrote an exposition of one such construction of a category of spectra following the ideas of Boardman. There seems to be little disagreement that the homotopy category resulting from Boardman's construction is the best. However some people, including those attempting to generalize group action notions from spaces to spectra, have found Boardman's category unsuitable for their purposes and another category of spectra resulting in the same homotopy category has been developed by May [LMS]. The objects of May's category are a special case of Ω-spectrum, and he refers to spectra as defined by Lima are referred to as "pre-spectra". My purpose in these notes is to discuss the contributions of Whitehead and Brown, and I shall not attempt to present constructions of any category of spectra, except peripherally.

Definition 13.3.1 A *spectrum* is a collection of pointed spaces $\{Z_n\}_{n \in \mathbb{Z}}$ together with pointed maps $e_n : SZ_n \to Z_{n+1}$.

Given a pointed space X, the *suspension spectrum* of X, denoted $S^\infty X$, is the spectrum with $(S^\infty X)_n = S^n X$ and $e_n = 1_{S^{n+1}X}$.

An Ω-*spectrum* is one in which the adjoint $e'_n : Z_n \to \Omega Z_{n+1}$ of the structure map is a homotopy equivalence for each n. (Spectra in May's category are Ω-spectra in which e'_n is a homeomorphism for each n.)

Let Z be a CW-spectrum (i.e., each space is a CW-complex) and let $e'_n : Z_n \to \Omega Z_{n+1}$ be the adjoint of e_n. Let T_n^Z be the infinite mapping telescope of the system

$$Z_n \xrightarrow{e'_n} \Omega Z_{n+1} \xrightarrow{\Omega e'_{n+1}} \Omega^2 Z_{n+2} \xrightarrow{\Omega^2 e'_{n+2}} \cdots$$

$$\longrightarrow \Omega^k Z_{n+k} \xrightarrow{\Omega^k e'_{n+k}} \Omega^{k+1} Z_{n+k+1} \longrightarrow \cdots.$$

Collapsing $Z_n \times [0,1]$ and using Proposition 13.1.2 gives a homotopy equivalence $T_n^Z \approx \Omega T_{n+1}^Z$ which turns T^Z into an Ω-spectrum. (A morphism of degree r from Z to Z' in the homotopy category of Boardman's category of spectra turns out to be equivalent to $\varinjlim_n [T_n^Z, T_{n+r}^{Z'}]$.) For a space X, the space $T_0^{S^\infty X} = \varinjlim_n \Omega^n S^n X$ is written $\Omega^\infty S^\infty X$ or $Q(X)$.

Let Z be an Ω-spectrum. Let $\mathcal{AB}$ denote the category of abelian groups. For each n, define a functor $Y^n : \mathcal{T}op_* \to \mathcal{AB}$ by $Y^n(X) = [X, Z_n]$, with natural transformations $\sigma_n : Y^n \to Y^{n+1}$ given by the composite

$$Y^n(X) = [X, Z_n] \xrightarrow{(e'_n)_\#} [X, \Omega Z_{n+1}] = [SX, Z_{n+1}] = Y^{n+1}(SX).$$

From the definitions, it is clear that $\{Y^n\}$ forms a reduced cohomology theory. We shall usually write $Z^*(\ )$ for this cohomology theory. To get the homology theory associated to Z, set $Z_n(X) = \varinjlim_n \pi_{n+k}(X \wedge Z_k)$. The fact that this yields a homology theory follows from Proposition 13.2.2.

4 Eilenberg-MacLane Spaces and Brown Representability

A space with one nonzero homotopy group is called an *Eilenberg-MacLane space*.

Proposition 13.4.1 *Let G be a group and n a positive integer. If either $n = 1$ or G is abelian then there exists an Eilenberg-MacLane space X with $\pi_n(X) = G$.*

Proof There are several possible constructions. For G abelian, one possibility is to take the geometric realization of the simplicial set whose kth degree is $(\mathbb{Z}S^n)_k \otimes G$, where S^n is the simplicial n-sphere and $\mathbb{Z}S^n$ is as in Chapter 12. In other words it is the simplicial set formed by tensoring the free abelian group on on $(S^n)_k$ with G for each k.

Another construction is to use a presentation of G as a quotient F/R of a free abelian group F with relations R. Begin with a wedge of spheres $X^{(n)} = \bigvee S^n$ corresponding to the generators of F. Therefore $\pi_n\left(X^{(n)}\right) = H_n\left(X^{(n)}\right) = F$. Next attach an $(n+1)$-cell for each relation in R by using an attaching map in $\pi_n\left(X^{(n)}\right)$ corresponding to the element, thereby constructing a space $X^{(n+1)}$ with $\pi_n\left(X^{(n+1)}\right) = H_n\left(X^{(n+1)}\right) = G$. Continue by attaching an $(n+2)$-cell for each element of $\pi_{n+1}\left(X^{(n+1)}\right)$ obtaining a space $X^{(n+2)}$ with the right homotopy groups through dimension $n+1$. Iterating the last step for each $k > n+1$ leads to a space X with the desired properties.

In the case $n = 1$, the space BG, where G is thought of as a topological group with the discrete topology, has the desired properties.

There is also a plan for dealing with higher n's using a classifying space approach, although it requires more details about classifying spaces than we have discussed. Specifically, in the case where G is abelian, the multiplication map $\mu : G \times G \to G$ is a group homomorphism and so induces a map of classifying spaces $B(G) \times B(G) \cong B(G \times G) \xrightarrow{B\mu} BG$. There are ways to replace the space BG by a homotopy equivalent space in such a way that this map creates a topological group structure whose classifying space becomes the desired space in the case $n = 2$. This procedure can then be iterated to deal with higher n's. $\qquad\square$

The proof of this proposition can be generalized to construct "representing" spaces for any cohomology theory Y satisfying the wedge axiom so that the preceding becomes the special case $Y = \tilde{H}^*(\ ;G)$. Think of the procedure described in the second proof as adding cells to a space so as to produce spaces $X^{(m+1)}$ with $[S^m, X^{(m+1)}]$ having the right value, $\tilde{H}^n(S^m;G)$, for larger and larger m. The generalization is as follows.

Theorem 13.4.2 (Brown Representability Theorem) *Let Y be a reduced cohomology theory satisfying the wedge axiom. Then there exist unique (up to homotopy equivalence) CW-complexes Z_n such that $Y^n(X)$ is naturally isomorphic to $[X, Z_n]$.*

Proof We start by constructing, by induction on k, a space $Z_n(k)$ and an element $\iota_n(k) \in Y^n(Z_n(k))$ such that the map $f \mapsto f^*(\iota_n(k))$ from $[S^m, Z_n(k)]$ to $Y^n(S^m)$ is an isomorphism for $m < k$ and an epimorphism for $m = k$. The induction begins with $k = -1$ where we set $Z_n(-1) = *$ and $\iota_n(-1) = 0$. Suppose that $Z_n(k)$ and $\iota_n(k)$ have been constructed. Let K be the kernel of $f \mapsto f^*(\iota_n(k))$ taking $[S^k, Z_n(k)]$ to $Y^n(S^k)$. Define $Z_n(k+1)$ to be the homotopy cofibre of the map $\bigvee_{x \in K} S^k \xrightarrow{\phi} Z_n(k) \vee \bigvee_{y \in Y^n(S^{k+1})} S^{k+1}$ where the restriction of ϕ to the summand labelled by $x \in K$ is the composite of $x : S^k \to Z_n(k)$ and the inclusion of the summand $Z_n(k)$ into the wedge. By the wedge axiom

$$Y^n \left(Z_n(k) \vee \bigvee_{y \in Y^n(S^{k+1})} S^{k+1} \right) \cong Y^n \big(Z_n(k) \big) \times \prod_{y \in Y^n(S^{k+1})} Y^n(S^{k+1})$$

and

$$Y^n \left(\bigvee_{x \in K} S^k \right) \cong \prod_{x \in K} Y^n(S^k)$$

By construction $\phi^* \big(\iota_n(k) \big) = 0$ and so from the exact sequence

$$Y^n \big(Z_n(k+1) \big) \xrightarrow{c^*} Y^n \left(Z_n(k) \vee \bigvee_{y \in Y^n(S^{k+1})} S^{k+1} \right) \xrightarrow{\phi^*} Y^n \left(\bigvee_{x \in K} S^k \right)$$

there exists $\iota_n(k+1) \in Y^n \big(Z_n(k+1) \big)$ such that

$$c^* \big(\iota_n(k+1) \big) = \left(\iota_n(k), \prod_{y \in Y^n(S^{k+1})} y \right).$$

Then $Z_n(k+1)$ and $\iota_n(k+1)$ have the desired properties, completing the induction.

Let $f_n(k) : Z_n(k) \to Z_n(k+1)$ be the composite

$$Z_n(k) \hookrightarrow Z_n(k) \vee \bigvee_{y \in Y^n(S^{k+1})} S^{k+1} \xrightarrow{c} Z_n(k+1)$$

and note that $f_n(k)^* \big(\iota_n(k+1) \big) = \iota_n(k)$. Let Z_n be the infinite mapping telescope of $\ldots \longrightarrow Z_n(k-1) \xrightarrow{f_n(k-1)} Z_n(k) \xrightarrow{f_n(k)} Z_n(k+1) \longrightarrow \ldots$ and let $\iota_n \in Y^n(Z_n)$ be an element restricting to $\iota_n(k)$ for each k, which exists by Theorem 13.1.3b. By construction, $(\iota_n)_X^* : [X, Z_n] \to Y^n(X)$ is an isomorphism when X is a sphere and so by the 5-Lemma it is an isomorphism for any finite complex. Since the preceding gives natural isomorphisms

$$[X, Z_n] \cong Y^n(X) \cong Y^{n+1}(SX) \cong [SX, Z_{n+1}] \cong [X, \Omega Z_{n+1}]$$

for any finite complex X, applying this to the finite subcomplexes of Z_n gives a compatible family of maps into ΩZ_{n+1} which therefore induce a map from their direct limit, Z_n, to ΩZ_{n+1}. Since this map induces an isomorphism on homotopy groups (spheres being finite complexes) it is a homotopy equivalence so $\{Z_n\}$ is an Ω-spectrum. We have a natural transformation $\{(\iota_n)^*\}$ from the cohomology theory represented by this Ω-spectrum to $Y^*(\ )$ which induces an isomorphism on all finite CW-complexes and thus induces an isomorphism on all CW-complexes by Theorem 13.1.7. Uniqueness follows from general categorical considerations. $\square$

Return now to the special case where $Y = \tilde{H}^*(\ ; G)$ and denote its nth representing space by $K(G, n)$. It is clear the spaces $K(G, n)$ are Eilenberg-MacLane spaces. Furthermore if X is any Eilenberg-MacLane space with $\pi_n(X) = G'$ then by the universal coefficient theorem and Hurewicz theorem

$$[X, K(G, n)] \cong H^n(X; G) \cong \mathrm{Hom}\big(H_n(X), G\big) \cong \mathrm{Hom}(G', G).$$

In particular, if $G' = G$ then $1_G \in \mathrm{Hom}(G, G)$ yields a map $X \to K(G, n)$ which induces a weak homotopy equivalence. We can choose $\iota_n \in H^n \big(K(G, n) \big)$ to correspond to 1_G under the above isomorphism. In summary we have the following.

Theorem 13.4.3 *Let G be an abelian group.*

a) *Up to weak homotopy equivalence, there is a unique Eilenberg-MacLane space $K(G,n)$ for each n.*

b) *For any space W, the map $[W, K(G,n)] \to H^n(W;G)$ given by $f \mapsto f^*(\iota_n)$ is an isomorphism.*

c) $[K(G',n), K(G,n)] \cong \mathrm{Hom}(G', G).$ □

CHAPTER 14

Cohomology Operations and the Steenrod Algebra

1 Primary Cohomology Operations

Definition 14.1.1 Let Y and Y' be reduced cohomology theories. A natural transformation from Y^n to $Y'^{n'}$ for some integers n and n' is called a *primary cohomology operation* from Y^n to $Y'^{n'}$. A sequence of natural transformations $\left(\theta(n) : Y^n \to Y'^{n+k}\right)_{n\in\mathbb{Z}}$ commuting with the suspension in the graded sense $(\sigma_X \circ \theta(n)_X = (-1)^k \theta(n+1)_{SX} \circ \sigma_X : Y^n(X) \to Y'^{n+k+1}(SX))$ is called a *stable primary cohomology operation* of degree k from Y to Y'.

Although we shall later introduce higher order cohomology operations, when there is no possibility of confusion we will often omit the word "primary" when referring to primary cohomology operation.

If Y and Y' are the reduced cohomology theories coming from Ω-spectra $\{Y_n\}$ and $\{Y'_n\}$ then primary cohomology operations from Y^n to $Y^{n'}$ are in bijective correspondence with the elements of $[Y_n, Y'_{n'}]$. According to the following proposition, a stable primary cohomology operation of degree k from Y to Y' corresponds to a sequence of homotopy classes $\hat{\theta}(n) : Y_n \to Y'_{n+k}$ such that $\hat{\theta}(n) = (-1)^k \Omega\left(\hat{\theta}(n+1)\right)$. (A stable primary operation of degree k corresponds to a morphism of degree k from $\{Y_n\}$ and $\{Y'_n\}$ in Boardman's homotopy category of spectra.)

Let $\sigma : SY_n \to Y_{n+1}$ be the structure map of the Ω-spectrum $\{Y_n\}$. The suspension isomorphism is induced by σ and will be written $\sigma_* : Y^n(X) \cong Y^{n+1}(SX)$.

Proposition 14.1.2 *Let θ be a stable cohomology operation represented by a sequence of maps $\left(\hat{\theta}(n) : Y_n \to Y'_{n+k}\right)$. Then $\hat{\theta}(n) = (-1)^k \Omega\left(\hat{\theta}(n+1)\right)$. Conversely a sequence of maps $\left(\hat{\theta}(n) : Y_n \to Y'_{n+k}\right)$ satisfying $\hat{\theta}(n) = (-1)^k \Omega\left(\hat{\theta}(n+1)\right)$ defines a stable cohomology operation.*

Proof Both statements are equivalent to the commutativity of the diagram

$$
\begin{array}{ccc}
SY_n & \xrightarrow{\;(-1)^k S\hat{\theta}(n)\;} & SY'_{n+k} \\
\downarrow{\scriptstyle\sigma} & & \downarrow{\scriptstyle\sigma} \\
Y_{n+1} & \xrightarrow{\;\hat{\theta}(n+1)\;} & Y'_{n+k+1}
\end{array}
$$

for all n. $\qquad\square$

Notice that a primary operation yields for every X a set function from $Y^n(X)$ to $Y'^{n'}(X)$ which need not be a group homomorphism; it is induced by a map $Y_n \to Y'_{n'}$ which need not be an H-map. However in the case of a stable operation, each of the corresponding maps is a loop map, so the function $Y^n(X) \to Y'^{n+k}(X)$ is automatically a group homomorphism.

A stable cohomology operation from Y to Y' can be composed with one from Y' to Y'' in the obvious way. For a (reduced) cohomology theory Y, the collection of all stable primary self-cohomology-operations forms an algebra, denoted $Y^*(Y)$. In the case where Y equals $H^*(\ ;\mathbb{Z}/p\mathbb{Z})$ for a prime p, this algebra is called the mod p Steenrod algebra and denoted $\mathcal{A}_p$, or simply $\mathcal{A}$ when p is understood.

2 The Steenrod Algebra as a Hopf Algebra

Let p be a prime. Throughout this section, $H^*(\ )$ will be used to denote cohomology with $(\mathbb{Z}/p\mathbb{Z})$-coefficients, $\mathcal{A}$ will denote $\mathcal{A}_p$, and we will write K_n for $K(\mathbb{Z}/p\mathbb{Z}, n)$. Note that $H^*(X \times Y) = H^*(X) \otimes H^*(Y)$ for all X and Y since we are working over a field. Let $\iota_n \in H^n(K_n)$ correspond to the identity map and let $\Delta_{i,j} : K_i \times K_j \to K_{i+j}$ correspond to $\iota_i \otimes \iota_j \in H^{i+j}(K_i \times K_j) \cong \left(H^*(K_i) \otimes H^*(K_j)\right)^{i+j}$.

By Ganea's theorem, there is a homotopy fibration $S(K_n \wedge K_n) \longrightarrow SK_n \xrightarrow{\ \sigma\ } K_{n+1}$, so the map $\sigma : SK_n \to K_{n+1}$ is $2n$-connected. Thus for a given k, $[K_n, K_{n+k}] = H^{n+k}(K_n)$ is independent of n provided n is sufficiently large. Therefore stable cohomology operations of degree k can be identified with $H^{n+k}(K_n)$ for large n.

Let θ be a stable cohomology operation of degree k. Choose large i and j (relative to k), set $n = i + j$, and let $\hat{\theta} \in H^{n+k}(K_n)$ represent the operation θ. In $H^{n+k}(K_i \times K_j)$ write $\hat{\theta}(\Delta_{i,j}) = \sum_{s+t=k} \hat{\theta}'_s \otimes \hat{\theta}''_t$ where $\hat{\theta}'_s$ belongs to $H^{i+s}(K_i)$ and $\hat{\theta}''_t$ belongs to $H^{j+t}(K_j)$. Note that $H^{m+q}(K_m) = 0$ for $q < 0$ so that $0 \leq s, t \leq k$ for all s and t. The stable cohomology operations represented by $\hat{\theta}'_s$ and $\hat{\theta}''_t$ are independent of i and j (given our choice of large i and j). We set $\psi(\theta) = \sum_{s+t=k} \theta'_s \otimes \theta''_t$ where θ'_s and θ''_t are the stable cohomology operations represented by $\hat{\theta}'_s$ and $\hat{\theta}''_t$.

Theorem 14.2.1 (External Cartan Formula) *Let $f \in H^n(X)$ and $g \in H^{n'}(Y)$ be cohomology classes and let θ be a stable cohomology operation of degree k. Then $\theta(f \otimes g) = \left(\psi(\theta)\right)(f \otimes g)$ in $H^{n+n'+k}(X \times Y)$. That is, $\theta(f \otimes g) = \sum_{s+t=k} (-1)^{tn'} \theta'_s(f) \otimes \theta''_t(g)$ where $\psi(\theta) = \sum_{s+t=k} \theta'_s \otimes \theta''_t$.*

Proof By naturality it suffices to consider the special case where $f = \iota_n$ in $H^n(K_n)$ and $g = \iota_{n'}$ in $H^{n'}(K_{n'})$. Since $\iota_n \otimes \iota_{n'}$ lies in $\operatorname{Im} c^* : H^{n+n'}(K_n \wedge K_{n'}) \to H^{n+n'}(K_n \times K_{n'})$ it suffices to consider the corresponding formula in $H^*(K_n \wedge K_{n'})$. Let $\bar{\Delta}_{i,j} : K_i \wedge K_j \to K_{i+j}$ be the map corresponding to $\iota_n \otimes \iota_{n'}$. The commutative diagram

$$
\begin{array}{ccc}
S^i K_n \wedge S^j K_{n'} & \xrightarrow{\ S^{i+j}\bar{\Delta}_{n,n'}\ } & S^{i+j} K_{n+n'} \\
\Big\downarrow{\sigma^i \wedge \sigma^j} & & \Big\downarrow{\sigma^{i+j}} \\
K_{n+i} \wedge K_{n+j} & \xrightarrow{\ \bar{\Delta}_{n+i,n'+j}\ } & K_{n+n'+i+j}
\end{array}
$$

along with Proposition 14.1.2 allows us to assume that n and n' are large in which case the theorem follows from the definition of $\psi(\theta)$. $\qquad\square$

By composing with the diagonal, $\Delta^* : H^*(X \times X) \to H^*(X)$, Theorem 14.2.1 gives

Corollary 14.2.2 (Cartan Formula) *Let $f \in H^n(X)$ and $g \in H^{n'}(X)$ be cohomology classes and let θ be a stable cohomology operation of degree k. Then $\theta(f \otimes g) = \sum_{s+t=k}(-1)^{tn'}\left(\theta'_s(f)\right)\left(\theta''_t(g)\right)$ where $\psi(\theta) = \sum_{s+t=k}\theta'_s \otimes \theta''_t$.* $\square$

Theorem 14.2.3 $\psi : \mathcal{A} \to \mathcal{A} \otimes \mathcal{A}$ *is a coalgebra structure which turns $\mathcal{A}$ into an associative coassociative Hopf algebra over $\mathbb{Z}/p\mathbb{Z}$.*

Proof Most of the conditions are trivial to check. Coassociativity follows from the commutative diagram

$$
\begin{array}{ccc}
K_i \times K_j \times K_k & \xrightarrow{\;K_i \times \Delta_{j,k}\;} & K_i \times K_{j+k} \\[4pt]
\downarrow{\scriptstyle \Delta_{i,j} \times K_k} & & \downarrow{\scriptstyle \Delta_{i,j+k}} \\[4pt]
K_{i+j} \times K_k & \xrightarrow{\;\Delta_{i+j,k}\;} & K_{i+j+k}
\end{array}
$$

The condition $\psi(\beta \circ \alpha) = \psi(\beta)\psi(\alpha)$ follows by applying the external Cartan formula to compute β on $\alpha(\iota_i \otimes \iota_j) \in H^*(K_i \times K_j)$. $\square$

3 Construction of the Steenrod Reduced Power Operations

Let W_* be the standard $(\mathbb{Z}/p\mathbb{Z})[\mathbb{Z}/p\mathbb{Z}]$-free resolution of $\mathbb{Z}/p\mathbb{Z}$ (the reduction modulo p of the resolution described in Section 4.3). As in Section 4.3, we will use the notation T for a generator of $\mathbb{Z}/p\mathbb{Z}$ and e_i for the generator of W_i and $N = 1 + T + \ldots + T^{p-1}$. We will write $w_i \in W^*$ for the dual of e_i with respect to the basis $\{T^j e_i\}$ of W_*.

Let the group $\mathbb{Z}/p\mathbb{Z}$ act on $S_*(X)^{\otimes p}$ by permuting the factors and act on $W_* \otimes S_*(X)$ by the standard action on W_* and trivial action on $S_*(X)$. By acyclic models (Theorem 5.5.1) there is a natural chain map (unique up to chain homotopy) $\theta_X : W_* \otimes S_*(X) \to S_*(X)^{\otimes p}$ extending the natural transformation $T^j e_0 \otimes x \mapsto x^{\otimes p}$ in degree 0. Recall that the method of acyclic models involves constructing chain maps and chain homotopies by choosing solutions to equations for each generator of the model chain complexes, which in this case are $W_* \otimes S_*(\Delta^n)$. By using the $(\mathbb{Z}/p\mathbb{Z})$-action to extend the choices for $e_i \otimes S_*(\Delta^n)$ to $T^j e_i \otimes S_*(\Delta^n)$, we can arrange for θ to commute with the action of $\mathbb{Z}/p\mathbb{Z}$. Set $D_i(x) = \tilde{\theta}(e_i \otimes x^{\otimes p}) \in S^{p|x|-i}$ where $\tilde{\theta} : W_* \otimes S^*(X)^{\otimes p} \to S^*(X)$ is the $(\mathbb{Z}/p\mathbb{Z})[\mathbb{Z}/p\mathbb{Z}]$-module map given by $\langle \tilde{\theta}(e_m \otimes y), a \rangle = (-1)^{m|y|}\langle y, \theta(e_m \otimes a)\rangle$. (See the end of Section 4.4 for the definition of the tensor product of a chain complex and a cochain complex.) Equivalently, $\theta^*(x^{\otimes p}) = \sum_{i=0}^q N w_i \otimes D_i(x)$ since $x^{\otimes p}$ is invariant under the action of $\mathbb{Z}/p\mathbb{Z}$. The operations $(D_i)_X : S^q(X) \to S^{pq-i}(X)$ are natural transformations. Since the composite $S_*(X) \hookrightarrow W_* \otimes S_*(X) \xrightarrow{\theta_X} S_*(X^{\otimes p})$ is chain homotopic to the composite $S_*(X) \xrightarrow{(\Delta^{p-1})_*} S_*(X^p) \xrightarrow{AW} S_*(X)^{\otimes p}$ (by acyclic models), $D_0(x)$ is cohomologous to x^p (i.e., they differ by a coboundary).

Theorem 14.3.1 *If y is cohomologous to z in $S^*(X)$ then $D_i(y)$ is cohomologous to $D_i(z)$ and so D_i induces a well defined map on cohomology.*

Note We do not yet know that D_i is a homomorphism, so it does not suffice to consider the case $z = 0$.

Proof Let I denote the chain complex with R-module basis $c \in I_1$, $b \in I_0$, $a \in I_0$ and differential given by $dc = b - a$. Thus I is the normalized chain complex of the simplicial set $\mathbb{Z}I$. The augmentation given by $\epsilon_I(c) = 0$, $\epsilon_I(b) = 1$, $\epsilon_I(a) = 1$ induces a chain homotopy equivalence $\epsilon_I : I \cong \underline{\mathbb{Z}/p\mathbb{Z}}$, where $\underline{\mathbb{Z}/p\mathbb{Z}}$ denotes the complex consisting of $\mathbb{Z}/p\mathbb{Z}$ in degree 0 and 0 elsewhere. Let J be the kernel of the chain homotopy equivalence $\epsilon_I^{\otimes p} : I^{\otimes p} \to \underline{\mathbb{Z}/p\mathbb{Z}}$. Define $\alpha : W \to W \otimes J$ by $\alpha(w) = w \otimes (b^{\otimes p} - a^{\otimes p})$. Since $W \otimes J$ has zero homology, in particular $\alpha_* = 0$ on $H_0(\ )$ and so by Theorem 4.3.3 $\alpha \simeq 0$ as $(\mathbb{Z}/p\mathbb{Z})[\mathbb{Z}/p\mathbb{Z}]$-module chain maps; let $s : \alpha \simeq 0$ be a $(\mathbb{Z}/p\mathbb{Z})[\mathbb{Z}/p\mathbb{Z}]$-module chain homotopy.

Given $x \in S^*(X)$ such that $\delta x = y - z$, define $\phi_x : S^{-|x|-1}I \to S^*(X)$ by $\phi_x(c) = (-1)^{|x|+1}x$, $\phi(b) = y$, and $\phi(a) = z$, where $S^m I$ denotes the mth suspension of the chain complex I and the cochain complex $S^q(X)$ is regarded as a negatively graded chain complex. Let $de_i = r_i e_{i-1}$, where $r_i \in (\mathbb{Z}/p\mathbb{Z})[\mathbb{Z}/p\mathbb{Z}]$ is either N or $(T - 1)$ depending upon the parity of i.

$$d\tilde{\theta}(W_* \otimes \phi_x^{\otimes p})se_i = \tilde{\theta}(W_* \otimes \phi_x^{\otimes p})dse_i$$

$$= \tilde{\theta}(W_* \otimes \phi_x^{\otimes p})(\alpha - sd)e_i$$

$$= \tilde{\theta}(W_* \otimes \phi_x^{\otimes p})(e_i \otimes b^{\otimes p} - e_i \otimes a^{\otimes p} - sr_i e_{i-1})$$

$$= \tilde{\theta}(e_i \otimes y^{\otimes p}) - \tilde{\theta}(e_i \otimes z^{\otimes p}) - r_i\tilde{\theta}(W_* \otimes \phi_x^{\otimes p})se_{i-1}$$

$$= D_i(y) - D_i(z) - r_i\tilde{\theta}(W_* \otimes \phi_x^{\otimes p})se_{i-1}.$$

However $(\mathbb{Z}/p\mathbb{Z})[\mathbb{Z}/p\mathbb{Z}]$ acts trivially (i.e., through its augmentation) on $S^*(X)$ and $\epsilon(N) = \epsilon(T - 1) = 0$ in $\mathbb{Z}/p\mathbb{Z}$ and so $D_i(y) - D_i(z) = d\tilde{\theta}(W_* \otimes \phi_x^{\otimes p})se_i$ regardless of the parity of i. $\qquad\square$

Theorem 14.3.2 *Let x, y belong to $S^*(X)$ and $S^*(Y)$ respectively. If $p > 2$ then*

$$D_{2n}(x \otimes y) = (-1)^{|x||y|(p-1)/2} \sum_{i+j=n} D_{2i}(x) \otimes D_{2j}(y)$$

and

$$D_{2n+1}(x \otimes y) =$$

$$(-1)^{|x||y|(p-1)/2}\left(\sum_{i+j=n} D_{2i+1}(x) \otimes D_{2j}(y) + (-1)^{|x|}D_{2i}(x) \otimes D_{2j+1}(y)\right).$$

If $p = 2$ then

$$D_n(x \otimes y) = \sum_{i+j=n} D_i(x) \otimes D_j(y).$$

Proof Suppose $p > 2$. Since $W_* \otimes W_*$ is an acyclic resolution of $(\mathbb{Z}/p\mathbb{Z}) \oplus (\mathbb{Z}/p\mathbb{Z})$, the diagonal map $\mathbb{Z}/p\mathbb{Z} \to (\mathbb{Z}/p\mathbb{Z}) \oplus (\mathbb{Z}/p\mathbb{Z})$ induces a $(\mathbb{Z}/p\mathbb{Z})[\mathbb{Z}/p\mathbb{Z}]$-module chain map $\mu : W_* \to W_* \otimes W_*$ (unique up to chain homotopy). Explicitly, one such map is given on $(\mathbb{Z}/p\mathbb{Z})[\mathbb{Z}/p\mathbb{Z}]$-module generators by

$$\mu(e_{2n}) \;=\; \sum_{i+j=n} e_{2i} \otimes e_{2j} \;+\; \sum_{i+j=n-1} \sum_{0 \le s < t < p} T^s e_{2i+1} \otimes T^t e_{2j+1}$$

and

$$\mu(e_{2n+1}) \;=\; \sum_{i+j=n} e_{2i} \otimes e_{2j+1} \;+\; \sum_{i+j=n} e_{2i+1} \otimes T e_{2j}.$$

For any spaces X and Y, let $\theta'_{X \times Y}$ be the natural composition

$$W_* \otimes S_*(X) \otimes S_*(Y) \xrightarrow{\mu \otimes S_*(X) \otimes S_*(Y)} W \otimes W \otimes S_*(X) \otimes S_*(Y) \xrightarrow{W \otimes \tau \otimes S_*(Y)}$$

$$W \otimes S_*(X) \otimes W \otimes S_*(Y) \xrightarrow{\theta_X \otimes \theta_Y} S_*(X)^{\otimes p} \otimes S_*(Y)^{\otimes p} \xrightarrow{\tau} \big(S_*(X) \otimes S_*(Y)\big)^{\otimes p},$$

where τ denotes maps which interchange factors (in the graded sense). Since $\theta'_{X \times Y}$ extends the same map in degree 0 that $\theta_{X \times Y}$ does, it is chain homotopic to $\theta_{X \times Y}$ by acyclic models. Therefore D_k can be computed on $S^*(X \times Y)$ using $\tilde{\theta}'$.

$$D_{2n}(x \otimes y) = \tilde{\theta}'\big(e_{2n} \otimes (x \otimes y)^{\otimes p}\big)$$

$$= (-1)^{|x||y|p(p-1)/2} \sum_{i+j=n} \tilde{\theta}_X\big(e_{2i} \otimes x^{\otimes p}\big) \otimes \tilde{\theta}_Y\big(e_{2j} \otimes y^{\otimes p}\big)$$

$$+ (-1)^{|x||y|p(p-1)/2} \sum_{i+j=n-1} \sum_{0 \le s < t < p}$$

$$\tilde{\theta}_X\big(T^s e_{2i+1} \otimes x^{\otimes p}\big) \otimes \tilde{\theta}_Y\big(T^t e_{2j+1} \otimes y^{\otimes p}\big)$$

$$= (-1)^{|x||y|(p-1)/2} \sum_{i+j=n} D_{2i}(x) \otimes D_{2j}(y)$$

$$+ (-1)^{|x||y|p(p-1)/2} \sum_{i+j=n-1} \sum_{0 \le s < t < p}$$

$$\tilde{\theta}_X\big(e_{2i+1} \otimes x^{\otimes p}\big) \otimes \tilde{\theta}_Y\big(e_{2j+1} \otimes y^{\otimes p}\big)$$

$$= (-1)^{|x||y|(p-1)/2} \sum_{i+j=n} D_{2i}(x) \otimes D_{2j}(y)$$

$$+ (-1)^{|x||y|p(p-1)/2} \sum_{i+j=n-1} \binom{p}{2}$$

$$\tilde{\theta}_X\big(e_{2i+1} \otimes x^{\otimes p}\big) \otimes \tilde{\theta}_Y\big(e_{2j+1} \otimes y^{\otimes p}\big)$$

$$= (-1)^{|x||y|(p-1)/2} \sum_{i+j=n} D_{2i}(x) \otimes D_{2j}(y)$$

using that $\mathbb{Z}/p\mathbb{Z}$ acts trivially on $S^*(X) \otimes S^*(Y)$, that $p(p-1)$ interchanges are required to go from $(x \otimes y)^{\otimes p}$ to $x^{\otimes p} \otimes y^{\otimes p}$, and that $(-1)^{pk} = (-1)^k$.

The other cases are similar. $\qquad\square$

We now introduce the upper notation, D^k, which has the advantage that its degree is given by the index k rather than depending on both the index and the degree of the element on which it operates. We also introduce a unit modulo p which eliminates the signs in the preceding coproduct formulas. Set

$$D^k(x) = (-1)^{[k/2]}\nu(|x|)D_{|x|(p-1)-k}(x) \in S^{|x|+k}(X)$$

where

$$\nu(q) = \begin{cases} 1 & \text{if } p = 2; \\ (-1)^{q(q-1)(p-1)/4}\left(\left(\tfrac{p-1}{2}\right)!\right)^q & \text{if } p > 2 \end{cases}$$

and $[t]$ is the greatest integer not exceeding t. We shall see later that $D^k(x) = 0$ if $k < 0$ and that if $p > 2$ then $D^k(x) = 0$ unless either $k \equiv 0 \pmod{2(p-1)}$ or $k \equiv 1 \pmod{2(p-1)}$. Since $\nu(a+b) = (-1)^{ab(p-1)/2}\nu(a)\nu(b)$, translating Theorem 14.3.2 into upper notation yields

Theorem 14.3.3 *Let x, y belong to $S^*(X)$ and $S^*(Y)$ respectively. If $p > 2$ then*

$$D^{2n}(x \otimes y) = \sum_{i+j=n} D^{2i}(x) \otimes D^{2j}(y)$$

and

$$D^{2n+1}(x \otimes y) = \sum_{i+j=n} D^{2i+1}(x) \otimes D^{2j}(y) + (-1)^{|x|}D^{2i}(x) \otimes D^{2j+1}(y).$$

If $p = 2$ then

$$D^n(x \otimes y) = \sum_{i+j=n} D^i(x) \otimes D^j(y).$$

$\square$

For each q, the natural map D^n is a cohomology operation $K_q \to K_{q+n}$.

Lemma 14.3.4 $D^0 : H^1(S^1) \to H^1(S^1)$ *is the identity operation.*

Proof The lemma is equivalent to the claim that D_{p-1} is multiplication by $(-1)^m m!$ on $H^1(S^1)$, where $m = (p-1)/2$.

Once again let I denote the chain complex with $(\mathbb{Z}/p\mathbb{Z})[\mathbb{Z}/p\mathbb{Z}]$-basis $c \in I_1$, $b \in I_0$, $a \in I_0$ and differential given by $dc = b-a$. Define $\epsilon : I \to I$ by $\epsilon(b) = \epsilon(a) = a$ and $\epsilon(c) = 0$. Define a chain homotopy $s : 1_I \simeq \epsilon$ by $s(a) = 0$, $s(b) = c$, and $s(c) = 0$. Define $S : I^{\otimes p} \to I^{\otimes p}$ by $S = \sum_{i=0}^{p-1} \epsilon^{\otimes i} \otimes s \otimes I^{\otimes(p-i-1)}$. In the following we will sometimes write xy for $x \otimes y$. The definition of S implies:

$$S(a^p) = 0; \tag{1}$$

$$S(b^p) = \sum_{i=0}^{p-1} a^i c b^{p-i-1}; \tag{2}$$

$$S(a^k cx) = 0 \quad \text{for } k \geq 0 \text{ and any } x; \tag{3}$$

$$S(b^t a^s cx) = \sum_{i=0}^{t-1} a^i c b^{t-i-1} a^s cx \quad \text{for } s \geq 0,\ t \geq 1 \text{ and any } x. \tag{4}$$

Sublemma 14.3.5 $dS + Sd = 1 - \epsilon^{\otimes p}$.

Proof Since $d\epsilon = \epsilon d = 0$ and $\epsilon(x) = 0$ when $|x| > 0$,

$$(dS + Sd)(x_1 x_2 \cdots x_p)$$

$$= \sum_{i=1}^{p} (-1)^{|x_1| + \ldots + |x_{i-1}|} (\epsilon x_1) \cdots (\epsilon x_{i-1})(ds x_i) x_{i+1} \cdots x_p$$

$$+ \sum_{i=1}^{p} \sum_{j>i} (-1)^{|x_1| + \ldots + |x_{j-1}|} (\epsilon x_1) \cdots (\epsilon x_{i-1})(s x_i) x_{i+1} \cdots x_{j-1}(dx_j) x_{j+1} \cdots x_p$$

$$+ \sum_{j=1}^{p} \sum_{i<j} (-1)^{|x_1| + \ldots + |x_{j-1}| + 1} (\epsilon x_1) \cdots (\epsilon x_{i-1})(s x_i) x_{i+1} \cdots x_{j-1}(dx_j) x_{j+1} \cdots x_p$$

$$+ \sum_{j=1}^{p} (-1)^{|x_1| + \ldots + |x_{j-1}|} (\epsilon x_1) \cdots (\epsilon x_{j-1})(sd x_j) x_{j+1} \cdots x_p$$

$$= \sum_{i=1}^{p} (-1)^{|x_1| + \ldots + |x_{i-1}|} (\epsilon x_1) \cdots (\epsilon x_{i-1})\big((1 - \epsilon)x_i\big) x_{i+1} \cdots x_p$$

$$= (1 - \epsilon^{\otimes p})(x_1 x_2 \cdots x_p).$$

$\square$

Define $\theta : W_* \otimes I \to I^{\otimes p}$ by:

$$\theta(e_0 \otimes a) = a^p;$$
$$\theta(e_0 \otimes b) = b^p;$$
$$\theta(e_j \otimes a) = \theta(e_j \otimes b) = 0 \quad \text{for } j > 0;$$

$$\theta(e_{2i} \otimes c) = i! \sum_{\alpha_0 + \beta_0 + \ldots + \alpha_i + \beta_i = p - 2i - 1} a^{\alpha_0} c b^{\beta_0} c a^{\alpha_1} c b^{\beta_1} c \cdots c a^{\alpha_i} c b^{\beta_i};$$

$$\theta(e_{2i+1} \otimes c) = i! \sum_{\alpha_0 + \beta_0 + \ldots + \alpha_i + \beta_i = p - 2i - 2} c a^{\alpha_0} c b^{\beta_0} c a^{\alpha_1} c b^{\beta_1} c \cdots c a^{\alpha_i} c b^{\beta_i}.$$

Sublemma 14.3.6 *If $|x| > 0$ then $S\theta dx = \theta x$.*

Proof It is immediate from the definitions that $\theta(e_i \otimes a) = S\theta d(e_i \otimes a) = \theta(e_i \otimes b) = S\theta d(e_i \otimes b) = 0$ for $i > 0$ and also $S\theta d(e_0 \otimes c) = S\theta(e_0 \otimes b - e_0 \otimes a) = S(b^p - a^p) = \sum_{r=0}^{p-1} a^r c b^{p-r-1} = \theta(e_0 \otimes c)$.

Since T acts trivially on I, using equation (3) gives

$$S\theta d(e_{2i+1} \otimes c) = S\theta\big((T - 1)e_{2i} \otimes c\big) - S\theta(e_{2i+1} \otimes dc)$$

$$= i! S(T - 1) \sum_{\alpha_0 + \beta_0 + \ldots + \alpha_i + \beta_i = p - 2i - 1} a^{\alpha_0} c b^{\beta_0} c a^{\alpha_1} c b^{\beta_1} c \cdots c a^{\alpha_i} c b^{\beta_i}$$

$$= i! \sum_{\substack{\alpha_0 + \beta_0 + \ldots + \alpha_i + \beta_i = p - 2i - 1 \\ \beta_i > 0}} c a^{\alpha_0} c b^{\beta_0} c a^{\alpha_1} c b^{\beta_1} c \cdots c a^{\alpha_i} c b^{\beta_i - 1}$$

$$= \theta(e_{2i+1} \otimes c).$$

Similarly for $i > 0$,

$$S\theta d(e_{2i} \otimes c) = S\theta(Ne_{2i-1} \otimes c) - S\theta(e_{2i} \otimes dc)$$

$$= (i-1)! SN \sum_{\alpha_0 + \ldots + \beta_{i-1} = p - 2i - 2} ca^{\alpha_0} cb^{\beta_0} ca^{\alpha_1} cb^{\beta_1} c \cdots ca^{\alpha_{i-1}} cb^{\beta_{i-1}}$$

$$= (i-1)! S \sum_{\alpha_0 + \ldots + \beta_{i-1} = p - 2i - 2} \sum_{j=0}^{i-1} \sum_{r=1}^{\beta_j}$$

$$b^r ca^{\alpha_j + 1} cb^{\beta_j + 1} c \cdots a^{\alpha_{i-1}} cb^{\beta_{i-1}} ca^{\alpha_0} cb^{\beta_0} c \ldots ca^{\alpha_j - 1} cb^{\beta_j - 1} ca^{\alpha_j} cb^{\beta_j - r}$$

$$= (i-1)! \sum_{\alpha_0 + \ldots + \beta_{i-1} = p - 2i - 2} \sum_{j=0}^{i-1} \sum_{r=1}^{\beta_j} \sum_{t=1}^{r-1}$$

$$a^t cb^{r-t-1} ca^{\alpha_j + 1} cb^{\beta_j + 1} c \cdots a^{\alpha_{i-1}} cb^{\beta_{i-1}} ca^{\alpha_0} cb^{\beta_0} c \ldots ca^{\alpha_j} cb^{\beta_j - r}$$

$$= (i-1)! \sum_{j=0}^{i-1} \phi(e_{2i} \otimes c)/i!$$

$$= \phi(e_{2i} \otimes c).$$

$$\qquad \square$$

Sublemma 14.3.7 θ *is a chain map.*

Proof From the definitions,

$$d\theta(e_0 \otimes a) = \theta d(e_0 \otimes a) = d\theta(e_0 \otimes b) = \theta d(e_0 \otimes b) = 0,$$

$$d\theta(e_1 \otimes a) = \theta d(e_1 \otimes a) = d\theta(e_1 \otimes b) = \theta d(e_1 \otimes b) = 0,$$

and

$$d\theta(e_0 \otimes c) = d \sum_{\alpha_0 + \beta_0 = p - 1} a^{\alpha_0} cb^{\beta_0}$$

$$= \sum_{\alpha_0 + \beta_0 = p - 1} a^{\alpha_0}(b - a) b^{\beta_0} = b^p - a^p = \theta d(e_0 \otimes c).$$

Therefore $d\theta(x) = \theta(x)$ for $|x| \le 1$. For $|x| > 1$,

$$d\theta x = dS\theta dx = (1 - \epsilon^{\otimes p} - Sd)\theta dx = \theta dx - \epsilon^{\otimes p}\theta dx - Sd\theta_I dx = \theta dx$$

using that $\epsilon^{\otimes p}\theta dx = 0$ since $|\theta dx| > 0$, and $d\theta dx = \theta ddx = 0$ by induction. $\quad \square$

It follows from Sublemma 14.3.7 that θ_I for the interval I can be chosen to be the θ above and thus by naturality θ can be used to compute $D_X(x)$ for any space X and any generator $x : I \to X \in S_1(X)$. Represent the generator of $H_1(S^1)$ by the quotient map $f : I \to I/\partial I = S^1$ and let v be a generator of $H^1(S^1)$ such that $\langle v, f \rangle = 1$. Then $f = f_*(c)$, since c can be thought of as $1_I \in S_1(I)$.

Let $D_{p-1}(v) = \gamma v$ and set $m = (p-1)/2$ as before. Then

$$\begin{aligned}
\gamma &= \langle \gamma v, f \rangle \\
&= \langle D_{p-1}(v), f \rangle \\
&= \langle w_{p-1}, e_{p-1} \rangle \langle D_{p-1}(v), f \rangle \\
&= (-1)^{p-1} \langle w_{p-1} \otimes D_{p-1}(v), e_{p-1} \otimes f \rangle \\
&= (-1)^{p-1} \langle (\theta_{S^1})^* v^p, e_{p-1} \otimes f \rangle \\
&= \langle v^p, \theta_{S^1}(e_{p-1} \otimes f) \rangle \\
&= \langle v^p, f_* \theta_I(e_{p-1} \otimes c) \rangle
\end{aligned}$$

$$= m! \left\langle v^p, f_* \left(\sum_{\alpha_0 + \beta_0 + \ldots + \alpha_m + \beta_m = p - 2m - 1} a^{\alpha_0} cb^{\beta_0} ca^{\alpha_1} cb^{\beta_1} c \cdots ca^{\alpha_m} cb^{\beta_m} \right) \right\rangle$$

$$= m! \left\langle v^p, f_* \left(\sum_{\alpha_0 + \beta_0 + \ldots + \alpha_m + \beta_m = 0} a^{\alpha_0} cb^{\beta_0} ca^{\alpha_1} cb^{\beta_1} c \cdots ca^{\alpha_m} cb^{\beta_m} \right) \right\rangle$$

$$\begin{aligned}
&= m! \langle v^p, f_*(c^p) \rangle \\
&= m! \langle v^p, f^p \rangle \\
&= (-1)^{p(p-1)/2} m! \langle v, f \rangle^p \\
&= (-1)^m m!
\end{aligned}$$

as desired. $\qquad\square$

The suspension isomorphism $\sigma : \tilde{H}^q(X) \cong \tilde{H}^{q+1}(SX)$ may be chosen to be $(-1)^{q+1}$ times the composite

$$\tilde{H}^q(X) \cong \tilde{H}^q(X) \otimes \tilde{H}^1(S^1) \cong \tilde{H}^{q+1}(X \wedge S^1) = \tilde{H}^{q+1}(SX).$$

Let v denote a generator of $H^1(S^1)$. Then for $x \in H^*(X)$, applying Theorem 14.3.3 gives

$$\begin{aligned}
D^n(\sigma(x)) &= (-1)^{|x|+1} D^n(x \otimes v) = (-1)^{|x|+1} D^n(x) \otimes D^0(v) \\
&= (-1)^{|x|+1} D^n(x) \otimes v = (-1)^{|x|+1+n+|x|+1} \sigma(D^n(x)) = (-1)^n \sigma(D^n(x)).
\end{aligned}$$

Thus D^n a stable cohomology operation of degree n and its coproduct is described by Theorem 14.3.3. In particular, $[K_q, K_{q+n}] = 0$ for $n < 0$ and so $D^n = 0$ for $n < 0$. Also, all elements of $[K_q, K_q]$ are multiples of the identity so D^0_X is multiplication by a constant which is independent of the space X. Considering that $D^0_{S^1}(v) = v$, this constant is 1, so D^0 is the identity operation on all spaces. Similarly D^1 corresponds to an element of $H^{n+1}(K_n) = \mathbb{Z}/p\mathbb{Z}$ so $D^1 = \lambda \beta$ for some $\lambda \in \mathbb{Z}/p\mathbb{Z}$, and λ can be found by evaluating D^1 on any space. It turns out that this coefficient is also 1 and more generally that $D^{2n+1} = \beta D^{2n}$, however we shall not need to use this.

Theorem 14.3.8 *If $p > 2$ then $D^{2i}(x) = 0$ unless either $i \equiv 0 \pmod{2(p-1)}$ or $i \equiv 1 \pmod{2(p-1)}$.*

This will appear as a corollary of results in the next section, so we will defer its proof until after Theorem 14.4.4. A direct proof is given in [EpS] by proving some facts about group homology, and in particular that $H_i^{\mathrm{alg}}(\mathbb{Z}/p\mathbb{Z}) \to H_i^{\mathrm{alg}}(S_p)$ is zero unless either $i \equiv 0 \pmod{2(p-1)}$ or $i \equiv 1 \pmod{2(p-1)}$, where S_p is the symmetric group on p symbols.

In the sequel, statements will sometimes be made in the form they have for $p > 2$ with the necessary modifications for $p = 2$ appearing in square brackets [ ].

Define $\mathcal{P}^s = D^{2(p-1)s}$. [When $p = 2$, define instead $\mathrm{Sq}^s = D^s$.] In statements which are indicated as holding for all primes, $\mathcal{P}$ should be interpreted as standing for Sq when the prime is 2.

In summary we have the following, where we have not yet proved statement (4) in the case $p > 2$, pending the proof of Theorem 14.3.8.

Theorem 14.3.9 *There exist stable cohomology operations* $\mathcal{P}^n$ *with the following properties:*

1) $\mathcal{P}^0(x) = x$;
2) $\mathcal{P}^n(x) = x^p$ if $2n = |x|$ [or $\mathrm{Sq}^{|x|} = x^2$ when $p = 2$];
3) $\mathcal{P}^n(x) = 0$ if $2n > |x|$ [or $\mathrm{Sq}^n = 0$ for $n > |x|$ when $p = 2$];
4) $\psi(\mathcal{P}^n) = \sum_{i+j=n} \mathcal{P}^i \otimes \mathcal{P}^j$. $\qquad\qquad\square$

4 Calculation of the Steenrod Algebra

We wish to describe the Steenrod algebra in terms of generators and relations. Since $H^*(\ ;\mathbb{Z}/p\mathbb{Z})$ is represented by the Ω-spectrum $\{K(\mathbb{Z}/p\mathbb{Z}, n)\}$, to do this we calculate $[K(\mathbb{Z}/p\mathbb{Z}, n), K(\mathbb{Z}/p\mathbb{Z}, n + k)] = H^{n+k}\big(K(\mathbb{Z}/p\mathbb{Z}, n)\big)$ for all n and k.

Suppose first that $p > 2$. Since $\beta^2 = 0$, any composition of Steenrod reduced power operations can be written in the form $\beta^{\epsilon_1} \mathcal{P}^{i_1} \beta^{\epsilon_2} \mathcal{P}^{i_2} \ldots \mathcal{P}^{i_k} \beta^{\epsilon_{k+1}}$, with ϵ_j equal to 0 or 1. The above element will sometimes be denoted $\mathcal{P}^{(I,E)}$, where $I = (i_1, i_2, \ldots, i_k)$ and $E = (\epsilon_1, \epsilon_2, \ldots, \epsilon_{k+1})$ and its length k will be denoted $\mathrm{Len}(I, E)$. The element $\mathcal{P}^{(I,E)}$ (and the pair of sequences (I, E)) will be called *admissible* if $i_j \geq p i_{j+1} + \epsilon_{j+1}$ for all j. If (I, E) is admissible, its *excess*, $e(I, E)$, is defined by $e(I, E) = \sum_{j=1}^{k}\big(2(i_j - p i_{j+1}) - \epsilon_{j+1}\big)$. If all of the ϵ's are zero we will simply write $\mathcal{P}^I$.

For $p = 2$, a composition $\mathrm{Sq}^{i_1} \mathrm{Sq}^{i_2} \ldots \mathrm{Sq}^{i_k}$ will be denoted $\mathrm{Sq}^{(i_1, i_2, \ldots, i_k)}$. The sequence I is called admissible if $i_j \geq 2 i_{j+1}$ for all i, and if I is admissible then its excess is defined by $e(I) = \sum_{j=1}^{k}(i_j - 2 i_{j+1})$.

Theorem 14.4.1 $H^*(K_n) = S[\{\mathcal{P}^{(I,J)}(\iota_n) \mid e(I, J) < n\}]$.

Proof For $p = 2$ and $n = 1$ we know that $K_1 = B(\mathbb{Z}/\mathbb{Z}/2\mathbb{Z}) = \mathbb{R}P^\infty$ so $H^*(K_1) = S[v]$ with $|v| = 1$ and thus the theorem holds in this case. For $p > 2$, the short exact sequence of topological groups $0 \longrightarrow \mathbb{Z}/p\mathbb{Z} \longrightarrow S^1 \xrightarrow{p} S^1 \longrightarrow 0$ (where $\mathbb{Z}/p\mathbb{Z}$ is given the discrete topology) yields a homotopy fibration $B(\mathbb{Z}/p\mathbb{Z}) \to \mathbb{C}P^\infty \to \mathbb{C}P^\infty$, and the Serre spectral sequence can be applied to the induced homotopy fibration $S^1 \to B(\mathbb{Z}/p\mathbb{Z}) \to \mathbb{C}P^\infty$ to calculate $H^*(K_1) = H^*\big(B(\mathbb{Z}/p\mathbb{Z})\big)$. This gives $H^*(K_1) = S[u, v]$ where $|u| = 1$, $|v| = 2$, and $\beta(u) = v$. Therefore the theorem holds for $k = 1$; suppose by induction that it holds for $k \leq n$. Since by Theorem 11.9.6 the transgression τ is induced by a map of spaces and the Steenrod operations are natural, all of the generators of $H^*(K_n)$ are transgressive and $\tau\big(\mathcal{P}^{(I,E)}(\iota_n)\big) = \mathcal{P}^{(I,E)}\big(\tau(\iota_n)\big) = \mathcal{P}^{(I,E)}(\iota_{n+1})$. Furthermore, since $\mathcal{P}^{|x|/2} = x^p$ [respectively $\mathrm{Sq}^{|x|} = x^2$], the p^tth powers of the even degree generators are also images of ι_n under compositions of Steenrod reduced power operations so they too are transgressive. Applying Borel's Theorem (11.9.14) gives that $H^*(K_{n+1})$ is a free commutative algebra on $\{\mathcal{P}^{(I,E)}(\iota_{n+1}) \mid e(I, E) \leq n\} \cup X \cup X'$ where X consists of those elements $\mathcal{P}^{(I,E)}(\iota_{n+1})$ such that $\mathcal{P}^{(I,E)}(\iota_n)$ is a p^tth power of one of the polynomial generators and $X' = \beta(X)$ for $p > 2$ while $X' = \emptyset$ when $p = 2$. It is easy to check that $X \cup X'$ consists precisely of the admissible sequences of excess n. $\square$

It follows from Theorem 14.4.1 that an additive basis for the Steenrod algebra consists of all admissible sequences $\mathcal{P}^{(I,E)}$ of Steenrod reduced power operations. There must therefore be formulas which equate an inadmissible composition of reduced power operations with a unique sum of admissible ones. These formulas are called the *Adem relations*. We will now show that each Steenrod operation is determined by its action on $(K_1)^n$ for sufficiently large n and this will give us a method of computing the Adem relations.

Let $v_1, \dots, v_n$ denote generators for the polynomial algebra

$$H^*\big((K_1)^n\big) = H^*\Big(\big(B(\mathbb{Z}/2\mathbb{Z})\big)^n\Big) = (\mathbb{Z}/2\mathbb{Z})[v_1, \dots, v_n]$$

where $|v_j| = 1$. Let $s_1, s_2, \dots, s_n$ be the elementary symmetric polynomials in $v_1, v_2, \dots, v_n$ and let s^I denote $s_1^{i_1} s_2^{i_2} \cdots s_n^{i_n}$ for $I = \{i_1, i_2, \dots, i_n\}$.

Theorem 14.4.2 *For any integer n, the map $\phi : \mathcal{A}_2 \to H^*\big((K_1)^n\big)$ given by* $\mathrm{Sq}^I \mapsto Sq^I(s_n)$ *is a monomorphism in degrees less or equal to n.*

Proof We shall write $\mathrm{sdeg}(m)$ (the "symmetric degree") for the order of a monomial $m \in (\mathbb{Z}/2\mathbb{Z})[s_1, s_2, \dots, s_n]$ to distinguish from its degree $|m|$ as an element of $H^*\big((K_1)^n\big)$.

It is clear that Sq^I takes symmetric polynomials to symmetric polynomials. Notice that if $j > 0$ then each monomial of $\mathrm{Sq}^j(s_t)$ has symmetric degree of at least 2 and that $\mathrm{Sq}^j(s_n) = s_j s_n$ for $j \leq n$. Let $I = (i_1, i_2, \dots, i_k)$ be admissible with $|I| \leq n$. Therefore $i_1 \leq n$ and so $\mathrm{Sq}^{i_1} s_n = s_{i_1} s_n$. Suppose by induction on k that $\phi\Big(\mathrm{Sq}^{(i_2, i_3, \dots, i_k)}\Big)$ has the form

$$s_n\big(s_{i_2} s_{i_3} \cdots s_{i_k} + (\text{terms } m \text{ having } \mathrm{sdeg}(m) > k - 1)\big).$$

Let $s_n m$ be a monomial in the preceding expression. Then $|m| = |\,\mathrm{Sq}^{(i_2, i_3, \dots, i_k)}\,| < i_1$ since I is admissible, so $\mathrm{Sq}^{i_1} m = 0$. Therefore,

$$\mathrm{Sq}^{i_1}(s_n m) = \sum_{q+r=i_1} (\mathrm{Sq}^q s_n)(\mathrm{Sq}^r m)$$

$$= \sum_{q+r=i_1} s_q s_n \, \mathrm{Sq}^r m$$

$$= s_{i_1} s_n m + s_n\big(\text{terms } m' \text{ having } \mathrm{sdeg}(m') \geq \mathrm{sdeg}(m) + 1\big).$$

Thus

$$\phi\big(\mathrm{Sq}^I\big) = s_n\big(s_{i_1} s_{i_2} s_{i_3} \cdots s_{i_k} + (\text{terms } m \text{ having } \mathrm{sdeg}(m) > k)\big).$$

Hence this expression holds for all admissible Sq^I with $|\,\mathrm{Sq}^I\,| \leq n$ and so they form a linearly independent set. $\square$

A similar result for odd primes is given in Theorem 14.4.4 below. Let $u, \dots, u_n$, $v_1, \dots, v_n$ denote generators for the free commutative algebra

$$H^*\big((K_1)^n\big) = H^*\Big(\big(B(\mathbb{Z}/p\mathbb{Z})\big)^n\Big) = E[u_1, \dots, u_n] \otimes (\mathbb{Z}/p\mathbb{Z})[v_1, \dots, v_n]$$

where $|u_j| = 1$ and $|v_j| = 2$. Let $\mathcal{P}[m]$ denote the composition $\mathcal{P}^{p^{m-1}} \mathcal{P}^{p^{m-2}} \dots$ $\mathcal{P}^p \mathcal{P}^1$. While the preceding proof for $p = 2$ used part (4) of Theorem 14.3.9 which is still unproven for $p > 2$, the proof of Theorem 14.4.4 uses the following lemma which follows from Theorem 14.3.3 and the first three parts of Theorem 14.3.9.

Lemma 14.4.3 *In $H^*\big(B(\mathbb{Z}/p\mathbb{Z})\big)$*

a) $\mathcal{P}[m](v) = v^{p^m}$ *and every other sequence (admissible or not) acts trivially on* v.

b) $\mathcal{P}[m]\beta(u) = v^{p^m}$ *and every other sequence (admissible or not) acts trivially on* u. $\square$

Theorem 14.4.4 *For any integer n, the map $\phi : \mathcal{A}_p \to H^*\big((K_1)^{2n}\big)$ given by $\mathcal{P}^{(I,E)} \mapsto \mathcal{P}^{(I,E)}(u_1 u_2 \cdots u_n v_{n+1} v_{n+2} \cdots v_{2n})$ is a monomorphism in degrees less or equal to n.*

Proof Suppose by induction that the theorem holds for integers less than n. Let $X_n = u_1 u_2 \cdots u_n v_{n+1} v_{n+2} \cdots v_{2n}$. Let

$$\mathcal{I}_q = \{(I, E) \mid (I, E) \text{ admissible and } |(I, E)| = q\}$$

for $q \leq n$. Suppose $\sum_{(I,E)\in\mathcal{I}_q} a_{(I,E)} \mathcal{P}^{(I,E)}(X_n) = 0$. We wish to show that $a_{(I,E)} = 0$ for all (I, E). Since there is a maximum possible length for elements of $\mathcal{I}_q$ in order for its degree not to exceed q, we may suppose by downward induction on m that $a_{(I,E)} = 0$ for all terms with $\mathrm{Len}(I, E) > m$. By part (b) of the lemma, terms in $\sum_{(I,E)\in\mathcal{I}} a_{(I,E)} \mathcal{P}^{(I,E)}(X_n)$ contributing to the coefficient of $v_n{}^{p^m} v_{2n}$ are in one-to-one correspondence with elements of $\mathcal{I}_q$ of length m ending with β. Equating this coefficient to zero yields an expression of the form $\sum a_{(I,E)} \mathcal{P}^{(I',E')}(X_{n-1}) = 0$ where (I, E) runs through all sequences of length m ending with β, and so these coefficients are 0 by the induction assumption on n. Similarly using part (a) of the lemma and examining the coefficient of $u_n{}^{p^m} v_{2n}$ shows that the rest of the coefficients of terms of length m are zero, completing the induction on m. $\square$

From the previous two theorems we see that all relations of $\mathcal{A}$ in degree q can be verified by evaluating the expression on $u_1 u_2 \cdots u_n v_{n+1} v_{n+2} \cdots v_{2n}$ [respectively $v_{n+1} v_{n+2} \cdots v_{2n}$] for sufficiently large n.

Proof of Theorem 14.3.8 Suppose that n is not congruent to either 0 or 1 $(\mathrm{mod}\, 2(p-1))$ and assume by induction that $D^{2k} = 0$ for all $k < n$ such that k is not congruent to either 0 or 1 $(\mathrm{mod}\, 2(p-1))$. Then Theorem 14.3.3 implies that $\psi(D^{2n}) = D^{2n} \otimes D^0 + D^0 \otimes D^{2n}$. For degree reasons, $D^{2n} \neq \mathcal{P}^{p^{q-1}} \mathcal{P}^{p^{q-2}} \ldots \mathcal{P}^p \mathcal{P}^1$, and $D^{2n} \neq \mathcal{P}^{p^{q-1}} \mathcal{P}^{p^{q-2}} \ldots \mathcal{P}^p \mathcal{P}^1 \beta$ for any q. Therefore D^{2n} acts trivially on the generators of $H^*\big((B(\mathbb{Z}/p\mathbb{Z}))^t\big)$ for all t and because of its coproduct this implies that D^{2n} acts trivially on all elements of $H^*\big((B(\mathbb{Z}/p\mathbb{Z}))^t\big)$. Therefore $D^{2n} = 0$ by Theorem 14.4.3. $\square$

Theorem 14.4.5 (Adem Relations) *For all p, if $a < pb$ then*

$$\mathcal{P}^a \mathcal{P}^b = \sum_{j=0}^{[a/p]} (-1)^{a+j} \binom{(p-1)(b-j) - 1}{a - pj} \mathcal{P}^{a+b-j} \mathcal{P}^j$$

and for $p > 2$, if $a \leq pb$ then

$$\mathcal{P}^a \beta \mathcal{P}^b = \sum_{j=0}^{[a/p]} (-1)^{a+j} \binom{(p-1)(b-j)}{a-pj} \beta \mathcal{P}^{a+b-j} \mathcal{P}^j$$

$$+ \sum_{j=0}^{[(a-1)/p]} (-1)^{a+j-1} \binom{(p-1)(b-j)-1}{a-pj-1} \mathcal{P}^{a+b-j} \beta \mathcal{P}^j.$$

Proof Consider the case $p > 2$. Set

$$c_j(a,b) = (-1)^{a+j} \binom{(p-1)(b-j)-1}{a-pj},$$

$$c'_j(a,b) = (-1)^{a+j} \binom{(p-1)(b-j)}{a-pj}, \quad \text{and}$$

$$c''_j(a,b) = (-1)^{a+j-1} \binom{(p-1)(b-j)-1}{a-pj-1}.$$

Notice that $c''_j(a,b) = c_j(a-1,b)$ and by Pascal's triangle ($\binom{m}{k} = \binom{m-1}{k} + \binom{m-1}{k-1}$), we see that $c'_j(a,b) + c''_j(a,b) = c_j(a,b)$. Let

$$R'(a,b) = \mathcal{P}^a \beta \mathcal{P}^b - \sum_{j=0}^{[a/p]} c'_j(a,b)\beta \mathcal{P}^{a+b-j} \mathcal{P}^j - \sum_{j=0}^{[(a-1)/p]} c''_j(a,b) \mathcal{P}^{a+b-j} \beta \mathcal{P}^j$$

and let $R(a,b)$ be the relation involving $\mathcal{P}^a \mathcal{P}^b$. Let

$$X_n = u_1 u_2 \cdots u_n v_{n+1} v_{n+2} \cdots v_{2n} \in H^{3n}\big((K_1)^{2n}\big).$$

By Theorem 4.4.3, to check the second Adem relation it suffices to check that $R'(a,b)(X_n) = 0$ for all n. Set $u = u_n$ and $v = v_{2n}$. By induction on n, it suffices to check that $R'(a,b)(uy) = 0$ and that $R'(a,b)(vy) = 0$ under the assumption that all Adem relations evaluate to 0 on y.

Consider the first of these equalities. Let $x = \beta(u)$.

$$\mathcal{P}^s \beta \mathcal{P}^t (uy) = \mathcal{P}^s \beta \big(x\mathcal{P}^t(y)\big) = \mathcal{P}^s \big(v\mathcal{P}^t(y) - u\beta \mathcal{P}^t\big)$$
$$= x^p \mathcal{P}^{s-1}\mathcal{P}^t(y) + x\mathcal{P}^s \mathcal{P}^t(y) - u\mathcal{P}^s \beta \mathcal{P}^t$$

and similarly $\beta \mathcal{P}^s \mathcal{P}^t(uy) = x\mathcal{P}^s \mathcal{P}^t(y) - u\beta \mathcal{P}^s \mathcal{P}^t(y)$. Therefore

$$R'(a,b)(uy) = x^p \left(\mathcal{P}^{a-1}\mathcal{P}^b(y) - \sum_j c''_j(a,b)\mathcal{P}^{a+b-1-j}\mathcal{P}^j(y) \right)$$

$$+ x\left(\mathcal{P}^a \mathcal{P}^b(y) - \sum_j (c'_j(a,b) + c''_j(a,b))\mathcal{P}^{a+b-j}\mathcal{P}^j(y) \right) - uR'(a,b)(y)$$

$$= x^p R(a-1,b)(y) + xR(a,b)(y) - uR'(a,b)(y) = 0.$$

For the second equality, by the same method we get

$$R'(a,b)(vy)$$
$$= vR'(a,b)(y) + v^p R'(a-p,b-1)(y) + v\Big(R'(a-1,b)(y) + R'(a,b-1)(y)$$
$$+ \sum_j (c'_j(a,b) + c'_{j+1}(a,b) - c'_j(a-1,b) + c'_j(a,b-1))\beta \mathcal{P}^{a+b-1-j}\mathcal{P}^j(y)$$
$$+ \sum_j (c''_j(a,b) + c''_{j+1}(a,b) - c''_j(a-1,b) + c''_j(a,b-1))\mathcal{P}^{a+b-1-j}\beta \mathcal{P}^j(y) \Big)$$

so it suffices to show that

$$c'_j(a,b) + c'_{j+1}(a,b) - c'_j(a-1,b) - c'_j(a,b-1) \equiv 0 \quad (\mathrm{mod}\ p)$$

and

$$c''_j(a,b) + c''_{j+1}(a,b) - c''_j(a-1,b) - c''_j(a,b-1) \equiv 0 \quad (\mathrm{mod}\ p).$$

For the first of these,

$$c'_j(a,b) + c'_{j+1}(a,b) - c'_j(a-1,b) - c'_j(a,b-1)$$

$$= (-1)^{a+j} \binom{(p-1)(b-j)}{a-pj} + (-1)^{a+j+1} \binom{(p-1)(b-j-1)}{a-pj-p}$$

$$- (-1)^{a+j-1} \binom{(p-1)(b-j)}{a-pj-1} - (-1)^{a+j} \binom{(p-1)(b-j-1)}{a-pj}$$

$$= (-1)^{a+j} \binom{(p-1)(b-j)+1}{a-pj} - (-1)^{a+j} \binom{(p-1)(b-j-1)}{a-pj-p}$$

$$- (-1)^{a+j} \binom{(p-1)(b-j-1)}{a-pj}$$

by Pascal's triangle. However this is congruent to zero by the mod p version of Pascal's triangle: $\binom{m}{k} = \binom{m-p}{k} + \binom{m-p}{k-p} \equiv 0 \ (\mathrm{mod}\ p)$. The second identity is also an application of Pascal's triangle and its mod p version.

The proof of the other Adem relation is similar (except a bit simpler), eventually reducing to an application of Pascal's triangle and Pascal's triangle mod p. $\qquad\square$

5 Dual of the Steenrod Algebra

Let $\mathcal{A}_*$ denote the dual $\mathrm{Hom}(\mathcal{A}, \mathbb{Z}/p\mathbb{Z})$ of the mod p Steenrod algebra. In the dual basis in $\mathcal{A}_*$ to the basis of admissible monomials in $\mathcal{A}$ set $\xi_k = (\mathcal{P}[k])^*$ for $k \geq 1$ and $\tau_k = (\mathcal{P}[k]\beta)^*$ for $k \geq 0$. This yields an algebra homomorphism $\phi : S[X] \to \mathcal{A}_*$ where $X = \{\tau_0, \xi_1, \tau_1, \dots, \xi_k, \tau_k, \dots\}$.

For a finite sequence $I = (e_0, r_1, e_1, r_2, \dots, e_k, r_k)$ with $r_j \geq 0$ and $e_j = 0$ or 1, we set $\omega(I) = \tau_0^{e_0} \xi_1^{r_1} \tau_1^{e_1} \cdots \tau_k^{e_k} \xi_k^{r_k} \in S[X]$. Order the variables in X by increasing degree, order the set $\mathcal{I}$ of monomials of $S[X]$ by the induced right lexicographical ordering, and order the sequences correspondingly. For $I \in \mathcal{I}$ of length k, define $\{s_k, s_{k-1}, \dots, s_1\}$ inductively (downward) by $s_j = ps_{j+1} + r_j + e_j$ and define a function $\theta : \mathcal{I} \to \mathcal{A}$ by $\theta(I) = \beta^{e_0} \mathcal{P}^{s_1} \beta^{e_1} \mathcal{P}^{s_2} \beta^{e_2} \cdots \mathcal{P}^{s_k} \beta^{e_k}$. Then θ is a bijection from the monomials of $S[X]$ to the basis of admissible monomials for $\mathcal{A}$.

Theorem 14.5.1 $\langle \theta(I), \omega(I) \rangle = \pm 1$ *and* $\langle \theta(I), \omega(J) \rangle = 0$ *if* $I < J$.

Proof Every inadmissible sequence acts trivially on $H^1(K_1)$ and $H^2(K_1)$ (Lemma 14.4.3) so cannot contain the term $\mathcal{P}[k]$ or $\mathcal{P}[k]\beta$ when written admissibly. Thus $\langle I, \xi_k \rangle = 0$ if (I, E) is inadmissible and similarly $\langle I, \tau_k \rangle = 0$ if (I, E) is inadmissible. Therefore by the Cartan formula, a term of the form $\mathcal{P}[k] \otimes M$ appears in $\psi(\theta(I))$ if and only if $\omega(I)$ is divisible by ξ_k, in which case $\mathcal{P}[k] \otimes \theta(I')$ appears, where $\omega(I') = \omega(I)/\xi_k$. Similarly $\mathcal{P}[k]\beta \otimes N$ appears if and only if $\omega(I)$ is divisible by τ_k in which case $\mathcal{P}[k]\beta \otimes \theta(I')$ appears where $\omega(I') = \omega(I)/\tau_k$. Suppose $I \leq J$. Write $\omega(J) = z\omega(J')$ where z is the highest order variable dividing $\omega(J)$. Consider first the case where $z = \tau_k$ for some k. $\langle \theta(I), \omega(J) \rangle = \langle \psi(\theta(I)), \tau_k \otimes \omega(J') \rangle$. Since only $\mathcal{P}[k]\beta$ evaluates nontrivially on τ_k this equals $\pm\langle\theta(I'),\omega(J')\rangle$ if $\omega(I) = \tau_k\omega(I')$ and 0 if $\omega(I)$ is not divisible by τ_k. Similarly if $z = \xi_k$ then

$$\langle\theta(I),\omega(J)\rangle = \begin{cases} \pm\langle\theta(I'),\omega(J')\rangle & \text{if } \omega(I) = \xi_k\omega(I'); \\ \\ 0 & \text{if } \omega(I) \text{ is not divisible by } \xi_k. \end{cases}$$

Unless $I = J$, repeated application of these two formulas will eventually give 0, and gives ± 1 if $I = J$. $\qquad\square$

Corollary 14.5.2 $\phi : S[X] \to \mathcal{A}_*$ *is an isomorphism.*

Proof The preceding theorem says the matrix for ϕ is triangular with ± 1's on the diagonal. $\qquad\square$

Having completed our description of the multiplication on the Hopf algebra $\mathcal{A}_*$, we now describe its comultiplication.

Theorem 14.5.3

1) $\psi\xi_k = \sum_{j=0}^{k} \xi_{k-j}^{p^j} \otimes \xi_j$;

2) *if* $p > 2$ *then* $\psi\tau_k = \sum_{j=0}^{k}(\xi_{k-j}^{p^j} \otimes \tau_j + \tau_k \otimes 1)$.

Write $H^*\big(B(\mathbb{Z}/p\mathbb{Z})\big) = E[u] \otimes (\mathbb{Z}/p\mathbb{Z})[v]$ if $p > 2$ or $H^*\big(B(\mathbb{Z}/2\mathbb{Z})\big) = (\mathbb{Z}/2\mathbb{Z})[v]$ when $p = 2$. Notice that for a monomial $\alpha \in \mathcal{A}$, $\alpha(v^{p^j}) = v^{p^m}$ if and only if $\alpha = \mathcal{P}^{p^{m-1}}\mathcal{P}^{p^{m-2}}\cdots\mathcal{P}^{p^{m-j}}$. The coproduct $\psi(\xi_k)$ is characterized by the fact that if α and γ are monomials in $\mathcal{A}$ then $\langle\alpha \otimes \gamma, \psi(\xi_k)\rangle \neq 0$ if and only if $\alpha\gamma(x)$ is a nonzero multiple of x^{p^k}. Therefore, $\langle\alpha \otimes \gamma, \psi(\xi_k)\rangle \neq 0$ if and only if there exists j such that $\alpha = \mathcal{P}^{p^{k-1}}\mathcal{P}^{p^{k-2}}\cdots\mathcal{P}^{p^j}$ and $\gamma = \mathcal{P}^{p^{j-1}}\mathcal{P}^{p^{j-2}}\cdots\mathcal{P}^1$. This implies that $\psi\xi_k = \sum_{j=0}^{k} \xi_{k-j}^{p^j} \otimes \xi_j$. The proof of part (2) is similar, using in addition that $\alpha(u) = v^{p^m}$ if and only if $\alpha = \mathcal{P}[m]\beta$. $\qquad\square$

Although the algebra generators of $\mathcal{A}_*$ were chosen from the dual basis to the basis of admissible sequences in $\mathcal{A}$, other monomials in this free commutative algebra need not be dual basis elements, so the basis of monomials gives a second basis for $\mathcal{A}_*$. The dual basis to this basis of monomials gives a second basis for $\mathcal{A}$, called the *Milnor basis*. The element in the Milnor basis which is dual to $w(e_0, r_1, e_1, r_2, e_2, \ldots, r_k, e_k)$ is usually written $\mathcal{P}^{\{e_0, r_1, e_1, r_2, e_2, \ldots, r_k, e_k\}}$. The elements $\mathcal{P}^{\{0,0,\ldots,1,0\}}$ (only nonzero entry is r_k) and $\mathcal{P}^{\{0,0,\ldots,0,1\}}$ (only nonzero entry is e_k) are sometimes written $\mathcal{P}^{\Delta_k}$ and $\mathcal{Q}^{\Delta_k}$ respectively. They are primitives in $\mathcal{A}$ and can be defined inductively by $\mathcal{P}^{\Delta_1} = \mathcal{P}^1$, $\mathcal{Q}^{\Delta_0} = \beta$, and $\mathcal{P}^{\Delta_k} = [\mathcal{P}^{p^k}, \mathcal{P}^{\Delta}_{k-1}]$ and $\mathcal{Q}^{\Delta_k} = [\mathcal{P}^{p^k}, \mathcal{Q}^{\Delta_{k-1}}]$.

6 Sample Applications of the Steenrod Algebra

Example 14.6.1 Suppose n is not of the form 2^t for any t. Then there does not exist a space X such that $H^*(X; \mathbb{Z}/2\mathbb{Z}) = (\mathbb{Z}/2\mathbb{Z})[x]/(x^3)$ with $|x| = n$.

Remark This is the famous "Hopf invariant 1" problem. The real, complex, quaternionic, and Cayley projective planes, which are the homotopy cofibres of the squaring map on S^1 and the three Hopf maps, provide examples of such spaces for $n = 1, 2, 4$, and 8. The fact that such a space does not exist for $n = 2^t$ with $t > 3$ is a theorem of Adams originally proved using higher order cohomology operations and then reproved more easily using K-theory operations.

Proof By inspection of the Adem relations in the mod 2 Steenrod algebra, Sq^n is decomposable into a sum of products of other operations, since n is not of the form 2^t. Therefore the existence of such a space X would yield a contradiction since $Sq^n(x) = Sq^{|x|}(x) = x^2$, but $H^q(X; \mathbb{Z}/2\mathbb{Z}) = 0$ for $n < q < 2n$. $\qquad\square$

Example 14.6.2 Observe that if n is odd then
$$H^*(S^N \mathbb{R}P^n) \cong H^*(S^N \mathbb{R}P^{n-1} \vee S^{N+n}).$$
Suppose n is not of the form $2^t - 1$ for any t. Then
$$S^N \mathbb{R}P^n \not\simeq S^N \mathbb{R}P^{n-1} \vee S^{N+n}$$
for any N.

Proof Recall that $H^*(\mathbb{R}P^n; \mathbb{Z}/2\mathbb{Z}) \cong (\mathbb{Z}/2\mathbb{Z})[x]/(x^{n+1})$ where $|x| = 1$. It follows from the properties of Steenrod operations that the action of the operations is given by $Sq^i x^j = \binom{j}{i} x^{i+j}$. Since the Steenrod operations are stable (commute with suspension) this shows that the action of Sq^i on $H^*(S^N \mathbb{R}P^n)$ takes the generator in degree $N + j$ to $\binom{j}{i}$ times the generator in degree $N + i + j$. If n is not of the form $2^t - 1$ then there exists $i > 0$ and j such that $i + j = n$ and $\binom{j}{i} \not\equiv 0 \pmod 2$ so that the generator in degree $N + n$ is in the image of Sq^i for some $i > 0$. This is not consistent with the action of the operations on $S^N \mathbb{R}P^{n-1} \vee S^{N+n}$. $\qquad\square$

Example 14.6.3 Let $\eta : S^{n+1} \to S^n$ denote the $(n-2)$nd suspension of the Hopf map, whose homotopy cofibre is $\mathbb{C}P^2$. Suppose n is congruent to either 0 or 1 $\pmod 2$. Then for no N does there exist a map $f : S^{N+n+1} \to S^N \mathbb{R}P^n$ such that the composite $S^{N+n+1} \xrightarrow{S^N f} S^N \mathbb{R}P^n \xrightarrow{S^N c} S^{N+n}$ is $S^N \eta$, where $c : \mathbb{R}P^n \to S^n$ is the pinch map onto the top cell.

Proof Suppose that such a map f exists and let X be the homotopy cofibre of X. Then $\tilde{H}^q(X; \mathbb{Z}/2\mathbb{Z}) \cong \mathbb{Z}/2\mathbb{Z}$ if $N+1 \le q \le N+n$ or $q = N + n + 2$ and 0 otherwise. There is an induced map $X \to S^{N+n-2}\mathbb{C}P^2$ which induces an isomorphism on $H^{N+n}(\ ; \mathbb{Z}/2\mathbb{Z})$ and $H^{N+n+2}(\ ; \mathbb{Z}/2\mathbb{Z})$, so naturality of the operations along with knowledge of their action on $H^*(\mathbb{C}P^2; \mathbb{Z}/2\mathbb{Z})$ imply that there is a Sq^2 connecting the top two generators in $H^*(X; \mathbb{Z}/2\mathbb{Z})$. Similarly, using the map $S^N \mathbb{R}P^n \to X$ and naturality gives that Sq^2 takes the generator in degree $N+n-2$ to $\binom{n-2}{2}$ times the generator in degree N, and this coefficient is nonzero mod 2 given our hypothesis on n. Since $H^{N+n+1}(X; \mathbb{Z}/2\mathbb{Z}) = 0$, the relation $Sq^2 Sq^2 = Sq^1 Sq^2 Sq^1$ derived from the Adem relations yields a contradiction. $\qquad\square$

Theorem 14.6.4 *The H-space S^7 (with multiplication induced from the Cayley numbers) is not homotopy associative.*

Proof For an H-space X, there exists a map $X * X \to SX$ given by $(x, t, y) \mapsto (xy, t)$; we will write $B_2 X$ for its homotopy cofibre. This map is called the *Hopf construction* on X and is natural with respect to H-maps and so induces a map between the corresponding homotopy cofibres. If X is homotopy associative, then the map $JX \to X$ given by $(x_1, x_2, \ldots, x_n) \to x_1 x_2 \cdots x_n$ is an H-map and so there is an induced map $JX * JX \to X * X$ whose restriction to $X * X$ is the identity map. However if X is a connected CW-complex then JX is homotopy equivalent to a topological group G (Theorem 7.9.2 and Theorem 9.2.4) and the map in the Milnor bundle $G \to G * G \to SG$ corresponds to the Hopf construction. (To make the maps equal one should use a modified action on $G * G$ which introduces an inverse in the action on one factor, or else use a modified Hopf construction $(x, t, y) \mapsto (xy^{-1}, t)$; however $x \to x^{-1}$ is a homotopy equivalence so these modifications do not affect the homotopy cofibres.) It follows from Chapter 9 (see the iterated Ganea construction) that the space denoted $B_2(G)$ above is the same as Milnor's $B_2(G)$. Therefore Milnor's bundle $G \to G * G * G \to B_2(G)$ shows that

$$\Omega B_2(JX) \approx JX \times \Omega(JX * JX * JX).$$

Suppose that S^7 were homotopy associative. Then the preceding gives a composite map

$$\Omega S^{23} = \Omega(S^7 * S^7 * S^7) \to \Omega(JS^7 * JS^7 * JS^7) \to \Omega B_2(JS^7) \to \Omega B_2(S^7)$$

which we will denote by γ. Let $\alpha : S^7 \to \Omega B_2(S^7)$ be a generator of the least nonvanishing homotopy group of $\Omega B_2(S^7)$.

Lemma 14.6.5 *Suppose Y_{k-1} is a simply connected space with $H^*(Y_{k-1}) = \mathbb{Z}[x]/(x^k)$, where $|x| = 8$. Let Y_k be the homotopy cofibre of some map $f : S^{8k-1} \to Y_{k-1}$. Then the following are equivalent:*

a) $H^*(Y_k) = \mathbb{Z}[x]/(x^{k+1})$;
b) $H^*(\Omega Y_k) \cong H^*\left(S^7 \times \Omega S^{8(k+1)-1}\right)$;
c) *there is a homotopy equivalence $\Omega Y_{k-1} \approx S^7 \times \Omega S^{8k-1}$ such that the composite $\Omega S^{8k-1} \xrightarrow{\Omega f} \Omega Y_{k-1} \approx S^7 \times \Omega S^{8k-1} \xrightarrow{\pi''} \Omega S^{8k-1}$ is a homotopy equivalence.*

Proof The equivalence of (a) and (b) is obtained easily by applying the Serre spectral sequence to the fibration $\Omega Y_k \to PY_k \to Y_k$.

Let F be the homotopy fibre of $Y_{k-1} \hookrightarrow Y_k$ and let G be the homotopy fibre of $S^7 \to \Omega Y_k$. Since $S^{8k-1} \xrightarrow{f} Y_{k-1} \hookrightarrow Y_k$ is null homotopic, there is an induced map $S^{8k-1} \to F$ that induces an isomorphism on $H_{8k-1}(\ )$, which is the least nonvanishing degree of F. Since the homotopy fibre of $Y_{k-1} \cup CF \to Y_k = Y_{k-1} \cup CS^{n-1}$ is $(8k + 6)$-connected by Ganea's theorem, the map $S^{8k-1} \to F$ induces a homology isomorphism through degree $8k + 4$.

Suppose now that (c) holds. From the Serre spectral sequence, to prove (b) it suffices to know that the second lowest nonvanishing homology group of ΩY_k occurs beyond degree $8k - 1$, or equivalently, that G is at least $(8k - 2)$-connected. However the diagram of homotopy fibrations

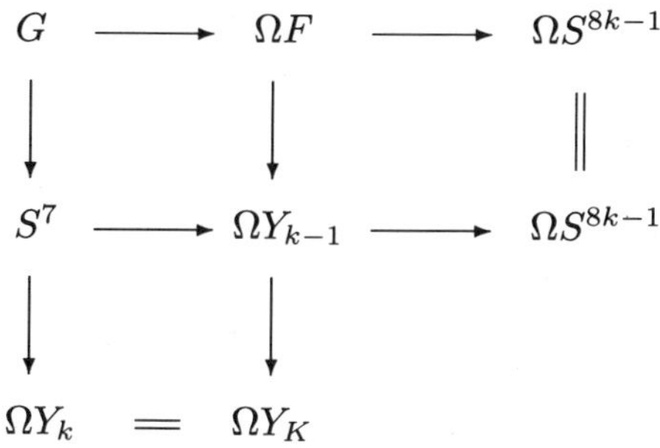

shows that G is at least $(8k + 2)$-connected.

Conversely, if (b) holds then G is $(8k+5)$-connected. We already know from the equivalence of (a) and (b) (applied with $k-1$ replacing k) that $H^*(\Omega Y_{k-1}) \cong H^*(S^7 \times \Omega S^{8k-1})$; we will construct a map of spaces inducing this isomorphism. The connectivity of G and the homotopy pullback square

$$
\begin{array}{ccc}
G & \longrightarrow & \Omega F \\
\downarrow & & \downarrow \Omega(\ ) \\
S^7 & \longrightarrow & \Omega Y_{k-1}
\end{array}
$$

show that $\Omega F \xrightarrow{\ \Omega(\ )\ } \Omega Y_{k-1}$ induces an isomorphism on $H^{8k-2}(\ )$, and therefore so does the composite $\Omega S^{8k-1} \xrightarrow{\ \Omega(\ )\ } \Omega F \xrightarrow{\ \Omega(\ )\ } \Omega Y_{k-1}$, which is $\Omega(f)$. Therefore it is easy to check that the composite

$$
S^7 \times \Omega S^{8k-1} \xrightarrow{\ (\ )\times\Omega(f)\ } \Omega Y_{k-1} \times \Omega Y_{k-1} \xrightarrow{\ \text{mult}\ } \Omega Y_{k-1}
$$

induces the homology isomorphism above and is the desired homotopy equivalence.
$\qquad\square$

Proof of Theorem 14.6.4 (cont.). Applying the Hopf construction to S^7 gives the Hopf map $\sigma : S^{15} \to S^8$. Since there is a homotopy fibration $S^7 \to S^{15} \xrightarrow{\ \sigma\ } S^8$, we get a decomposition $\Omega S^8 \approx S^7 \times \Omega S^{15}$ with $\Omega S^{15} \xrightarrow{\ \Omega\sigma\ } \Omega S^8 \approx S^7 \times \Omega S^{15} \xrightarrow{\ \pi''\ } \Omega S^{15}$ homotopic to the identity. Applying the lemma with $k=2$ gives $H^*\big(B_2(S^7)\big) \cong \mathbb{Z}[x]/(x^3)$ where $|x| = 8$. (Thus the Hopf map does indeed have Hopf invariant 1!) The lemma now also gives $H^*\big(\Omega B_2(S^7)\big) \approx H^*(S^7 \times \Omega S^{23})$. Using the construction of γ and naturality it is easy to check that $(\Omega\gamma)_*$ is an isomorphism on $H^{22}(\ )$ and using this it follows as in the proof of the lemma that the composite

$$
S^7 \times \Omega S^{23} \xrightarrow{\ \alpha\times\gamma\ } \Omega B_2(S^7) \times \Omega B_2(S^7) \xrightarrow{\ \text{mult}\ } \Omega B_2(S^7)
$$

induces a homology isomorphism and is thus a homotopy equivalence. Now applying the lemma with $k=3$ shows that cohomology of the homotopy cofibre of γ, C_γ, is given by $H^*(C_\gamma) \cong \mathbb{Z}[x]/(x^4)$ where $|x| = 8$. However there does not exist a space with this cohomology ring because reducing modulo 3 and applying the Steenrod operation $\mathcal{P}^4$ gives $\mathcal{P}^4(x) = \mathcal{P}^{|x|/2}(x) = x^3$ which contradicts the Adem relation $\mathcal{P}^1\mathcal{P}^3 = \mathcal{P}^4$ since $\mathcal{P}^3(x)$ belongs to $H^{20}(C_\gamma; \mathbb{Z}/3\mathbb{Z}) = 0$. $\qquad\square$

Remark 1 It may not be immediately clear from what was presented here that the fibre of the Hopf map $S^{15} \to S^8$ is S^7, which is equivalent to the decomposition of ΩS^8. It is true in general that the homotopy fibre of the Hopf construction $X * X \to SX$ is X, (for a connected CW-H-space X), but we have demonstrated this only in the case where X has the homotopy type of loop space where we can use Milnor's construction. In the case of S^7 an ad hoc verification can by done by checking that $\Omega S^{15} \xrightarrow{\ \Omega(\ )\ } \Omega S^8$ induces an isomorphism on $H^{14}(\ )$ and this can be done by applying naturality to the H-map $S^7 \to K(\mathbb{Z}, 7)$, taking advantage of the fact that $K(\mathbb{Z}, 7)$ has the homotopy type of a loop space.

Remark 2 The fact that homotopy associativity of an H-space allows one to build the first two stages of Milnor's classifying construction is part of a larger theory developed by Stasheff which relates higher order homotopy associativities with existence of higher levels of Milnor's construction. See [Sta].

7 Secondary Cohomology Operations

This section takes a brief glimpse at secondary operations and illustrates their use with a typical example. Throughout the section, F^f and C_f will denote respectively the homotopy fibre and cofibre of f.

Let $\alpha : W \to X$, $\beta : X \to Y$, and $\gamma : Y \to Z$ be maps such that $\beta \circ \alpha \simeq *$ and $\gamma \circ \beta \simeq *$. Since $\beta \circ \alpha \simeq *$, there exists an extension $\bar{\beta} : C_\alpha \to Y$ of β. The restriction of $\gamma \circ \bar{\beta}$ to X is $\gamma \circ \beta \simeq *$ and so $\gamma \circ \bar{\beta}$ extends to a map denoted $\{\gamma, \beta, \alpha\} : SW \to Z$ from SW, the homotopy cofibre of $X \to C_\alpha$. By varying the choices of the extensions, $\{\gamma, \beta, \alpha\} \in [SW, Z]$ may be changed by addition of some element of $\gamma_\#[SW, Y] + S\alpha^\#[SX, Z]$, so we will regard $\{\gamma, \beta, \alpha\} \in [SW, Z]$ as an element of $[SW, Z]/(\gamma_\#[SW, Y] + S\alpha^\#[SX, Z])$. This construction is called the *secondary composition* or *Toda bracket* of α, β, and γ. $(\gamma_\#[SW, Y] + S\alpha^\#[SX, Z])$ is called the *indeterminacy* in the secondary composition.

The dual construction formed by lifting β to $\hat{\beta} : X \to F_\gamma$ and then lifting $\hat{\beta} \circ \alpha$ to ΩZ gives an element of $[W, \Omega Z]/(\Omega\gamma_\#[W, \Omega Y] + \alpha^\#[X, \Omega Z])$ which we shall denote $\langle \gamma, \beta, \alpha \rangle$. In the "stable range" (i.e., given sufficient connectivity assumptions) $\langle \gamma, \beta, \alpha \rangle$ is the adjoint of $\{\gamma, \beta, \alpha\}$.

An equivalent formulation is as follows. Begin by choosing lifts $\hat{\beta} : X \to F^\gamma$ and $\hat{\alpha} : W \to F^\beta$ of β and α respectively. By Proposition 7.6.1, the homotopy pullback of $\hat{\beta} : X \to F^\gamma$ and $\partial : \Omega Z \to F^\gamma$ is F_β so there is an induced map $\tilde{\beta} : F^\beta \to \Omega Z$. Then $\langle \gamma, \beta, \alpha \rangle = \tilde{\beta} \circ \hat{\alpha}$. The other construction can be done similarly using a homotopy pushout.

Let Y, Y', and Y'' be reduced cohomology theories represented by Ω-spectra (Y_n), (Y'_n) and (Y''_n). Suppose that for some n and k we are given q pairs of cohomology operations $\theta_i \in [Y_n, Y'_{n+|\theta_i|}]$ and $\tau_i \in [Y'_{n+|\theta_i|}, Y''_{n+k}]$ for $i = 1, \dots, q$, such that $\sum_{i=1}^q \tau_i \sigma_i = 0$ (where $|\tau_i| + |\theta_i| = k$ for all i). Set $\beta = (\theta_i)_{i=1}^q \in \left[Y_n, \prod_{i=1}^q Y'_{n+|\theta_i|}\right]$ and $\gamma = \sum_{i=1}^q \tau_i \in [\prod_{i=1}^q Y'_{n+|\theta_i|}, Y''_{n+k}]$. Then $\gamma \circ \beta = \sum_{i=1}^q \tau_i \sigma_i = 0$ so we can choose a lift $\hat{\beta} : Y_n \to F^\gamma$ which in turn yields a map $\Phi = \tilde{\beta} : F^\beta \to \Omega Y''_{n+k} = Y''_{n+k-1}$. Such a choice of Φ will be called a *secondary cohomology operation* associated to the relation $\sum_{i=1}^q \tau_i \sigma_i = 0$ among primary cohomology operations. For any space X, the secondary operation Φ is a function from $\bigcap_{i=1}^q \operatorname{Ker} \theta_i \subset Y^n(X)$ to $Y''^{n+k-1}(X) \Big/ \left(\sum_{i=1}^q (\Omega\tau_i)^* \left(Y''^{n+k-1-|\tau_i|}(X) \right) \right)$ defined as follows. Suppose that $x \in Y^n(X)$ is a cohomology class in a space X having the property that $\theta_i(x) = 0$ for all i. The composition $X \xrightarrow{x} Y_n \xrightarrow{\beta} \prod_{i=1}^q Y'_{n+|\theta_i|}$ is trivial so there exists a lift $\hat{x} : X \to F^\beta$ of x. Then $\Phi(x)$ is defined as the element of $Y''^{n+k-1}(X) \Big/ \left(\sum_{i=1}^q (\Omega\tau_i)^* \left(Y''^{n+k-1-|\tau_i|}(X) \right) \right)$ represented by $\Phi \circ \hat{x}$. Thus $\Phi(x)$ represents the secondary composition $\langle \sum_{i=1}^q \tau_i, (\theta_i)_{i=1}^q, x \rangle$ in $Y''^{n+k-1}(X)$ with smaller indeterminacy due to the fixed choice of $\Phi = \tilde{\beta}$ in forming the composition.

Example 14.7.1 Write K_q for $K(\mathbb{Z}/2\mathbb{Z}, q)$. Let Φ be a secondary cohomology operation associated to the Adem relation $Sq^2 Sq^2 + Sq^3 Sq^1 = 0$ on K_n. Let $\eta_n : S^{n+1} \to S^n$ denote the attaching map of the top cell in $S^{n-2}\mathbb{C}P^2$ (which is the $(n-2)$nd suspension of the Hopf map). Let X be the homotopy cofibre of $\eta_n \circ \eta_{n+1} : S^{n+2} \to S^n$. Then $\tilde{H}^q(X; \mathbb{Z}/2\mathbb{Z}) = \mathbb{Z}/2\mathbb{Z}$ if $q = n$ or $n+3$ and 0 otherwise. Let $x \in H^n(X; \mathbb{Z}/2\mathbb{Z})$ and $y \in H^{n+3}(X; \mathbb{Z}/2\mathbb{Z})$ denote generators. Note that $\mathrm{Sq}^2(x) = 0$ and $\mathrm{Sq}^1(x) = 0$ for degree reasons so $\Phi(x)$ is defined. Since $\mathrm{Sq}^2 : H^{n+1}(X; \mathbb{Z}/2\mathbb{Z}) \to H^{n+3}(X; \mathbb{Z}/2\mathbb{Z}) = 0$ and $\mathrm{Sq}^3 : H^n(X; \mathbb{Z}/2\mathbb{Z}) \to H^{n+3}(X; \mathbb{Z}/2\mathbb{Z}) = 0$ the indeterminacy is 0 in this case. We will show that $\Phi(x) \neq 0$ in $H^{n+3}(X; \mathbb{Z}/2\mathbb{Z})$ and in particular that $\eta_n \circ \eta_{n+1} \neq 0$ in $[S^{n+1}, S^n]$.

As above, set $\beta = (\mathrm{Sq}^2, \mathrm{Sq}^1) : K_n \to K_{n+2} \times K_{n+1}$ and $\gamma = \mathrm{Sq}^2 + \mathrm{Sq}^3 : K_{n+2} \times K_{n+1} \to K_{n+4}$. The methods of Chapter 7 yield a homotopy cofibration sequence $S^{n-1}\mathbb{C}P^2 \xrightarrow{i} X \xrightarrow{j} S^{n-2}\mathbb{C}P^2$. Let $X \xrightarrow{j} S^{n-2}\mathbb{C}P^2 \xrightarrow{f} S^n\mathbb{C}P^2$ be the induced fibration. In the diagram

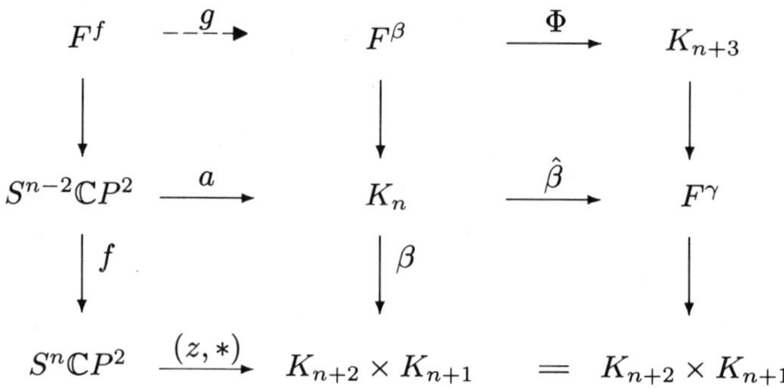

in which the columns are homotopy fibrations and $a : S^{n-2}\mathbb{C}P^2 \to K_n$ and $z : S^n\mathbb{C}P^2 \to K_{n+2}$ are nontrivial, the bottom left square homotopy commutes since

$$f^* : H^{n+2}(S^n\mathbb{C}P^2; \mathbb{Z}/2\mathbb{Z}) \to H^{n+2}(S^{n-2}\mathbb{C}P^2; \mathbb{Z}/2\mathbb{Z})$$

is an isomorphism and $\mathrm{Sq}^2(a) \neq 0$. Let $g : F^f \to F^\beta$ be an induced map of homotopy fibres. The lift $\hat{x} : X \to F^\beta$ of x can be chosen to be $g \circ \hat{j}$ where $\hat{j} : X \to F^f$ is a lift of j. Then $\Phi(x)$ is represented by the composition $\Phi \circ \hat{x}$. Since $i^*\big(\Phi(x)\big) \in H^{n+3}(S^{n-1}\mathbb{C}P^2; \mathbb{Z}/2\mathbb{Z})$ is represented by the composition

$$S^{n-1}\mathbb{C}P^2 \longrightarrow \Omega S^n\mathbb{C}P^2 \xrightarrow{(\Omega z, *)} K_{n+3} \times K_{n+2} \xrightarrow{\pi'} K_{n+3}$$

which equals $\mathrm{Sq}^2(z) \neq 0$, we see that $\Phi(x) \neq 0$. It follows that $\eta_n \circ \eta_{n+1} \not\simeq 0$ since if it were trivial then X would be homotopy equivalent to $S^n \vee S^{n+3}$, and it is easy to see that Φ acts trivially on $H^n(S^n \vee S^{n+3}; \mathbb{Z}/2\mathbb{Z})$. $\qquad\square$

Bibliography

[A] Adams, J. F. [1972], *A student's guide to the literature*, London Math Soc. Lecture Notes Series **4**, Cambridge University Press.

[A2] Adams, J. F. [1974], *Stable Homotopy and Generalized Homology*, Chicago University Press.

[A3] Adams, J. F. [1990], *On the non-existence of elements of Hopf invariant* 1, Ann. of Math. **72**, 20–104.

[AH] Adams, J. F. and Hilton, P. J. [1956], *On the chain algebra of a loop map*, Comm. Math. Helv. **30**, 305–330.

[An] Anick, D. J. [1989], *Hopf algebras up to homotopy*, Jour. AMS **2**, 417–453.

[BK] Bousfield, A. K. and Kan, D. M. [1972], *Homotopy Limits, Completions and Localization*, Lecture Notes in Math **304**, Springer-Verlag.

[BL] Baues, H. J. and Lemaire, J.-M. [1977], *Minimal models in homotopy theory*, Math. Ann. **225**, 219–242.

[C] Curtis, E. [1971], *Simplicial Homotopy Theory*, Advances in Math. **6**, 107–209.

[D] Dugundji, J. [1966], *Topology*, Allyn and Bacon Series in Advanced Mathematics.

[EilS] Eilenberg, S. and Steenrod, N. [1952], *Foundations of Algebraic Topology*, Princeton Math Series **15**, Princeton University Press.

[EM] Eilenberg, S. and Moore, J. C. [1966], *Homology and fibrations*, Comment. Math. Helv. **40**, 199–236.

[EpS] Epstein, D. B. A. and Steenrod, N. [1962], *Cohomology Operations*, Princeton Math Series **50**, Princeton University Press.

[GH] Greenberg, M. and Harper, J. [1981], *Algebraic Topology: A First Course*, Benjamin.

[HS] Hilton, P. J. and Stammbach, U. [1970], *A Course in Homological Algebra*, Graduate Texts in Mathematics **4**, Springer-Verlag.

[H] Husemuller, D. [1966], *Fibre Bundles*, McGraw-Hill.

[J] James, I. [1955], *Reduced product spaces*, Ann. of Math. **62** 170–197.

[K] Kane, R. K. [1988], *The Homology of Hopf Spaces*, North-Holland Mathematical Studies **40**, North-Holland.

[LMS] Lewis, G., May, J. P. and Steinberger, M. [1986], *Equivariant Stable Homotopy*, Springer Lecture Notes in Math. **1213**, Springer-Verlag.

[Mac] Mac Lane, S. [1963], *Homology*, Grundlehren series **114**, Springer.

[Mas] Massey, W. [1991], *A Basic Course in Algebraic Topology*, Graduate Texts in Mathematics **127**, Springer-Verlag.

[Mas1] Massey, W. [1952], *Exact couples in algebraic topology, I, II*, Ann. of Math. **56**, 363–396; **57** [1953], 248–286.

[Mas2] Massey, W. [1954], *Products in exact couples*, Ann. of Math. **59**, 558–569.

[May] May, J. P. [1967], *Simplicial objects in algebraic topology*, Van Nostrand.

[May1] May, J. P. [1970], *A general algebraic approach to Steenrod operations*, The Steenrod algebra and its applications, Springer Lecture Notes in Math. **168**, 153–231.

[May2] May, J. P. [1975], *Classifying Spaces and Fibrations*, Memoirs of the Amer. Math. Soc. **155**, AMS.

[Mi1] Milnor, J. [1956], *Constructions of universal bundles, I*, Ann. of Math **63**, 272–284.

[Mi2] Milnor, J. [1956], *Constructions of universal bundles, II*, Ann. of Math **63**, 430–436.

[Mi3] Milnor, J. [1957], *The geometric realization of a semi-simplicial complex*, Ann. of Math **65**, 357–362.

[Mi4] Milnor, J. [1958], *The Steenrod algebra and its dual*, Ann. of Math **67**, 150–171.

[Mi5] Milnor, J. [1959], *On spaces having the homotopy type of a CW-complex*, TAMS **90**, 272–280.

[Mi6] Milnor, J. [1962], *On axiomatic homology theory*, Pac. J. Math **12**, 337–341.

[Miy] Miyazaki, H. [1952], *Paracompactness of CW-complexes*, Tohoku Math J. **4**, 309–313.

[MM] Milnor, J. and Moore, J. [1965], *On the structure of Hopf algebras*, Ann. of Math **81**, 211–264.

[MS] Milnor, J. and Stascheff, J. [1974], *Characteristic Classes*, Princeton Math Series **76**, Princeton University Press.

[MT] Mosher, Tangora [1968], *Cohomology Operations and Applications in Homotopy Theory*, Harper & Row.

[Mu] Munkres, J. R. [1975], *Topology: A First Course*, Prentice-Hall.

[RS] Rothenberg, M. and Steenrod, N. [1965], *The cohomology of classifying spaces of H-spaces*, Bull. AMS **71**, 872–875.

[Se1] Serre, J.-P. [1951], *Homologie singulièr des espaces fibré*, Ann. of Math **54**, 425–505.

[Se2] Serre, J.-P. [1953], *Cohomologie modulo 2 des complexes d'Eilenberg-MacLane*, Comm. Math. Helv. **27**, 198–231.

[Se3] Serre, J.-P. [1953], *Groupes d'homotopie et classes de groupes abeliens*, Ann. of Math. **58**, 258–294.

[Se4] Serre, J. P. [1965], *Lie algebras and Lie groups*, Benjamin.

[Sm] Smith, L. [1969], *Lectures on the Eilenberg-Moore Spectral Sequence*, Lecture Notes in Math **134**, Springer-Verlag.

[Sp] Spanier, E. H. [1981], *Algebraic Topology*, Springer-Verlag.

[Sta] Stascheff, J. D. [1970], *H-spaces from a homotopy point of view*, Lecture Notes in Math **161**, Springer-Verlag.

[St] Steenrod, N. E. [1951], *The Topology of Fibre Bundles*, Princeton Math Series **14**, Princeton University Press.

[St1] Steenrod, N. E. [1967], *A convenient category of topological spaces*, Michigan Math. J. **14**, 133–152.

[T] Toda, H. [1962], *Composition Methods in the Homotopy Groups of Spheres*, Princeton Math Series **49**, Princeton University Press.

[W] Whitehead, G. W. [1978], *Elements of Homotopy Theory*, Graduate Texts in Mathematics **61**, Springer-Verlag.

[Z] Zabrodsky, A. [1976], *Hopf Spaces*, North-Holland Mathematical Studies **22**, North-Holland.

Index